D1170901

Under Siege

Under Siege

Man, Men, and Earth

Kenneth A. Wagner
Ferris State College

Paul C. Bailey
Birmingham-Southern College

Glenn H. Campbell
Ferris State College

Intext Educational Publishers New York and London

Library of Congress Cataloging in Publication Data

Wagner, Kenneth Allan, 1919-
Under siege.

Includes bibliographies.
1. Ecology. 2. Man—Influence on nature. 2. Man—Influence of environment. I. Bailey, Paul Clinton, joint author. II. Campbell, Glenn, joint author. III. Title.
[QH541.W23 1973] 301.31 73-166
ISBN 0-7002-2434-3

Intext Educational Publishers
257 Park Avenue South
New York, New York 10010

Contents

Preface

We have attempted to discuss the more urgent environmental problems confronting man: their causes, current status, and the consequences of alternative solutions. We have tried not to be alarmists, but to be realistic in assessing the seriousness of these problems and in presenting some options that appear to be reasonable "outs" for man. We hope our readers will develop an awareness of their environment that will enable them to recognize new problems before they reach the crisis stage, and that this awareness will be converted into sensible citizen involvement leading to solutions.

Environmental problems are basically biological in nature and can be interpreted only through an understanding of the biological setting in which they occur. Therefore, this text presents biological concepts as they apply to the problems of environmental and cultural deterioration. The approach to these concepts is nontraditional. They are not organized systematically but are discussed where they relate to specific environmental problems. For example, human reproduction is described in conjunction with the discussion of human population problems, and pollution is seen as the introduction of foreign materials into balanced ecosystems. Throughout, the social and cultural implications of our environmental crisis are considered.

The book is divided into five sections. Part I offers a broad overview of man's current environmental problems. We attempt to provide sufficient background so that the reader will realize that the subject matter of this book applies to him. We return to the general case in Part V, which summarizes the principal areas of stress on our planet's ecosystems and points to hopeful indications that progress is being made. Specific suggestions for individual and community action include addresses of agencies that can help.

Part II, "Essentials," discusses the major factors that make up the physical part of our heritage. It considers the consequences of man's technology in those areas upon which life most intimately depends: air, water, and soil. To this list is added the radiation hazards that have resulted from the Twentieth Century development of atomic energy.

Part III, "Stresses: Ecological and Personal," covers a series of topics involving the complex relationships between man and other organisms. The role of disease in the balance of nature is reviewed from an ecological point of view and is followed by a more detailed consideration of those diseases which have the greatest impact on young people today—the venereal diseases. Factual information on the causes and results of drug use and abuse is intended to encourage students to make an objective analysis of social and individual drug problems.

The realistic note of pessimism in telling the story of man's intensified conflict with those organisms with which he competes for food, fibers, and other resources (Chapter 9, Pests, Plagues, and Poisons) is tempered with an appraisal of alternatives to a poisoned earth policy. Man, in his restless wanderings and in his efforts to subdue Earth by rebuilding ecosystems more to his liking, has displaced and misplaced many species of plants and animals. Some of this rearrangement of the "biological furniture" was intentional, some was not. The results have been disastrous in many cases but beneficial in others. The final chapter of Part III analyzes man's role, past and present, in accelerating the extinction of many species.

Part IV, "Assessing the Population Problem," is a consideration of one of the most critical problems facing modern man: the control of his own population before it reaches a state where it is controlled for him through disease, starvation, or atomic holocaust brought on by population stresses. The reader is introduced to the statistical facts of population in relation to the capacity of the earth to support man and other life forms.

We have tried, then, to assess man's position in the web of life, explore the problems that he has created, and present the relevant biological principles. We hope this approach will place the reader in a position to make sound judgments on critical environmental issues compatible not only with survival but also with an enhanced quality of life. By concentrated efforts stemming from a new and informed environmental ethic, the siege of Earth may yet be lifted.

The authors appreciate the assistance of a number of individuals who have helped make this book possible. Professor C. E. McCoy helped with editorial and organizational matters and Professor Richard McNeill provided material for one of the chapters. Professor Thomas J. Crowley participated in the original course revision that led to this book.

Under Siege

Hartmann's mountain zebra lives in the arid mountains of southwest Africa. It ranges from the seacoast to 50, and sometimes 100, miles inland along the western edge of the Namib Desert. Several factors have caused a steady decline in population. Among these are game-proof fences which restrict normal seasonal movement and cut off herds from access to the limited sources of water. *(Courtesy Lion Country Safari, Laguna Hills, California.)*

FOREWORD

This picture symbolizes the general state of the planet Earth. *(Courtesy U.S. Department of the Interior, Bureau of Reclamation.)*

1

Observations on the State of the Planet

Most crucial problems now facing man have a biological basis. The quality of the environment with which we interact determines the success or failure of the complex processes that sustain our lives. The similarity of these processes in all living things shatters the secure feeling generated by man's estimate of his role and capabilities on the fragile planet Earth. No organism exists in a biological vacuum. Conditions that threaten one life form may menace many more.

This chapter reviews the current status of some worldwide environmental problems: population, living space, waste accumulation, and depletion of natural resources. The technology that created these problems and which is being employed in an effort to solve them has direct and indirect effects far beyond its intended role. Mercury compounds applied to seeds to control fungi killed hogs that were fed the seeds and blinded children that ate contaminated pork. Insecticides directed toward the control of Dutch elm disease killed robins but did not save the elms (Fig. 1-1). Increased food production is made possible by building dams that impound water for irrigation of more land. In India and in Egypt, population increase during the time required to build dams more than offset the increased food production. In Mexico food production increased 80 percent over a 15-year period—and population increased 80 percent during the same period of time. People were still just as hungry.

Dutch elm disease

dams

Fig. 1-1. Massive spray programs were launched to destroy the beetle that carries the fungus causing Dutch elm disease. The failure of this effort is evidenced by the dead elms along American roads and streets.

Some Side Effects of Technological Progress

Problem	*Solution*	*New Problems Created by the Solution*
World food shortage	United States develops high-yield crops for export	Increased fertilizer use contaminates drinking water and ages lakes; holders of small acreages are put out of business.
Egyptians starving	Aswan High Dam for more irrigation	Mediterranean sardine fisheries are wiped out; blood fluke infection spreads; need for fertilizer to replace minerals from the Nile floods develops; concurrent population gain offsets increased food production.
Malaria and typhus	DDT	Population increase overtaxes world food supply; people starve.
Dutch elm disease	DDT	Pesticide kills robins and other songbirds; has little effect on elms.
Crop pests	Synthetic poisons	Destruction of insect predators and other natural controls occurs; ultimate increase in pest population takes place.
Fossil fuel shortage	Nuclear power	Thermal pollution results; disposal of radioactive wastes increases radiation hazards.
Smoke stacks	Scrubbers	Disposal of collected matter pollutes water.
Tree shortage	Synthetic substitutes	Substitutes are not biodegradable; incineration pollutes air.
Crowded highways	More freeways	Even more cars and more air pollution result; more drain on fossil fuel reserves occurs.
Great Lakes shipping	Welland Canal to bypass Niagara Falls	Lamprey and alewife invade the Great Lakes.
Pollution	Dilution	Saturation of the air, water, and soil with toxic materials; ecosystem undergoes destruction.
In general:	Problem + Solution = Two new problems	

Some areas of conservation are now accepted as government, business, or personal responsibilities. No longer must we labor to make a case for clean air, clean water, and the control of pesticide poisons. Too many people are being hurt by what they breathe, drink, and eat. In other areas, however, particularly those involving personal freedoms, people are less inclined to take a firm stand on the issues that may take away their right to reproduce, their right to exploit their land as they see fit, or their right to use any natural resources they can pay for.

personal freedom

In the October–November 1969 issue of *National Wildlife,* the National Wildlife Federation published its first National Environmental Quality Index

Fig. 1-2. A growing awareness of environmental problems is evident in this Earth Day march on the U.S. Department of the Interior building, Washington, D.C., April 22, 1970. *(Courtesy U.S. Department of the Interior, National Park Service.)*

EQ Index

(EQ) in an effort to inform citizens on environmental issues. The October–November 1971 issue of *National Wildlife* carried the third of these annual summaries and clearly showed what the fight to save Earth had accomplished. We look at their conclusions and at other data on the status of several environmental problems that have a biological basis. This is necessarily a brief summary as these problems are treated in detail in later chapters. It is strongly recommended that students compare the most recent National Wildlife Federation EQ Index with the situation as described in the sections that follow and evaluate current trends.

POPULATION

population

The concept of population can be applied at various geographic levels. The level chosen depends on the immediate problem under consideration. A population may be thought of as comprising all the individuals of a species throughout its world range, or we may discuss the population of one kind of plant or animal in a specific locality. In a broader sense, more than one kind of organism can be included in a population: for example, all the insects in an orchard.

In the thinking of many people, progress depends on an increasing human population. The presence of more people requires more of everything—more houses, more cars, more electric power, and more food; more of all the things people want as well as the things they need. In the industrialized nations in particular, an expanding economy with its rising standard of living relies on population increase to sustain its momentum. The efforts to meet these demands for goods and services have made possible the "American way of life" but at the same time have created the present environmental problems.

progress

Man's society operates much as a living organism. Both utilize material and energy sources, and both produce wastes. An organism that exhausts its resources, or cannot escape its own waste, perishes. In the past man has been able to find new resources by moving into new territory or, as in the Nile Delta, by having the necessary resources moved to him. Also, in the past a relatively low population density and crude technology resulted in a waste production level that natural systems could recycle. At present, however, by sheer numbers and technological advances man is rapidly depleting resources that cannot be renewed, and society's metabolic wastes are poisoning the air and the oceans. An increasing human population can only mean a lower quality of environment for man and for other living things.

Whether a population grows in number, declines, or remains constant during a particular period of time depends on the relationship between the number born and the number that die during that period. For plants and for animals other than man, survival of the species depends on a reproduction rate that greatly exceeds the carrying capacity of the habitat. Yet in undisturbed areas these populations may remain more-or-less constant over long periods of time because of a correspondingly high death rate. Millions are born so that dozens may live. Survival of the species is accomplished at the expense of the individual, and the odds against the newborn reaching maturity may be very high.

natural populations

One species, man, has learned to beat the odds. He has found ways to lower the death rate but has not offset this action by lowering the birthrate.

United States population

The United States population continues to grow, although the rate of increase has been declining since 1957. Estimates made in 1966 indicated that the United States population might reach 360 million individuals by A.D. 2000. However, with the continued drop in growth rate, the population by the year 2000 is now expected to be about 266 million. The present United States annual population increase of 1 percent will double the number of people in 70 years. This is critical from a world point of view because the United States standard of living takes a high toll of important resources and produces a disproportionate amount of pollution. A baby born in the United States potentially places 30 to 50 times more stress on the environment than a baby born in India.

zero population growth

A few countries have reduced their population growth rates even more. Ireland has achieved a balanced birthrate and death rate and now has zero population growth. At least four nations, Great Britain, France, Sweden, and Japan, have an annual population growth of less than 1 percent. Barbados, an island in the Caribbean, has the lowest birthrate in Latin America —an annual increase of 1.3 percent. This is remarkable when compared with other Latin American countries. The populations of Mexico and Central American countries increase at annual rates ranging from 2.9 (Guatemala) to 3.7 (Costa Rica). Barbados has their same problems—high population density (500 per square mile) and low income ($440 annually per capita). Costa Rica has about the same per capita income and the same annual death rate—8 per 1000 individuals. The major difference lies in the annual birthrate: 21 per 1000 in Barbados and 45 per 1000 in Costa Rica. All Central American countries have birthrates of about 40 per 1000.

Costa Rica

Earth was "blessed" with a net gain of 73 million people in 1970. The greatest increases were in countries where people were already the hungriest—in Latin America and Asia. These two regions, plus Africa, contributed 86 percent of the world's population growth during the 5-year period from 1966 to 1970.

optimum population

Determination of the optimum human population size for Earth would involve evaluating subjective criteria which differ widely in various parts of the world. By using experimental animals, and thus simpler standards and fewer variables, an optimum population size can more easily be calculated. In rat colonies, populations in excess of the optimum result in abnormal behavior which is detrimental to the colony. Pregnant females build inadequate nests or none at all. The percent of aborted embryos increases. The mothers fail to care for young that are born alive. Male rats become homosexual or withdraw from interaction with the rest of the colony. Our newspapers are overburdened to report even a small fraction of similar behavior in man. On such a basis the United States long ago passed its optimum population size. On a more pragmatic basis—the ability of natural systems to clean up man's pollution—a population of 150 million in the United States might be much more desirable than our present 207.1 million.

Fig. 1-3. The expanding of urban areas can be accomplished and still leave some trees. *(Courtesy U.S. Department of Agriculture, Soil Conservation Service.)*

LIVING SPACE

Americans are still moving to the cities. Half of the counties that were sparsely populated continue to lose people. Half of the United States population lives within 50 miles of either the Atlantic or the Pacific coast. Two of every three Americans live in urban areas. Predictions indicate that, by the year 2000, 70 percent of the population will live on 10 percent of the land in 12 giant metropolitan areas. The Atlantic seaboard will reach a population density of 1000 people per square mile at this rate. Such a shift in population distribution compounds present problems—air pollution, traffic jams, noise, antisocial behavior, drug abuse, poor housing, and substandard schools. Plans for sound land use are essential to provide the urban dweller with something resembling a good life. Cities need protected green space around and within them (Fig. 1-3).

RECYCLING WASTES

Solid waste disposal overwhelms our cities and stipples the landscape with the pox marks of a sick society. No other nation throws away as much as the United States. Each American's share of discarded garbage and paper amounts to at least 1 ton per year.

automobile tires

Many of the things we throw away could be reused or turned into something useful (Figs. 1-4 and 1-5). Junk cars now provide more than half the steel for new cars. Seven million junked cars a year become available for recycling. Automobile tires can be ground up and added to asphalt paving. The resulting road surface is crack-resistant and lasts four times longer than conventional surfacing. Ground-up bottles can also be used in asphalt paving. Aluminum cans are worth $200 per ton as salvage, and 1 ton of garbage, under heat and pressure treatment, can yield 1 barrel of crude oil.

Fig. 1-4 (left) and Fig. 1-5 (right). The past approach to waste disposal and the current efforts to solve several problems by recycling, respectively. *(Both photographs courtesy Michigan Department of Natural Resources.)*

Solid Waste Management

Americans consume 35 percent of the earth's industrial raw materials while accounting for only 6 percent of the world's population. The use of these large amounts of raw materials results in enormous quantities of solid wastes. Until recently, we were not greatly concerned with environmental problems associated with the collection and disposal of trash, garbage, or other solid wastes. In a country with low population density and seemingly unlimited natural resources, the most convenient disposal method—usually an open dump—seemed adequate. There appeared no reason to reuse wastes since virgin materials were abundant and usually cheaper than reclaimed materials. Today, however, this view has been replaced by a genuine concern, not only for improved disposal methods but for the recovery and reuse of the valuable and often irreplaceable resources that form a large part of the discards of our high-production, high-consumption society.

The solid wastes produced in the United States now total 4.3 billion tons per year. Of this amount 360 million tons are household, municipal, and industrial wastes. In addition, there are 2.3 billion tons of agricultural wastes and 1.7 billion tons of mineral wastes. Of this annual total 190 million tons, or 5.3 pounds per person per day, are collected and hauled away for disposal at a cost of 4.5 billion dollars per year.

Most present disposal methods pollute either land, air, or water. Three-fourths of the dumps contribute to air pollution and half of them are so located that their drainage aggravates the pollution of rivers and streams. Almost all municipal incinerators are obsolescent in terms of today's needs and technology.

By 1980 it is expected that waste collection will amount to over 340 million tons per year, or 8 pounds per person per day. It is estimated, in fact, that our solid waste load is presently increasing at twice the rate of our population increase.

Today, a new concept of solid waste management is evolving. It assumes that man can devise a sociotechnological system that will wisely control the quality and characteristics of wastes, efficiently collect wastes that must be removed, creatively recycle wastes that can be reused, and properly dispose of wastes that have no further use.

Recycling and reuse of wastes holds considerable potential for solving the problems of waste disposal. If enough of these materials can be made reusable, those that cannot be recycled may not constitute an insurmountable problem. The Owens-Illinois plant in Portland, Oregon, has demonstrated a successful program for recycling glass. The general public in the Portland area brings 150,000 pounds of glass to the factory each week, and industry sends another 90,000 pounds. These 240,000 pounds of glass do not become litter, and the space that would have been consumed in the sanitary landfill is given an extension of time. Western Kraft of Albany, Oregon, has had similar success in recycling paper and cardboard. In 1968 it reclaimed 5000 tons. By 1970 reclamation reached 18,000 tons of material—enough to fill 500 boxcars or cover a football field to a depth of 60 feet. By 1973 they expect to reconvert 30,000 tons annually. This paper recycling involves the addition of recycled fibers (approximately 15 percent) to virgin fiber during the process of paper manufacturing. Metal containers represent another solid waste now successfully recycled.

continued

The burning of crop wastes such as straw and corncobs has been strongly criticized by ecologists. A new process being developed by the U.S. Department of Agriculture converts these wastes to feed by adding sodium hydroxide and heating under pressure. These waste products are high in carbohydrates but are poorly digestible. The new process makes them more digestible for cattle and sheep. Straw, for example, has a digestibility of 30 to 40 percent before treatment. When heated with water at a pressure of 400 pounds per square inch, the cooked straw becomes 50 to 60 percent digestible. The digestibility can be increased to 70 to 80 percent when sodium hydroxide is added to the water and the mixture heated at a pressure of 400 pounds per square inch.

paper

We now reuse 20 percent of the annual 60 million tons of paper produced, but we could reprocess more. Germany, short on trees for paper pulp, reprocesses 35 percent of its annual paper production.

The recycling of waste is essential to our survival. Available landfill space is rapidly disappearing as rural communities refuse to become garbage pits for city pollution. Incineration cannot keep pace with the pile-up and destroys materials that should be converted to usable forms. Then, too, incineration constitutes a major source of air pollution.

The key to recycling is demand for the product. Recycling will increase only as the market for recycled materials increases. The 25 percent of our national product that is recycled involves very little in the way of household waste. At present, the U.S. Environmental Protection Agency is involved in studies of ways to stimulate recycling.

SOIL

The United States has some of the best soils in the world, mostly because it has not been farmed as long as other areas. No other country has as high a portion of its soil in classes 1 and 2 (see Chapter 4 for soil classification).

dust storms

Soil deterioration has been a recognized problem much longer than air pollution and water quality (Fig. 1-6). The dust storms of the 1930s triggered an avalanche of research on soil conservation. In the United States we farm about 60 percent of the land that could be farmed with presently available equipment and good management practices. At present, about 336 million acres are under cultivation. Each year 2 million acres are taken out of cultivation. Half of this is used for better outdoor living: wildlife refuges, parks, and other recreational areas. The other half is buried under concrete or water. Urban development consumes 420,000 acres each year, and a like amount goes for reservoirs and flood control. In many cases the latter serves a recreational function. Finally, 160,000 acres are used for

Fig. 1-6. A few years ago these two fields in Madison County, North Carolina, were equally gullied. The owner of the field at the left entered a Tennessee Valley Authority test demonstration program and by proper fertilization produced cover which checks erosion and permits livestock production. (*Courtesy Tennessee Valley Authority.*)

loss of farmland

airports and roads. However, 1 million acres of new cropland must go into production each year or be offset by increased yields to feed the annual increase in the United States population. These 2 million people, as adults, require the meat from one million hogs and 400,000 steers, the milk from one-half million cows, and the bread from 133,000 acres of wheat. Although we have a surplus of land at the present and do export large quantities of food, the land is under the stress of two forces. Expansion of housing areas, airport construction, and new roads take land out of cultivation each year. At the same time more and more cultivated land is needed to feed America's expanding population while maintaining our food exports to hungry nations. Figure 1-7 shows how much land needs better management.

We now export the crops from one-fourth of our cultivated land. In the future this figure may increase sharply. Not many observers believe that the world can remain half starving and half well-fed.

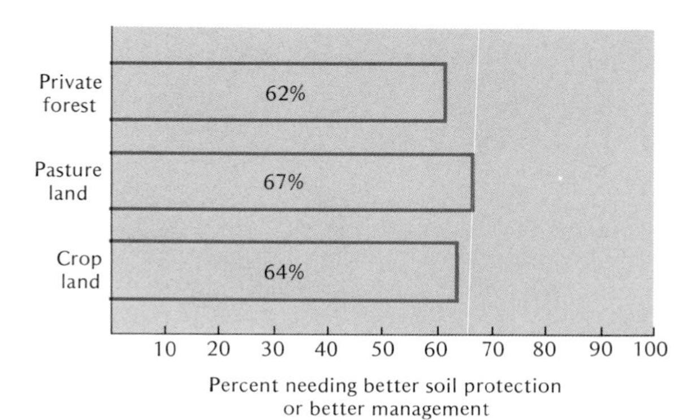

Fig. 1-7. Percent of land that needs better management. (Based on data from *National Wildlife,* 9(6):32, 1971.)

TIMBER

During 1970 the timber harvest on U.S. Forest Service lands was 87 percent of the allowable cut. Timber growth appears to have exceeded timber cut for this year. Better sawmill practices have resulted in more usable lumber from each log, but we should be able to use more that is now considered waste (slabs, edgings, and sawdust).

A federal goal of 3 million new housing units each year will require 7 billion board feet more from the national forests than is now being cut (Fig. 1-8). Can these demands, which amount to 60 percent more lumber from all sources, be met without destroying the multiple-use features of our public lands? Do we have the timber capacity for sustained yields at these levels?

fire

Fire continues to take a high toll of timber; 1970, in which 3.2 million acres burned, was the worst of 36 consecutive years. Replanting burned and cut-over areas is now about 5 million acres behind schedule.

tree breeding

On the positive side, new genetic stocks of trees are under development and some show excellent results. New cottonwoods grow 8 to 10 feet per year, twice the normal rate. New strains of pine mature in 25 years instead of 40. One hundred million of these pine seedlings, now in nurseries, will grow into trees 50 to 100 percent faster.

Fig. 1-8. Pulp wood stacking operation in Michigan. (*Courtesy Michigan Department of Natural Resources.*)

Fig. 1-9. Abandoned mine shaft houses such as this one in Michigan's upper peninsula remind us that mineral resources are depleted in many areas.

MINERALS

The energy squeeze threatens to reduce people's demands to their actual needs. In 1900 coal furnished 89 percent of the energy used in the United States; oil and natural gas supplied 7.8 percent. In 1971 coal accounted for only 28 percent, as energy from gas and oil rose to 68 percent of the total. Nuclear power and other sources amounted only to 4 percent.

The standard of living enjoyed by Americans boosts the energy-use figure. We make up 6 percent of the world's population but consume 30 percent of the world's energy output. The per capita share of the United States use of oil is 2310 gallons. Our net population gain each year is 2 million, but each year we add 4 million cars.

oil

Several other minerals important to our economy are in short supply (Fig. 1-9). Present use rates will see the known reserves of copper, lead, petroleum, uranium, tin, and zinc exhausted within 4 decades. The estimated reserves depend, however, on what is economically profitable to exploit. Then, too, substitution may play an important role. Recycling, forced upon us for other reasons, may relieve natural shortages.

WATER

The Little Deschutes River rises in the lava beds on the eastern slopes of Oregon's Cascade Mountains. It joins the Deschutes and together they run clear and cold through Bend, Oregon. The bottom of the riverbed is visible at depths of 6 to 8 feet. Some of Florida's short rivers that arise from underground caverns, such as the Wakulla and the Florida section of the St. Marks, carry so little sediment that they actually erode the floodplain during high water; most rivers drop sand and silt when they overflow their banks. Visibility in most major rivers of the United States, particularly those running through farmland, is 1 to 2 feet at low-water levels and essentially zero during flood stages.

river sediment

For more than 2 centuries we have been dumping wastes into our waters. Streams, rivers, and lakes have served as reservoirs for agricultural wastes, industrial chemicals, raw and partly treated sewage, oil, garbage, and trash. The problem of excess organic material in natural waters lies in the amount of dissolved oxygen used in the breakdown of the organic matter. Industries, especially those involving petroleum, steel, paper production, and organic chemicals, discharge three to four times as much oxygen-demanding waste as all home sewers combined. Industrial waste comprised 60 percent of water pollution in 1970 and 65 percent in 1971, and the industrial demand for water is increasing.

oxygen demand

feedlots

The raising of large numbers of cattle on giant feedlots may be a solution to the overgrazing of the West, but the treatment of feedlot animal waste as sewage, rather than as fertilizer (Fig. 1-10) to be returned to the land, compounds water pollution problems. Figure 1-11 shows the equivalency of animal and human waste.

oil spills

Giant ocean tankers, averaging a total of 100 oil spills each year, account for about 10 percent of the total oil spills reported. The cases that go unreported are unknown, of course, but a new law imposes a $10,000 fine for not reporting an oil spill. On a world basis 5 million tons of oil were accidently or intentionally dumped during 1971. This is enough to provide 16 gallons of gasoline for every car in America.

Less than one-third of the 1700 million gallons of water industry uses annually receives treatment before discharge. Application of today's technology could remove 85 percent of the pollutants. By 1980 we may need to remove 95 percent of the pollutants.

Several hopeful signs indicate that the water quality index may have reached a low point in 1971 and that we can look for gradual improvement in the years ahead. Many new sewage plants have been constructed and others improved. Somewhat belatedly and reluctantly industry has started to clean up its share of water pollution. Plant age and size are no longer accepted as excuses for pollution.

public drinking water

The safety of public drinking water has generally been assumed. Out-of-town visitors may not prefer the sulfur water of some areas but rarely do they question the bacterial count or the lead content. But how safe is our drinking water? Figure 1-12 shows the results of a survey of 969 city water systems in the United States. Nearly 500 were supplying water of inferior quality; the water of 87 was actually dangerous to drink.

Fig. 1-10. This newly constructed disposal lagoon handles 5000 gallons a day, the waste from 2500 hogs. The lagoon has ½ acre of surface area and a depth of 8 feet. The liquid is pumped periodically and spread on cropland. *(Courtesy U.S. Department of Agriculture, Soil Conservation Service.)*

Fig. 1-11. Equivalent waste production of farm animals and people.

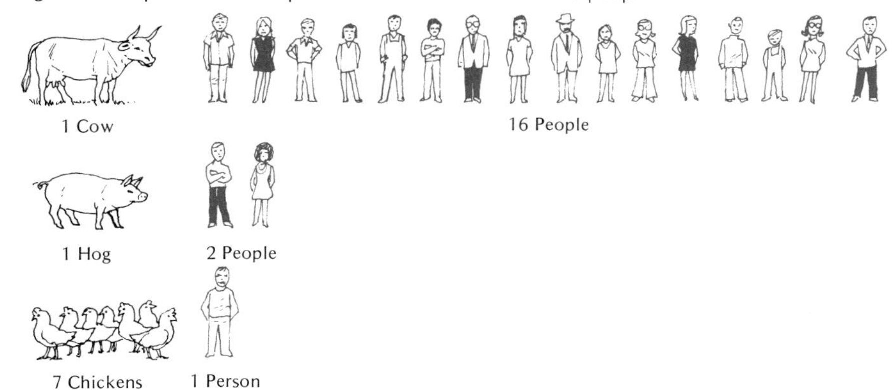

No annual inspection

Poorly trained water plant operators

Inferior-quality water

Bacteria or chemicals exceed safe limits

Potentially dangerous

Fig. 1-12. The safety of public drinking water. (Based on data from *National Wildlife*, 9(6): 25–40, 1971, on 969 public water systems.)

AIR

Transportation contributes 42 percent of all air pollution. The private automobile is responsible for most of the carbon monoxide, nitrogen oxides, and hydrocarbons. The goal for automobiles of no more than 10 percent of the 1970 emission level by 1975 is claimed to be an unrealistic standard for the short time allowed. Even if it is achieved, more than 1 million older cars will still be on the road.

Coal burning places second in fouling the air, principally by producing sulfur oxide and particulate matter. Some government agencies have

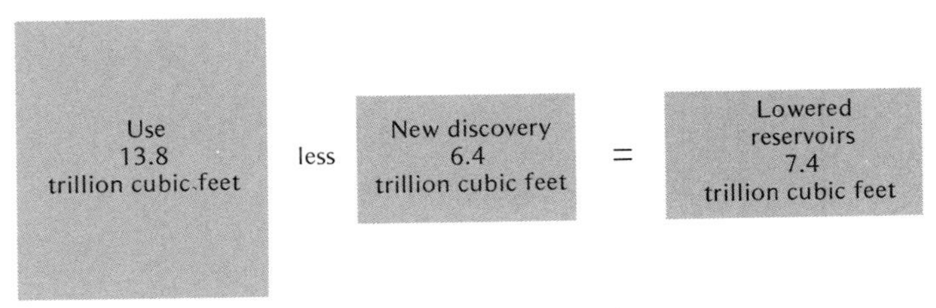

Fig. 1-13. Natural gas use versus explorations for 1971.

coal versus natural gas

recommended that cities switch to natural gas, which is low in sulfur. This would require a 300-percent increase in natural gas consumption in New York City and enough of an increase in other areas to raise the national need by 15 percent. Of all our mineral resources, natural gas is in the shortest supply and may be rationed by 1976. We now consume three times as much as in 1950. The quantity used in 1971, 13,800 billion cubic feet, was more than twice the volume of new sources found during the year (6400 billion cubic feet). The effect on reserve supplies is shown in Fig. 1-13. At the present use rate, the known reserves will be exhausted by 1984. Rationing could extend this date and higher prices may encourage more exploration. Nevertheless, we may be very near the end of our natural gas supply. More than likely, methane, the principal ingredient in natural gas, will be manufactured from coal.

carbon monoxide

Carbon monoxide levels in congested traffic are critical. A day of exposure to 10 parts per million of carbon monoxide slows a person's reactions and produces mental dullness. This degree of exposure increases the probability of auto accidents. Heavy traffic frequently results in carbon monoxide concentrations of 80 to 100 parts per million. A policeman directing traffic at a busy intersection may have his judgment impaired by carbon monoxide poisoning.

Death from respiratory diseases has increased by 50 percent for the nation as a whole and is especially high in heavily polluted areas (Fig. 1-14). Emphysema, a rare disease 30 years ago, now kills more than 25,000 persons per year in the United States.

Fig. 1-14. Air quality control is badly needed in all major cities. Factories, such as this sugar processing plant at Billings, Montana, mean jobs and prosperity but also pollute the air and make heavy demands on the water supply. The Great Western sugar plant at Billings uses 10,000 gallons of water per minute in processing sugar beets. One hundred train-car loads of beets can be processed in a 24-hour period. *(Courtesty Department of the Interior, Bureau of Reclamation.)*

On the positive side, combustion of all kinds has not reduced the oxygen content of the air. The results of a 3-year study show that the amount of oxygen in the air is the same now as it was more than 60 years ago. Photosynthesis in green plants returns oxygen to the air. The amount of carbon dioxide in the air is often the limiting factor for the rate of photosynthesis. An increase in carbon dioxide in the air, because of increased combustion, might have the compensating effect of increasing the rate of return of oxygen to the air.

WILDLIFE

Pesticides, poison bait, loss of habitat, and overhunting continue to put pressure on endangered species. Of these, loss of a place to live ranks first. Bulldozers and draglines "improve the site" on about 1 million acres per year, to the detriment of wildlife.

DDT

Pesticides, particularly dichlorodiphenyltrichloroethane (DDT) and dieldrin, continue to plague birds of prey. The former causes thin eggshells which break before hatching; the latter destroys the central nervous system. Eagles have been seen to fall during flight, wracked by convulsions and tremors, and die within 3 hours from dieldrin poisoning.

pelicans

Thousands of pelicans still live along the California coast, but only one chick hatched during the 1970 breeding season (Fig. 1-15). In Louisiana pelicans (the state bird) were restocked but might be expected to succumb to the same fate that destroyed previous flocks.

osprey
Cooper's hawk

Ospreys show a 12-percent annual decline (Fig. 1-16), and Cooper's hawks lose ground at a rate of 25 percent a year. The canvasback duck population has dropped by 25 percent in 5 years.

condor

The California condor, with a wingspread of 9 feet, is North America's largest bird. Once widespread in the Southwest, it now survives only in the Los Padres National Forest in California. Sixty to 70 condors still live, but

Fig. 1-15. DDT concentrated in pelican food has severely reduced successful reproduction. Eggshells are too thin to survive incubation. *(Courtesy Department of the Interior, Bureau of Sport Fisheries and Wildlife.)*

Fig. 1-16. Similar to several other kinds of fish-eating birds, the osprey is a victim of thin eggshells as a result of the effect of DDT on calcium metabolism. *(Courtesy Michigan Department of Natural Resources.)*

only 12 breeding pairs nest. Fifty percent fewer condors were sighted in the 1970 census than in previous years. This may have been due to the weather and not to an actual decline in population, however. The United States Gypsum Company is currently seeking to obtain a stripmining lease in the Upper Sespe watershed of the Los Padres National Forest in California. Phosphate ore in the area might support a stripping operation for 100 years —the ore is plentiful. Access roads and a processing plant employing 60 to 70 people would be involved. The problem lies in the location of the mining site—right in the center of all known nesting sites of the California condor. Successful breeding of the condor depends on complete isolation and the absence of disturbance.

mercury

Mercury has long been an ingredient of many pesticides. Methylated mercury is used for fungus control in such crops as wheat, oats, rye, barley, corn, rice, linseed (flax), sorghum, cotton, millet, various fruits, nuts, cucumbers, pumpkins, squash, and potatoes. The United States enacted pesticide regulations in 1947 placing mercury in the zero-tolerance category. This means that no residue is allowed on food materials shipped interstate. Considerable work has been done in Sweden on methyl mercury, the most deadly of the mercury compounds. Tests made in 1964 revealed that more than half of a sample of 200 game birds had mercury contents in excess of 2 parts per million. In several other tests game birds having 200 parts per million were recorded. Over a 100-year span, 1865 to 1965, the mercury level in bird feathers increased by a factor of 10.

The average amount of mercury in rainwater is 0.01 parts per billion. Rain falling through polluted urban air may contain 48 times this amount. With mercury in game birds, birds of prey, swordfish, tuna, and freshwater fish, how safe is man?

whales

The prospects of some wildlife may have taken a turn for the better. Eight species of whales have been declared endangered. Importation of whale products into the United States has been stopped. A United States commercial whaling fleet no longer exists. More land is being set aside for wildlife refuges. The one-half million acres added in 1970 brought the total to 30,725,000. Artificial insemination of birds that do not breed in captivity has increased the numbers of some critically low species. "Stealing" eggs from natural breeding grounds and hatching them in incubators has proved successful at Patuxent Wildlife Research Center in Maryland. This offsets losses that occur in the wild and has also been used to establish a captive population for breeding to restock natural areas. Twelve whooping cranes now live at Patuxent, but numbers sufficient for restocking are still a few years away. The natural population of whooping cranes, which breeds in Canada and winters in Louisiana, is now at 57 from a low at one time of 15.

whooping cranes

black duck

The black duck, seriously declining along the East Coast for numerous reasons, suffers a high mortality from predators because it builds its nest on the ground. Attempts are being made to condition black ducks to nest in boxes off the ground. If this can be accomplished and if the young reared in such boxes return to nest in them, the decline in the black duck population might be reversed.

Much of the work to save endangered species is a gamble, but some is paying off. Man has the opportunity to bring back from the edge of extinction species that his own undisciplined acts in the name of progress have so nearly exterminated. What he learns on this mission of mercy may also play a role in his own survival.

Suggested Readings

Ehrlich, P. R. *The Population Bomb*. New York: Ballantine Books, 1968.

Ehrlich, P. R. and J. P. Holdren. "Impact of Population Growth." *Science*, 171 (3977):1212–1217, 1971.

Goldman, M. I. "The Convergence of Environmental Disruption." *Science*, 170 (3953):37–42, 1970.

Hinrichs, N. *Population, Environment, and People*. New York: McGraw-Hill, 1970.

Moncrief, L. W. "The Cultural Basis for Our Environmental Crisis." *Science*, 170 (3857):508–512, 1970.

Paddock, W. and P. Paddock. *Famine 1975*. Boston: Little, Brown, 1967.

Rosen, W. G. "The Environmental Crisis: Through a Glass Darkly." *BioScience*, 20(22):1209–1211, 1216, 1971.

Strohm, H. "EQ Index." *National Wildlife*, 9(6):25–40, 1971.

The addax is a unique animal. It is the only large mammal that derives all its moisture from the plants it eats. Formerly, it ranged throughout the Sahara Desert from Algeria to the Nile River. The addax became extinct in Egypt prior to 1900; it has been exterminated in southern Algeria and Tunisia. Herds were machine-gunned by Italian military patrols during World War II. The last refuge of the addax is a waterless, uninhabited area in the western Sahara. Here it is hunted by three nomadic tribes which make forays into the desert. *(Courtesy Lion Country Safari, Laguna Hills, California.)*

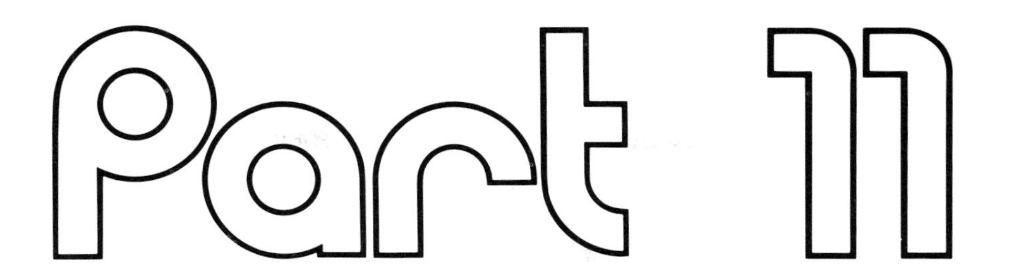

ESSENTIALS

Burning solid waste in the Ralls, Texas, area. This form of waste disposal, a common practice in many small communities, contributes to air pollution. Sanitary landfill is a better solution. *(Courtesy U.S. Department of Agriculture, Soil Conservation Service.)*

Air

PHYSICAL ASPECTS OF AIR

Composition of Air

The gaseous components of the air include two elements (nitrogen and oxygen) and one compound (carbon dioxide) that are involved in life support cycles.

Cellular respiration characterizes all living things. For land animals the atmosphere must serve as the oxygen source for the oxidation processes involved in respiration. In daylight photosynthesis in green land plants produces more oxygen than they use, but at night they too depend on atmospheric oxygen.

composition

Critical problems arise because living organisms are not automatically shielded from other materials in the air that may be harmful. The composition of the air has changed drastically since the time the earth's solid crust was first formed. Most of this change has been brought about by the activities of living organisms. Methane and ammonia, so abundant in the early atmosphere, are now present only as traces. They are part of a fraction (0.02 percent) that includes hydrogen, neon, xenon, krypton, helium, radon, ozone, hydrogen sulfide, sulfur dioxide, and other compounds. Oxygen, which now accounts for 20.93 percent of the air (by volume), has reached this concentration largely through the photosynthetic activity of plants. The remaining major component of air is nitrogen. It accounts for 78.09 percent of our atmosphere and is relatively inert in its molecular form (N_2). Plants use nitrogen, combined with hydrogen in ammonia or with oxygen in nitrates, by coupling it with carbohydrates to synthesize amino acids and then proteins, the building blocks of living systems.

At present, man's activities are bringing about local and worldwide changes in the composition of the atmosphere, which have both short- and long-term effects. Some obvious effects are seen when large city schools periodically curtail or eliminate physical education classes to avoid deep breathing or when the death rate increases during a smog-producing thermal inversion. Four thousand deaths were attributed to the London smog of December 1952.

An understanding of the air crisis confronting man depends upon our understanding the properties of air and the organs of the human body that utilize air. The physical properties diffusion, pressure, and temperature have direct bearing on problems of air pollution.

Diffusion

diffusion

Individual molecules of air (and all other substances, for that matter) are in constant motion. The direction of this motion is random and changes with each collision between molecules. The actual velocity of a molecule depends on its mass and the temperature and pressure under which it exists at a given moment. A molecule of water at sea level and at 81°F, if uninterrupted by other molecules, travels along a straight line a distance of 650 meters per second. Roughly, this amounts to 1450 miles per hour. Approximately 10 billion other molecules of water are in the way of this 1-second journey, however, and they also are in rapid motion. With each collision the direction of each molecule changes. This simply produces an erratic, nondirectional motion. Such motion in any mixture, gas or liquid, tends to distribute the components evenly throughout. This is diffusion. Diffusion plays a role in the distribution of air pollutants, but gravity, wind, convection currents, and rain exert a more pronounced effect.

Pressure

pressure

Gravity affects molecules of air in spite of their supposed random motion. More molecules of air per unit of volume are present near the earth's surface than at higher altitudes. At sea level a column of air 1 inch square and extending to the upper limit of the atmosphere weighs 14.7 pounds. This is also the weight of a 1-square-inch column of mercury 29.95 inches (76 centimeters) high and a 1-square-inch column of water 33 feet high. One atmosphere of pressure, then, is 14.7 pounds per square inch. Half the material of the earth's atmosphere occurs below 17,000 feet and three-fourths of it below 34,000 feet.

Air pressure has practical applications in survival as well as in everyday living. The total pressure equals the sum of the individual pressures of each component of the air. Of the 76 centimeters of mercury total air pressure at sea level, 23.14 percent (the percent by weight rather than by volume) is due to oxygen. Oxygen therefore accounts for a pressure equivalent to the weight of 17.59 centimeters of mercury at sea level.

One factor, a most important one, lies in the relationship between diffusion rate and pressure. To achieve the same rate of diffusion of oxygen into the bloodstream at an altitude of 17,000 feet, the oxygen pressure still must be 17.59 centimeters of mercury. Because the total air pressure at 17,000 feet is 1/2 atmosphere, an oxygen pressure equal to 17.59 centimeters of mercury can be achieved without artificially increasing the air pressure if the air mixture contains 64.28 percent oxygen. At 34,000 feet, however, the total air pressure is only 19 centimeters of mercury. At this altitude 92.6 percent oxygen is required to achieve an oxygen uptake equivalent to that at sea level.

Temperature

temperature

Temperature indicates the degree of hotness or coldness compared with an arbitrary scale. It is a measure of the average velocity of the molecules of a substance, the average kinetic energy. Heat refers to the total kinetic energy of a substance and thus is a measure of the quantity of energy.

The temperature range believed to exist within the universe extends from absolute zero (−273.15°C, −459.67°F) to millions of degrees. Some forms of life, primarily microorganisms in a dormant state, can survive −200 to +100°C. The amount of moisture makes a difference in heat tolerance. Certain mosses that grow on exposed rock and sand can survive temperatures of 100 to 115°C. In contrast, those that live in cool, damp habitats generally die at temperatures above 32 to 35°C.

Temperature fluctuation may be more stimulating than constant temperature. The eggs, larvae, and pupae of the codling moth develop 7 to 8 percent faster under conditions of variable temperature than when held at the average temperature of the fluctuations. Results of temperature variation are even more pronounced in grasshopper nymphs (12-percent increase in growth rate) and grasshopper eggs (38 percent faster development). The stimulatory effect of temperature variation on plants, animals, and with human cultures found in the Temperate Zone is a well-established ecological principle. This may explain why some laboratory experiments are not directly applicable to field conditions. In a deliberate effort to control variable factors and test only one of these factors at a time, many laboratory experiments are conducted at constant temperatures.

Considerable temperature differences may exist within a very short vertical or horizontal distance. Thus microclimates exist which allow

microclimate

plants and animals to survive under what would appear to be intolerable conditions. Small rodents known as voles (*Microtus*) live through the severe winters of the Alaskan tundra by staying under the snow. When the air temperature immediately above the surface of the snow is −70 to −60°F, the temperature 2 feet down, at the soil surface, may be +20°F. This difference of 80° or more produces a local climate typical of latitudes much farther south. The voles have relatively thin fur compared with animals that live on top of the snow.

THE HUMAN RESPIRATORY SYSTEM

The most direct effect of air pollution on the body involves the organs of breathing. A review of the human respiratory system is included here to provide a basis for understanding human responses to specific air pollutants and the hazards to health that are involved.

The exchange–transport mechanism for oxygen and carbon dioxide is based upon the physical principle of diffusion, regardless of the type of respiratory organ or system involved. The processes of aerobic metabolism within a cell create a decrease in oxygen tensions (or pressures) and an increase in carbon dioxide tensions. As blood containing relatively high oxygen and low carbon dioxide tensions circulates near a cell, an exchange of these gases occurs because oxygen diffuses from the region of its higher partial pressure (the blood) into the cell, and carbon dioxide moves from the region of its higher partial pressure (the cell) into the blood. As blood then circulates through the organ where the exchange of oxygen and carbon dioxide with the atmosphere occurs, oxygen from the atmosphere diffuses into the blood, and carbon dioxide from the blood diffuses into the atmosphere through the same pressure gradient that caused the exchange between the cell and the blood to occur. *Respiratory pigments* (such as *hemoglobin* in the erythrocytes of man) are primarily responsible for the transportation of oxygen by the bloodstream. They have a remarkable affinity for oxygen which enables blood to transport about 60 times as much oxygen as plasma, which holds the gas only in simple solution.

hemoglobin

Mammals, birds, reptiles, and adult amphibians possess lungs and, in addition, passageways through which air passes to and from the lungs. These passageways consist of the *nasal cavity, pharynx, trachea, bronchi,* and *bronchioles* (Fig. 2-1). As an animal inhales or draws air into its lungs, the air passes through the passageways into the microscopic air sacs or *alveolei* of the lungs. As the animal exhales or forces air from its lungs, the air passes back to the outside through these same passageways. Air enters the nasal cavities by way of the nostrils or *external nares*. As it passes through the nasal cavity, it is warmed, moistened, and filtered of dust and foreign particles by the mucus membrane which lines the nasal cavity. The *receptor cells* for the sense of smell are also located in the mucus membrane of the olfactory epithelium and, if chemically stimulated, send nerve impulses to the brain to be interpreted as smell. The nasal cavities communicate with the pharynx through two *internal nares*. The pharynx is a com-

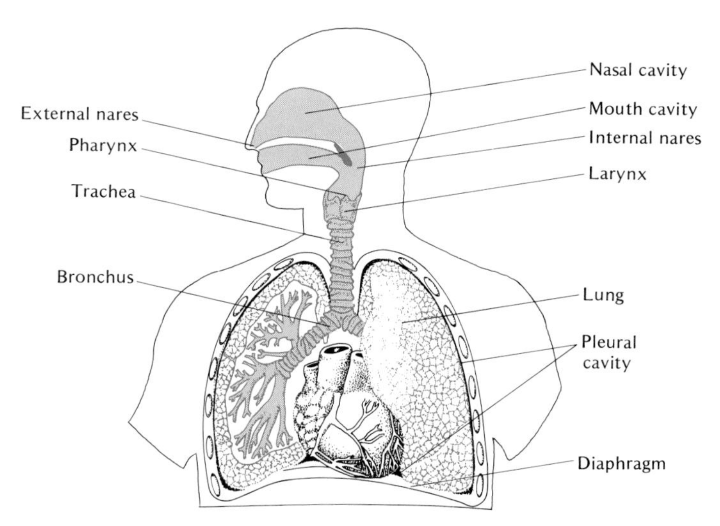

Fig. 2-1. The respiratory system of man.

mon passageway for both food and air, food moving from the pharynx into the esophagus, and air being directed into the larynx through an opening called the *glottis*. The glottis is normally open, allowing air to pass freely into the larynx. When we swallow, the larynx moves up so that the glottis is closed by the shelflike *epiglottis,* thus preventing the entry of food and water into the larynx and trachea. The *larynx,* located at the upper end of the trachea, is a chamber surrounded by a complex of cartilages. The *vocal folds,* one on either side, are stretched across the opening in the larynx. When the vocal folds are vibrated by the passage of air currents between them, sound results. The *trachea* or windpipe is a tubular structure about 4 to 5 inches long in man and has a diameter of about 1 inch. It is located in front of the esophagus and extends from the larynx through the neck and into the chest cavity, where it divides into the right and left bronchi or *bronchial tubes*. The wall of the trachea contains a series of C-shaped *cartilage rings* which prevent the trachea from collapsing and thus maintain an open passageway for air at all times. The right and left bronchi enter the right and left lungs, respectively, and begin to branch almost immediately into progressively smaller and smaller passageways. Although the larger branches of the bronchial tubes possess cartilage rings, the rings disappear as the branches become smaller. The small tubules which lack the cartilage rings are called *bronchioles* and are less than 1 millimeter in diameter. Each bronchiole continues to branch and become smaller until it finally opens into one of the microscopic air sacs or *alveoli* that constitute the functional units of the lungs (Fig. 2-2). The total number of alveoli in both lungs of a human being is estimated at nearly 1 billion. In an adult human these small sacs have about 100 square meters of exposed surface, or 50 times the body surface. Because the alveolar wall is composed of a single layer of squamous epithelial cells surrounded by a rich network of capillaries, it is well suited to function as a respiratory membrane.

trachea

bronchi

As animals developed lungs during the course of evolution, they also developed the ability to *breathe,* or to pull air into the lungs and to expel it. These mechanical processes are termed *inspiration* and *expiration,*

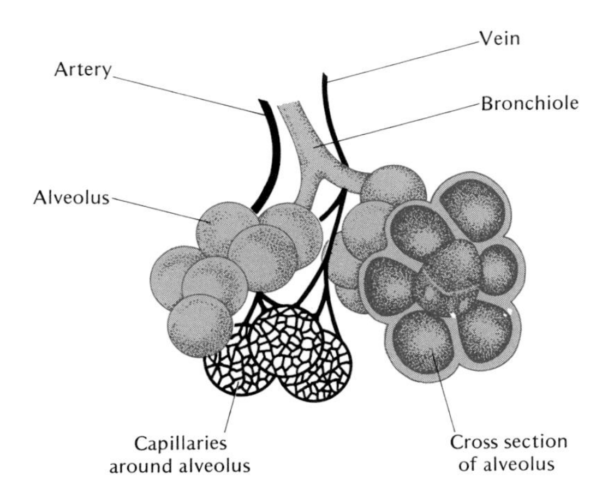

Fig. 2-2. Diagram showing alveoli in relationship to bronchioles and capillary networks.

Filtering the Air

Except for irritation to the eyes and skin, air pollutants must enter the body to cause harm. The respiratory tract provides a common route of entry. The normal breathing rate for man (15–20 times per minute) circulates about 3600 gallons of air through the lungs each day. A gallon of urban air in many areas contains as many as 8 million particles of foreign matter. This amounts to 28 billion particles per day entering the respiratory system. What defense, then, does the body have against these particles?

Nose hairs, the first obstruction to dirt in the air, collect dust particles and debris larger than 1/2000 inch in diameter. Mucus secreted by the lining of the nasal cavity removes almost all remaining particles larger than 1/5000 inch. The mucus also absorbs some gases and traps bacteria. This loaded mucus moves to the back of the throat and is swallowed or discharged.

Air then enters the trachea (windpipe) which is coated throughout its entire length with mucus and is lined with millions of cilia. The mucus and cilia screen particles down to 1/10,000 inch. Beating of the cilia (about 12 times per minute) moves the mucus and its pollution load to the top of the windpipe where it is swallowed or discharged.

Particles with a diameter of less than 1/10,000 inch enter the air sacs (alveoli) of the lungs. Those under 1/100,000 inch are absorbed into the bloodstream. Those that are unabsorbed (between 1/10,000 and 1/100,000 inch) remain in the alveoli where they may be engulfed by dust cells (white blood cells which trap dust and bacteria) and then expelled. However, dust cells are immobilized by nitrogen dioxide (from automobile exhausts) and the action of cilia that expels mucus is inhibited by nicotine. Also, dust cells cannot handle coal dust, soot, mineral dust, asbestos, metal, fiber glass or synthetic fibers. These become wrapped with fibrous connective tissue and form minute cysts. As a result, the walls of the air sacs thicken and oxygen absorption decreases.

respectively, and they occur at a rate of about 15 to 20 times per minute in an adult human at rest. Breathing is essentially involuntary; that is, it is performed automatically and usually unconsciously. Since the process depends upon the contraction of the diaphragm and the intercostal muscles, it is under the control of the nervous system. Breathing is possible because the lungs are located in a closed cavity, the size of which can be increased or decreased through muscular action. In man, as the *intercostal muscles* and the muscles of the *diaphragm* contract, the size of the *chest cavity* is increased and the inside pressure is therefore decreased. Because this decrease results in a pressure lower than atmospheric pressure, air rushes into the lungs. As these muscles relax, the size of the chest cavity is decreased, the air pressure within the lungs is increased to a pressure greater than atmospheric pressure, and air rushes to the outside. The pressure changes that occur within the chest cavity and the interior of the lungs are quite small, usually amounting to no more than a pressure of about 3 millimeters of mercury.

rate

The average amount of air drawn into the lungs during inspiration at any one time is about 500 milliliters, and an equal amount is exhaled. Of this air only about 350 milliliters actually reach the lungs, since approximately 150 milliliters remain in the air passages leading to the lungs. Breathing is not an emptying and filling of the lungs but is instead a partial emptying and a partial refilling, as some air remains in the lungs even after forced expiration.

capacity

The only essential difference in composition between inspired and expired air is a decrease in the oxygen concentration from approximately 21 to 16.3 percent and a corresponding increase in the amount of carbon dioxide to approximately 4.5 percent. Thus almost 20 percent of the total amount of oxygen in the air is removed for use during intracellular respiration.

TABLE 2-1
Some Sources of Air Pollution[a, b]

Source	*Carbon monoxide*	*Sulfur oxides*	*Nitrogen oxides*	*Hydro-carbons*	*Partic-ulate matter*	*Total*
Combustion of fuel						
Transportation	64	1	8	16.6	1.2	90.8
Other fuel use	1.7	24	10	1	8.9	45.6
Mills (pulp, paper, iron, steel)	2	9	3	1	3	18.0
Other industrial processes	7.7	[c]	[c]	3.6	4.5	15.8
Solid waste disposal	8.8	1.1	1.6	2.6	2.1	16.2
Miscellaneous	15.1	[c]	[c]	4.9	6.5	26.5
Total	99.3	35.1	22.6	29.7	26.2	212.9

[a] Quantities expressed as millions of tons per year, in the United States.
[b] Based on data from P. R. Ehrlich and A. H. Ehrlich, *Population, Resources, Environment,* San Francisco: Freeman, 1970; and from other sources.
[c] Less than 1 million tons; ignored in totals.

POLLUTION OF THE AIR

Fig. 2-3. Forest fires contribute to air pollution. *(Courtesy Michigan Department of Natural Resources.)*

Table 2-1 lists some sources of the pollution affecting the air we breathe. The orange sky over Gary, Indiana, the white sky over Los Angeles, California, the soot on the windowsills of Ohio Valley homes, and the labored breathing of an emphysema victim all tell us that something is wrong. Virtually all large cities throughout the world have serious air pollution problems.

Certain of the data shown in Table 2-1 warrant a closer look. The combustion of fuel accounts for most of the air pollution—about 64 percent. As we exhaust the readily accessible fossil fuel reserves and come to realize that oil is more valuable for something besides burning, cleaner air may be an inevitable result.

Quantities beyond comprehension have a certain shock value in the propaganda campaigns employed to direct human action. In emergencies—air pollution may well be one—this approach might be justified. The reaction to the suggestion that 213 million tons of poisons (Fig. 2-3) are added to our air each year, however, could stimulate responses that would create more problems than they solve. The 213 million tons amounts to about 1 ton per year for every person in the United States—year after year. An alert student might ask, however, how this quantity is diluted by the body of air over the United States.

dilution

The air mass over the continental United States weighs about 3 quadrillion tons—3×10^{15}. If the pollutants were evenly dispersed throughout this air, each 15 million tons of air would have 1 ton of pollutant in it. Stated in terms more often applied to pollutants, the pollutants we add to the air amount to less than 1 part per million per year (0.067 part per million).

This reassuring analysis does have a basic assumption which should be checked—"If the pollutants were evenly dispersed. . . ." The contrast between clean air and that shrouding our cities strongly suggests that complete mixing does not occur. Further investigation reveals that little of the surface air rises beyond 12,000 feet and many pollutants remain within the first 2000 feet (Fig. 2-4). Manmade and natural barriers reduce circulation. These facts help explain the commonly observed air pollution concentrations in the 10 to 50 parts per million range and occasionally at much higher levels.

Air pollution is not confined to cities. Smog has been seen in arctic regions and over the oceans. Man has been able through progress to foul the air of the entire planet. As more and more countries strive for increased industrialization, the problem may become insolvable. Even now, polluted air corrodes steel, blackens paint, rots windshield wipers, disintegrates synthetic fibers, obscures the landscape, damages crops, and kills people.

Fig. 2-4. Dust pollution picked up from a beach by winds of 35 miles per hour. *(Courtesy U.S. Department of the Interior, Bureau of Reclamation.)*

Thermal Inversions

thermal inversions

Under conditions that may be described as normal, the temperature of the air decreases with increased altitude to a height of 10 kilometers (5.4 nautical miles). Over this range the temperature drops at a fairly uniform rate, 6.8°C per kilometer (12.5°C per nautical mile). From this point at 10 kilometers (the lower edge of the tropopause), the temperature is more or less constant to a height of 30 kilometers (16.2 nautical miles). At lower elevations, however, temperature inversions take place which drastically affect the underlying atmosphere. An inversion occurs when a warm layer of air, often about 2000 feet above the countryside, is sandwiched between cooler air masses (Fig. 2-5). The warmer air layer greatly

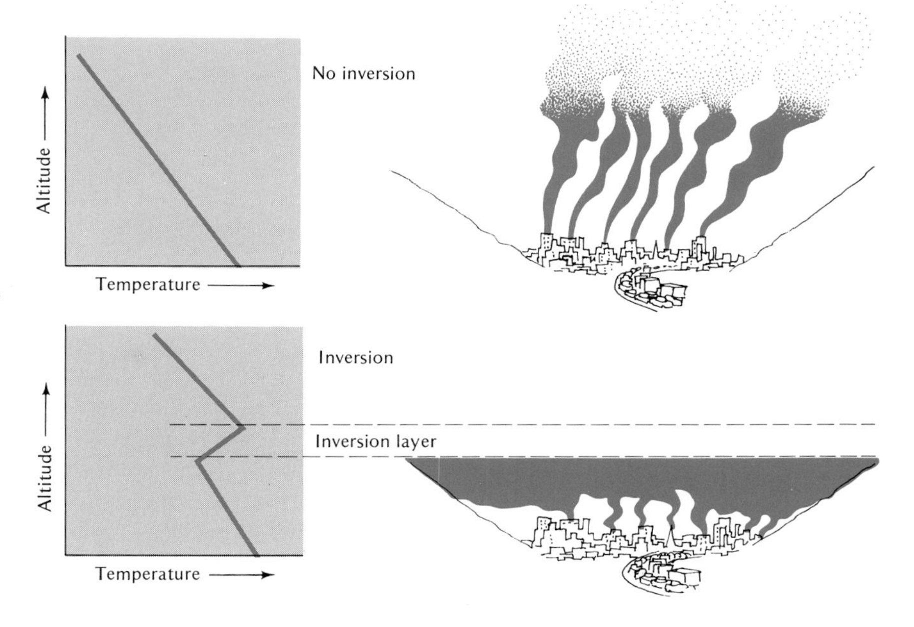

Fig. 2-5. "Clear" day in a factory situation (above) compared with intense air pollution produced by an inversion layer (below). The layer of warm air prevents dissipation of the smoke and other contaminants. (Redrawn from P. R. Ehrlich and A. H. Ehrlich, *Population, Resources, Environment,* San Francisco: Freeman, 1970.)

reduces the mixing of the air below with that above. Consequently, pollution-loaded air may be held over a restricted area for days at a time.

Valleys are particularly susceptible to the formation of temperature inversions. The ring of mountains around Los Angeles, coupled with the wind patterns in the eastern Pacific, produces a thermal inversion over the basin 7 days out of 16. During these periods smoke and other air pollutants are trapped within the basin.

Thermal inversions are not new. Four hundred years ago what is now San Pedro Bay was named "the Bay of Smokes" by Juan Cabrillo as he watched the effect of a thermal inversion on Indian campfire smoke back of the bay. The smoke rose only a few hundred feet and spread a blanket of haze over the basin.

Photochemical Smog

The term smog originally referred simply to a combination of smoke and fog such as that occurring over London in winter. When the residents of Los Angeles became harassed by an increasingly hostile atmosphere which burned their eyes, caused shortness of breath, and blighted plants they applied the name smog although neither fog nor smoke was involved.

efforts in Los Angeles

Los Angeles smog, which began during World War II, results from reactions in mixtures of oxygen, hydrocarbons, and nitrogen oxides in the presence of light. This is photochemical smog. The Air Pollution Control District operating in Los Angeles County has effectively reduced or eliminated many sources of air pollution by setting rigid emission control standards for refineries, power plants, and other sources. Fifty-seven open burning dumps, 12 large city-operated incinerators, most of the incinerators in commercial buildings, and one and one-half million residential incinerators have been eliminated. Burning of coal is illegal, and the sulfur content of fuel oil is regulated. Controls extend even to vapor escaping from gasoline loading and storage tanks. Emission control devices were required on all cars sold in California 2 years before Federal regulations were enacted.

automobile exhausts

Early efforts in California to find the cause of photochemical smog eventually singled out the automobile as the worst offender. Without an effort to reduce harmful emissions, the burning of 1000 gallons of gasoline in the internal combustion engine of an automobile discharges into the air about 3200 pounds of carbon monoxide, 200 to 400 pounds of hydrocarbons, 20 to 75 pounds of nitrogen oxides, 18 pounds of aldehydes, 17 pounds of sulfur, 2 pounds of ammonia, 2 pounds of organic acids, and other dangerous substances. The development of crankcase and tailpipe emission control devices greatly reduced the contribution of each automobile, but the increase in the number of automobiles in the area has largely offset the gains. This simply means that pollution levels stay about the same. Essentially, the air over Los Angeles has not improved since 1960. Without these efforts at control, however, the city might not be able to function at all by now.

Fig. 2-6. Air pollution over Billings, Montana, in 1969. *(Courtesy U.S. Department of the Interior, Bureau of Reclamation.)*

Lack of an effective mass transit system creates a barrier to any solution to the problem of air pollution in Los Angeles. The rapid growth of the city occurred during the reign of the automobile, and public transport systems were neglected in favor of extensive freeways. Thus each family must operate at least one car to move about the city. As a result, residents breathe an atmosphere which contains acids, ethers, ketones, alcohols, ozone, carbon monoxide, many compounds of nitrogen, and other poisons.

Coal Refuse Banks

The mining and preparation of coal results in an accumulation of waste at the mine site. This waste, consisting of bonded coal and rock, rock excavated during mining, carbonaceous shale, pyrites, and often broken mine timbers, discarded mine equipment, grease-soaked rags, and discarded oil containers was long considered an inevitable product of coal mining and was largely ignored. The problem was compounded in many places as local residents used refuse bank areas to dump household wastes. Even when the banks ignited and smoldered for years, producing acrid, offensive fumes, little heed was paid. Extinguishing a burning coal refuse bank was considered virtually impossible and to be a prohibitively expensive undertaking.

refuse banks

Refuse pile ignition starts in several ways. Spontaneous combustion accounts for two-thirds of refuse fires. As air circulates through the pile, oxidation occurs. Oxidation results in the production of heat. As the heat

spontaneous combustion

increases, the rate of oxidation increases and more heat is produced. Actual ignition temperatures will be reached if the air flow is insufficient to carry off the accumulating heat. Other causes of refuse pile ignition include forest fires, careless burning of trash near or on a pile, unextinguished campfires, and intentional ignition by man. In the past refuse piles have been burned to produce ash known as "red dog." This ash is used to surface streets, secondary roads, and parking areas. It is also used for landfill in low-lying areas.

Environmental Impact

emissions

A 1968 survey included a study of the emissions of burning coal refuse piles and the effects of these emissions on vegetation and on people. The most toxic gases produced include carbon monoxide, ammonia, hydrogen sulfide, sulfur dioxide and trioxide, sulfuric acid, and oxides of nitrogen. The 292 burning refuse piles produced 1.2 million tons of carbon monoxide in 1968. In highly industrialized areas near the anthracite and bituminous coal regions of Pennsylvania, studies showed that sulfur dioxide in the air averaged more than 1 part per million, with peaks reaching above 4.5 parts per million.

The synergistic effect of several pollutants creates a problem when efforts are made to determine the environmental impact of each separately. However, the following analysis gives a partial picture of the dangers involving sulfur compounds.

Sulfur dioxide in an annual mean concentration of as low as 0.03 part per million in the air produces chronic plant injury and excessive leaf

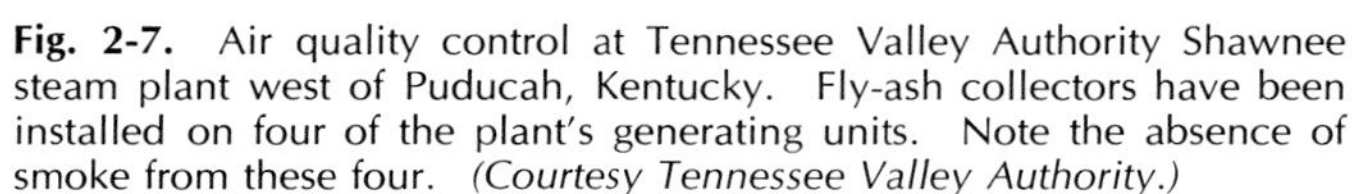

Fig. 2-7. Air quality control at Tennessee Valley Authority Shawnee steam plant west of Puducah, Kentucky. Fly-ash collectors have been installed on four of the plant's generating units. Note the absence of smoke from these four. *(Courtesy Tennessee Valley Authority.)*

fall. A case was reported in which a tree planted 3 miles away was adversely affected by gases emitted from a burning refuse pile. Somewhat higher quantities, 0.11 to 0.19 part per million, were correlated with increased absenteeism from work and increased hospital admissions for respiratory disease. The major effect was on older persons. A smoke concentration of 750 milligrams per cubic meter and a sulfur dioxide concentration of 0.25 part per million were accompanied by an actual increase in death rate.

sulfur dioxide

Hydrogen sulfide fumes cause darkening of lead-based paints by producing a lead sulfide film on the surface. The darkening of new or old paint in the vicinity of burning refuse piles warns of hazardous fumes. Danger also occurs in explosions of piles and in the resulting landslides. People have died as a result of breaking through the thin crust produced as the smoldering fire hollowed out the inside of the pile.

The seriousness of the problem stems from the fact that 260 of the 292 burning piles surveyed were within 5 miles of a community of at least 200 people. Thirty-one of the piles were adjacent to communities of 10,000 or more. Forty-five percent (131) of the sites were within 1 mile of such a community. Adverse effects on people may occur at distances up to 5 miles from the burning pile.

Other Sources of Air Pollution

Poisoned air has injured and killed plants for many years. Fluoride compounds, oxidants, and sulfur dioxide in particular have accounted for loss of vegetation in extensive areas. Although the most severe injuries occur close to the source, air pollution can affect plants many miles away. The ozone originating in Los Angeles has injured ponderosa pine 75 miles east in the San Bernardino Mountains.

Fluorides

The rapid growth of the aluminum industry during and after World War II, and other manufacturing processes which also emit fluorides into the air, marked the beginning of widespread fluoride injury to plants. Hydrogen fluoride and silicon tetrafluoride are emitted when minerals containing fluorides are heated. This occurs at brick plants, potteries, refineries, iron-enameling plants, steel-producing plants, phosphate fertilizer manufacturing plants, and aluminum reduction processing plants.

fluorides

Gaseous fluorides cause the most trouble, although some fluoride solids enter the air as particulate matter. After the gases enter the leaf stomata (Fig. 2-8), fluorides dissolve in the water film lining the substomatal chamber or adjacent air spaces. From here they are transported to the leaf margin. Accumulation of fluorides, often in high concentrations, kills marginal tissue in broad-leaved species or needle tips, usually, in pines. A large billboard[1] along U.S. Highway 30, east of Portland, Oregon,

[1] The billboard was removed after settlement of a court suit in favor of the rancher.

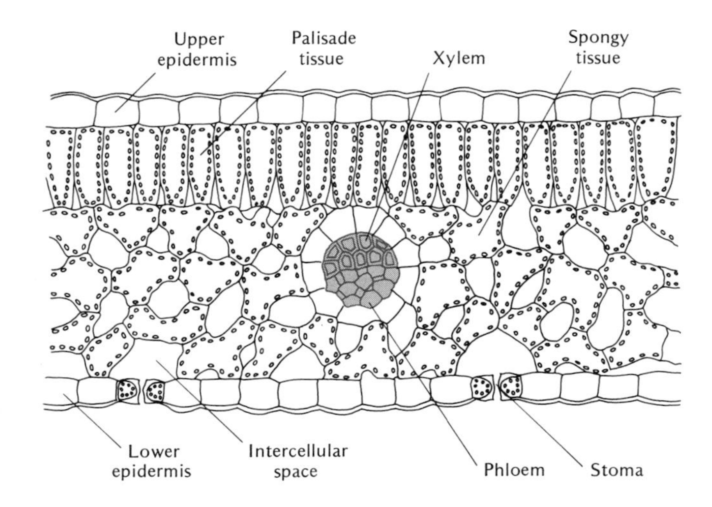

Fig. 2-8. The porous anatomy of leaves allows air pollutants to come in contact with internal cells.

lamented the death of cattle from eating poisoned grass on a ranch downwind from an alleged fluoride-emitting manufacturing plant. In recent years the installation of effective emission controls has greatly reduced fluorides in the air. Most problems that still exist involve local conditions where controls are relaxed.

Oxidants

Two oxidizing agents, ozone (O_3) and peroxyacetylnitrate (PAN), are produced in the formation of photochemical smog and therefore stem mainly from air pollution by automobiles. Ozone also forms naturally in the atmosphere at high altitudes by the action of electric discharge and ultraviolet light on diatomic oxygen (O_2). Ozone, an unstable, poisonous, powerful bleaching and oxidizing agent, probably has always caused some damage to vegetation as air turbulence carried it down to ground level. In recent years, however, more ozone and also PAN damage symptoms have been found near metropolitan areas having Los Angeles-type (photochemical) smog.

ozone

Ozone affects broad-leaved trees as well as conifers. In conifers, such as pine, low concentrations of ozone or high resistance in the tree results only in mottled yellow tissue patches on the needles. Acute damage occurs when the ozone concentration is high, or when the plants are particularly sensitive. Sensitivity varies considerably among different trees of the same species. In cases in which damage is acute, the needle tip, or often the entire needle, turns brown and dies. In broad-leaved species red, purple, or brown flecks on the upper surfaces of the leaves are the most common signs of ozone injury. The lower leaf surface seldom shows these symptoms.

Some persons are more sensitive to ozone than others. Generally, the first noticeable effect on people occurs when ozone reaches 0.3 part per million. At this low level nose and throat irritation begins. A 2-hour exposure to 1 to 3 parts per million of ozone may produce lack of coordination and extreme fatigue. Excessive accumulation of fluid in the lungs results from a 2-hour exposure to 9 parts per million of ozone.

Natural and synthetic textiles as well as rubber are adversely affected by even small amounts of ozone. In the presence of 0.01 to 0.02 part per million, rubber under slight tension produced cracks within 65 minutes. At 2 percent ozone (20,000 parts per million), cracks formed in less than 1 second. Ozone attacks these organic polymers at points in the molecule where carbon-to-carbon double bonds occur. The more of these double bonds in a substance, the more susceptible the substance is. When long-chain organic molecules are broken, often in several places, the substance loses strength and simply becomes more fluid. The other response to ozone, cross-coupling of parallel carbon chains, reduces elasticity and causes the material to become brittle.

Sulfur Dioxide

sulfur dioxide

Injury to plants from atmospheric sulfur dioxide (SO_2) has been known for the past 75 years. Earlier reports centered around the destruction of vegetation near ore smelters. The Ducktown and Copper Hill areas in Tennessee (Fig. 2-9) are classic examples. At present, ore smelters (particularly zinc, nickel, copper, and lead) contribute only 7 percent of the sulfur dioxide content of the air (Fig. 2-10).

Fig. 2-9. Destruction of vegetation by fumes from a copper smelter resulted in severe erosion at Ducktown, Tennessee.

Fig. 2-10. Air pollution from a zinc smelter near Amarillo, Texas. *(Courtesy U.S. Department of the Interior, Bureau of Reclamation.)*

Sixty percent of the sulfur dioxide in the air comes from the burning of coal. The electric power industry is the chief consumer of coal, although heating of houses still accounts for a sizable quantity. Refining and combustion of petroleum products produce 21 percent of the sulfur dioxide in the air, and minor quantities come from the production and use of sulfur and sulfuric acid, and from natural gas production.

Sulfur dioxide enters leaves through the stomata. Pine needles turn yellow or brown and may fall prematurely. Sometimes seedlings are dwarfed. In broad-leaved species injury most often occurs between the veins. Thus patches of ivory to brown alternate with the green vein areas.

Minor Air Pollutants that Damage Plants

Several additional materials in our atmosphere have long been recognized as damaging to plants. Ethylene from automobile exhausts, the manufacture of polyethylene and illuminating gas, petroleum refining, and combustion of most organic materials may produce premature leaf fall, flower injury, yellow leaves, and general abnormal growth. Ethylene gas is used commonly to ripen certain fruits after harvest.

chlorine

Chlorine (Cl_2) and hydrogen chloride (HCl) gases are products of burning polyvinyl chloride plastics, glassmaking, and oil refining. Some additional pollution comes from accidental spills. These gases cause stipples on the upper leaf surface similar to ozone injury. Ammonia, discharged from automobile engines and incinerators, or escaping from large refrigeration units and fertilizer tanks, produces a bronze or red upper leaf surface. Combustion at high temperatures when oxygen and nitrogen are present produces oxides of nitrogen. The internal combustion engine accounts for most of the nitrogen oxides in the air, but many sources exist. The manufacture of paint, roofing, nylon, soap, rubber, and nitric acid, as well as the incineration of organic wastes and the burning of fuel in power and heat production, all contribute a share to the atmospheric load. Symptoms of plant injury from nitrogen oxides are similar to those of damage produced by fluorides.

ammonia

AIR AND HEALTH

In 1969, 60 faculty members of the Medical School of the University of California at Los Angeles endorsed a recommendation directed to the residents of San Bernardino, Riverside, and Los Angeles, which stated that air pollution had become a major health hazard to most of the communities for most of the year. Those who did not really have to stay were advised to move to avoid bronchitis and emphysema. Each year Los Angeles doctors suggest to about 10,000 patients that they leave the Los Angeles Basin area.

Carbon Monoxide

carbon monoxide

Carbon monoxide combines with hemoglobin in the blood and thus displaces oxygen. Even in small amounts, carbon monoxide causes an increase in breathing rate and adds to the work load on the heart as tissues become deficient in oxygen. Eight hours exposure to 80 parts per million of carbon monoxide reduces the oxygen-carrying capacity of the blood by 15 percent. This is equivalent to the loss of 1 pint of blood. The loss of 1 quart of blood endangers life; those persons subject to exposure to 80 parts per million of carbon monoxide during a working day should not make blood donations the same day. The critical issue here is: How many people know what their carbon monoxide exposure really is? In snarled traffic, carbon monoxide counts often reach 400 parts per million. During traffic tie-ups, symptoms of carbon monoxide poisoning can be seen in increased headaches, loss of vision, reduced muscle coordination, nausea, and abdominal pains. Under these conditions driving ability suffers (Fig. 2-11). Nitrogen oxide has an effect similar to carbon monoxide.

Oxides of Sulfur

sulfur

Sulfur oxides are harshly irritating to the respiratory passages and cause coughing and choking. Those who have chronic lung diseases experience severe attacks when subjected to sulfur oxides in the air. Most smog deaths are probably due to sulfur. Sulfur dioxide from burning coal adheres to coal dust particles and thus spreads through the air. In the lungs it becomes sulfuric acid.

Severe cases of chronic asthma often result from contact with sulfur dioxide. Asthma is an allergic supersensitivity of the bronchial tubes.

Fig. 2-11. Battered center barrier indicates that drivers sometimes make errors in judgment. *(Courtesy Michigan Department of Natural Resources.)*

During asthmatic attacks circular muscles constrict the smaller branches of the bronchial "tree." The victim can inhale (with effort) but cannot exhale with sufficient force to clear the lungs. Carbon dioxide builds up in the lungs, and the patient's oxygen supply is lowered. While not a prime cause of death, asthma nevertheless kills several hundred Americans each year.

Two other respiratory diseases, bronchitis and emphysema, are substantially increased by air pollutants. Bronchitis is an inflammation of the bronchial tubes. Victims experience difficulty in expelling foreign matter by coughing; muscles of the bronchial tubes weaken and mucus accumulates. A progressive loss of breathing ability occurs.

Emphysema, in contrast, is a disease of the small air sacs, the alveoli. Small clusters of air sacs fuse to form one large pocket. This greatly reduces the surface area and so lowers oxygen absorption into the bloodstream. Generally, emphysema develops gradually over the years, sometimes as a consequence of smoking "two packs a day." The victim eventually dies of suffocation.

Evidence of Health Hazards

Several problems indicate the difficulty in analyzing the health hazards of air pollution. What happens to an individual or a group of people may be the result of the interaction of many factors. Often, any single relationship is difficult to prove and vested interests can make a good case for reasonable doubt. Pollutants are numerous and varied, and perhaps some critical ones are not yet suspected. Many are difficult to detect.

We have few long-term records, and these are badly needed to reveal the delayed effects of air pollution. Concentrations of pollutants vary geographically and differ at different times in the same locality. Coupled with inadequate monitoring techniques, the determination of precise exposure to specific pollutants is practically impossible. Also, various substances may interact synergistically. This means that the danger from two pollutants at the same time can be greater than the sum of the dangers of each separately. The relationship between smoking and asbestos, dual causes of lung cancer, is an example. Asbestos particles inhaled by nonsmokers are carried out of the lungs by the ever-moving sheet of mucus in the bronchial tubes and trachea. Cilia in these tubes propel the mucus, but smoking interferes with ciliary action. Thus smoking increases asbestos-induced lung cancer.

carcinogens

The length of exposure to airborne carcinogenic hydrocarbons (benzpyrene and others) is a factor in the development of lung cancer from these agents. Sulfur dioxide, like nicotine, interferes with the action of cilia in the trachea. Thus, for persons exposed to benzpyrene and to sulfur dioxide, the hazard is much greater than that calculated simply by adding the separate probabilities of lung cancer from the two.

The following facts are not presented as proof of cause but they are strong indicators of the effect of air quality.

Fig. 2-12. Exhaust emission from an airplane at National Airport, Washington, D.C. *(Courtesy U.S. Department of the Interior.)*

England has a higher level of air pollution than the United States and, among men, has twice our annual lung cancer death rate. The lung cancer rate on smoggy Staten Island, New York, is 55 per 100,000 persons; a few miles away where smog is not as frequent or persistent, the rate is 40 per 100,000.

An increase in emphysema parallels a rise in air pollution. Among smokers the emphysema rate in smog-ridden St. Louis, Missouri, is four times that of a comparable group in Winnipeg, Canada. The death rate from pneumonia is higher in more highly polluted areas. British postmen working in areas of high pollution have more serious chronic bronchitis than those who work in pollution-free areas. ("Pollution-free" is a relative term.) Ten years after the 1948 Donora, Pennsylvania, smog disaster, those who reported severe effects in 1948 had a higher death rate—they were dying before the expected time.

Some Air Particle Concentrations

A simple demonstration can reveal the large number of particles that may be present in what seems to be clean air. Release of moisture-laden air into the low pressure of a partial vacuum results in condensation of water droplets on particles in the air. Particles as small as 2×10^{-6} inch can be detected by this method. When properly calibrated and known volumes of air used, a light beam passing through the vacuum chamber and activating a photoelectric cell can give direct readings in numbers of particles per cubic centimeter.

Most of the small particles that reach the lungs stay there. In one demonstration a person inhaled from a bag in which the air contained 126,000 particles per cubic centimeter. Exhaled air was collected in another bag and shown to contain 42,000 particles per cubic centimeter. In this case about 66 percent of the small particulate matter remained in the lungs.

Air over the open oceans and polar regions contains about 200 particles per cubic centimeter. This may be considered the global background count. Locally, in areas of atmospheric pollution, the count goes much higher. The air at a height of 14 floors above 51st Street in New York City contained 170,000 particles per cubic centimeter on December 3, 1970. Another case was reported in which a person driving about 1 mile behind a pickup recorded 50,000 particles per cubic centimeter in an area where the off-road air (the local count) contained less than 2000 particles per cubic centimeter. The air in the Squirrel Hill and Pitt Tunnels (Pittsburgh, Pennsylvania) reaches 10^7 particles per cubic centimeter when the area count is only 5600. Airplane cabin air may reach a particulate level of 80,000 to 1 million per cubic centimeter during engine warm-up. As the plane climbs after takeoff, the particulate count may drop to 300 to 500 per cubic centimeter.

Fireplaces using hard maple can burn 32 pounds of wood in 1 hour and produce 30 trillion (3×10^{13}) smoke particles during that hour. These measure about 0.3 microns in diameter. Thirty-two pounds is also the approximate weight of the gasoline that an average car burns in 1 hour while traveling 60 miles. The automobile, however, produces particles 1/10 the size of the smoke particles and 100 times as many.

Up to 1973, people living in nonurban, low-pollution areas who did not smoke appeared not to be in danger from breathing. In metropolitan areas, however, with the inevitable smokestacks, cars, trucks, dirty streets, and extensive areas of concrete and asphalt, the respiratory system becomes overloaded. Complete coating of the mucus of the respiratory passages means that no more particles can be trapped, and large particles then go into the lungs.

HEARING

The capacity of the air to function in the transmission of sound forms the basis for still another type of pollution—too much sound and sound that is too loud. A look at the mechanics of hearing shows why sound pollution is a problem.

sound

The sensation of sound depends on pressure changes in the air. The external ear channels these pressure changes to the eardrum. The vibrations of the eardrum produced by the pressure changes set in motion three connected bones of the middle ear. These in turn start vibrations in the fluid of the inner ear because of the connection of the innermost middle ear bone to the oval window, an oval membrane covering part of the large end of the cochlea. The bones of the middle ear, because of the way in which they are joined, reduce the amplitude of the eardrum vibrations but at the same time intensify them (Fig. 2-13).

The cochlea resembles a snail shell. Running the length of the spiral, the organ of Corti contains rows of hair cells. The tiny hairs projecting from the cells into the basilar membrane respond to vibrations in the fluid

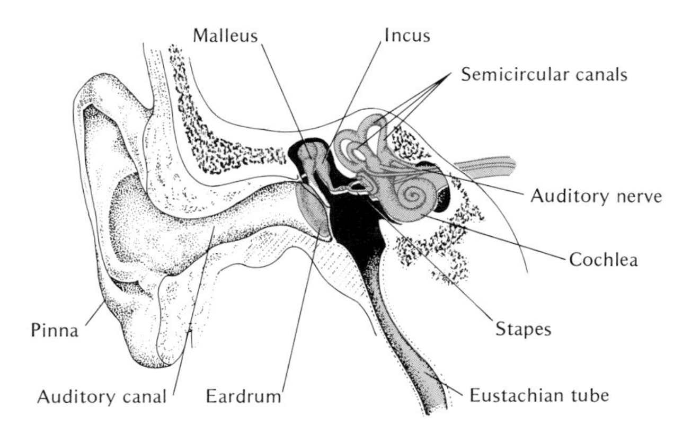

Fig. 2-13. Diagram of the internal structure of the human ear.

Fig. 2-14. The Tennessee long-eared bat. Bats navigate in total darkness by the detection of reflected high-frequency sounds which they emit.

the inner ear

of the inner ear—or perhaps to the vibrations of the basilar membrane as caused by vibrations in the fluid—and the hair cells initiate nerve impulses as a consequence. The short hair fibers at the large end of the cochlea are stimulated by higher notes. The longer fibers at the farther end respond to low notes. The nerve impulses are conducted to the hearing centers in the temporal lobe of the cerebrum by the auditory nerve.

The normal range of hearing for a healthy adult ear lies between 20 and 16,000 vibrations per second. Children may hear over a greater range. Bats hear sound with a frequency of 60,000 vibrations per second and even higher (Fig. 2-14).

The Energy of Sound

decibels

Decibels are small numbers which represent quantities of sound energy. The reference point, zero, is the weakest sound that healthy young ears can detect.

Differences in sound energy are perceived logarithmically. Each additional 10 decibels involves a sound energy increase of one order of magnitude. Ten decibels of sound have 10 times the energy of zero decibels; 20 decibels have 100 times as much, and 30 decibels 1000 times. Extremes of noise may reach 150 decibels, an energy increase of 10^{15} times the threshold intensity for a person of normal hearing.

At the proper acoustic energy and frequency, sound can crack plaster, break windows, and even destroy rock formations. On the positive side, sound can be used to mix paint and wash dishes. Sound has brewed a cup of coffee in 7 minutes.

Such energy also represents a threat to people. The eardrum may be broken by explosions, although it usually heals with little or no permanent injury. Ordinary factory noise, even though prolonged, has not been known to rupture the eardrum. However, damage to some of the 24,000 hair cells located on a membrane in the cochlea of the inner ear results from sudden, very loud noises or prolonged exposure to high noise levels. Generally, hair cells are destroyed in small patches and so noise seldom produces complete loss of hearing. These hair cells activate the branches of the auditory nerves and thus are directly concerned with the conversion of sound energy into nerve impulses. When the hair cell is destroyed, the connecting nerve endings shrink. The body does not repair or replace lost hair cells or the nerves with which they synapse. This type of hearing loss is permanent and cannot be corrected by surgery.

Decibels

Some examples of noise levels expressed in decibels:

Source	Decibels
Threshold of hearing	0
Rustling of leaves	20
Ordinary conversation	60
Vacuum cleaner at 10 feet	69
Loud orchestra in a large room	82
Beginning of hearing damage, if prolonged	85
Trailer truck on a highway, at 20 feet	90
Heavy city traffic	92
Inside a Los Angeles motor bus	96
Amplified musical instruments	90–105
Lawn mower	92–105
Punch press	100
New York City subway	102
Air hammer	104
Whistle on ocean liner, at 75 feet	112
Boeing 707 jet aircraft	
Ground level 1 mile beyond end of runway, climb rate 500 feet per minute	125
2000 Feet on either side of runway	100
Projected for the SST	
Ground level 1 mile beyond end of runway, climb rate 500 feet per minute	111
6000 Feet on either side of runway	100
13-Mile stretch under landing and takeoff pattern, at least	100
4-Mile stretch, 2 miles wide astride runway	100

effects

Noise may have effects other than on the ear itself. Sounds have been shown to raise blood cholesterol levels in rats and rabbits. Loud noise causes release of the adrenal hormone and this intensifies tension. As reported at the 1969 meeting of the Acoustical Society of America (Baron, 1970), high noise levels may aggravate high blood pressure, hardening of the arteries, and coronary disease.

Suggested Readings

Air Pollution and Trees. U.S. Department of Agriculture, Forest Service, Southeastern Area, Atlanta, Georgia. 1971.

Baron, R. A. *Tyranny of Noise.* New York: St. Martins Press, 1970.

Bazell, R. J. "Lead Poisoning: Combating the Threat from the Air." *Science,* 174(4009):574–576, 1971.

Burns, W. *Noise and Man.* Philadelphia: Lippincott, 1969.

Charlson, R. J. and L. F. Radke. "Atmospheric Aerosol: Does a Background Level Exist?" *Science,* 170(3955):315–317, 1970.

Ehrlich, P. R. and A. H. Ehrlich. *Population, Resources, Environment.* . San Francisco: Freeman, 1970.

Guild, S. R. "Effects of Sound: Pathological—Including Hearing Loss." *In* C. P. McCord, *The Acoustical Spectrum,* Ann Arbor: University of Michigan Press, 1952.

Harrington, D. and J. H. East, Jr. "Burning Refuse Dumps at Coal Mines." *Bureau of Mines Information Circular* 7439, 1948.

Hoffman, A. I. "Nationwide Inventory of Air Pollutant Emissions." *National Air Pollution Control Administration Publications AP-73,* August 1970.

Knudsen, V. O. "The Acoustical Spectrum." *In* C. P. McCord, *The Acoustical Spectrum,* Ann Arbor: University of Michigan Press, 1952.

Landsberg, H. E. "Man-Made Climatic Changes." *Science,* 170(3964):1265–1274, 1970.

Lansford, H. "The Supercivilized Weather and Sky Show." *Natural History,* 79(7): 92–97, 1970.

McNay, L. M. "Coal Refuse Fires, an Environmental Hazard." *Bureau of Mines Information Circular* 8515, 1971.

Morgan, B. B., G. Ozolins, and C. E. Tabor. "Air Pollution Surveillance Systems." *Science,* 170(3955):289–296, 1970.

Newell, R. E. "The Global Circulation of Atmospheric Pollutants." *Scientific American,* 224(1):32–42, 1971.

Shurcliff, W. A. *S/S/T and Sonic Boom Handbook.* New York: Ballantine Books, 1970.

Strahl, R. W. "Survey of Burning Coal-Mine Refuse Banks." *Bureau of Mines Information Circular 8209,* 1964.

Sussman, V. H. and J. J. Mulhern. "Air Pollution from Coal Refuse Disposal Areas." *Journal Air Pollution Control Association,* 14(7):279–284, 1964.

Weinstock, B. and H. Niki. "Carbon Monoxide Balance in Nature." *Science,* 176(4032):290–292, 1972.

Good-quality water is essential for all living organisms. *(Courtesy U.S. Department of the Interior, Bureau of Reclamation.)*

Water

Water, a simple compound containing only two elements, affects life and the nonliving world at every turn. Water plays many roles in living organisms and in their environment because

properties of water

1. Many things dissolve in water.
2. Water is liquid within the temperature range of life.
3. Protoplasm functions as a colloidal suspension in water.
4. Water buffers organisms and the environment against abrupt temperature changes because of its high heat capacity.
5. Water reaches its greatest density a few degrees above freezing; as a result, ice floats.
6. The splitting of water molecules during photosynthesis is the first step in the building of organic compounds, a process that feeds the world.
7. Water is the source of the oxygen that green plants return to the air.

THE NEEDS FOR WATER

Water for Food Production

Plants remove considerable amounts of water from the soil. Some is used in the processes of photosynthesis, but a much greater amount evaporates from stem and leaf surfaces. A reservoir covered with floating plants loses water faster than one that is plant-free. Water evaporation from plant surfaces (transpiration) forms one phase of the water cycle. This water is not immediately recoverable by man but eventually returns to the earth's surface in the form of rain or snow.

transpiration

Table 3-1 lists the amounts of water used in the production of various foods. These figures represent the total water use unless otherwise specified. The production of beef, for example, requires water to grow forage plants that cattle eat, the water they drink, and the water used in meat processing.

for food production

TABLE 3-1
Some Water Requirements[a]

Unit	*Water (gallons)*
In food production	
1 pound of meat[b]	2500–6000
1 pound of wheat[b]	60
1 pound of rice[b]	200–250
1 quart of milk[b]	1000
1 ear of corn (10–16 ounces)	50
1 acre of corn (100–250 bushels)	480,000–1,200,000
Daily drinking water	
Work horses	12
Dairy cattle	7–22
Range cattle	4.2–8.4
Pigs (200–380 pounds)	1.4–3.6
Sheep on dry pasture	0.6–1.6
Sheep on good pasture	little or none

[a] Based on data from D. K. Todd, *The Water Encyclopedia,* Port Washington, New York: Water Information Center, 1970.
[b] Includes water for processing.

Water for Personal Use

Figure 3-1 indicates the water use for an average family. The graph clearly shows that most water use involves personal hygiene. These data were developed from studies of home water use in the Akron, Ohio, area. The presence of modern conveniences in the home makes a considerable difference in personal water consumption. In rural areas, for example, the per capita domestic water use per day (1965 data) where families had running water ranged from a low of 50 gallons (in more than half of the states) to a high of 100 gallons (in Arizona, Florida, Idaho, Nevada, and North Carolina). Where families did not have running water, the per capita daily water use was about 10 gallons.

for personal use

The source of water also makes a difference in personal water consumption. Phoenix, Arizona, which uses reservoirs on the Salt River, has a daily personal water consumption of 200 gallons per person. Tuscon, Arizona, which depends on deep wells, uses only 160 gallons per person.

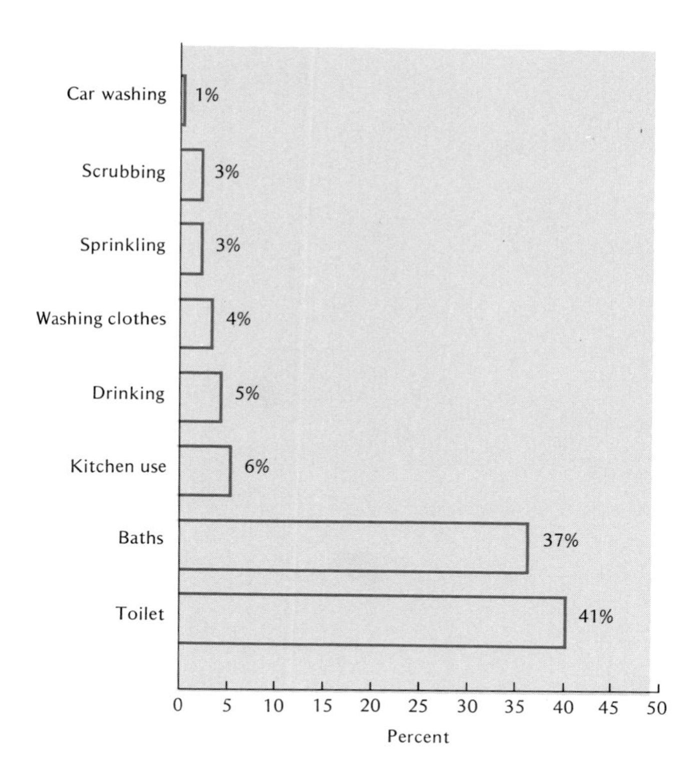

Fig. 3-1. Water use in an average home. (Based on data from D. K. Todd, *The Water Encyclopedia,* Port Washington, N.Y.: Water Information Center, 1970.)

Both these figures are high, however, compared with Madras, India, which has a daily consumption of 16 gallons per person, and Argentina where the average is 2 gallons per person per day.

Water for Industry

Industry accounts for most of the water use in the United States. As early as 1959, industrial cooling cycled 15.7 billion gallons of water each day. Processes other than cooling discharged 8.1 billion gallons into natural waters. Of the latter amount 3.7 billion gallons were untreated. The current and projected shares per person of water use in the United States are

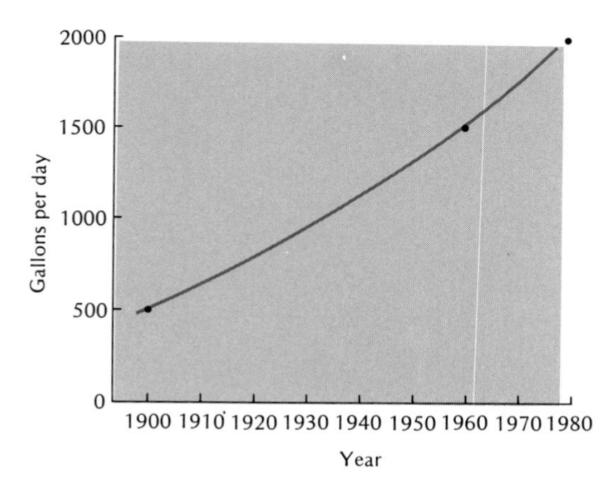

Fig. 3-2. Per capita share of total daily United States water use. 1900, 525 gallons; 1960, 1500 gallons; 1980, 2000 gallons. About 50 percent is used for irrigation. (Based on data from P. R. Ehrlich and A. H. Ehrlich, *Population, Resources, Environment,* San Francisco: Freeman, 1970.)

Fig. 3-3. Condensers for converting steam to water at Palisades nuclear power plant. *(Courtesy Consumers' Power Company, Michigan.)*

shown in Fig. 3-2. These data do not include rainwater used by crops. About half of this water, however, is water for irrigation.

Turbines, powered by liquid water or by steam turn the generators that produce almost all our electrical power. At hydroelectric dams the kinetic energy of moving water turns the turbines; steam-run power plants convert the potential energy of fossil fuels or atomic energy into electrical energy.

Although hydroelectric power is almost pollution-free, it supplies only about 5 percent of the total power now used in the United States. Production of power by this method is expected to drop to 4 percent of the total by 1980. The most suitable dam sites in the United States have already been utilized.

dams

In North America the developed capacity for hydroelectric power was 76,000 megawatts in 1967. Complete damming of all the rivers could raise this figure to 313,000 megawatts. In 1963, throughout the world, 341 high dams (over 250 feet) were in operation, 102 were under construction, and 443 were planned. Most of those under construction at that time are now in operation.

longevity of impoundments

Deposits of sediments in the impoundments behind dams seriously affect their useful lives. Eventually, silt completely fills the lake and the dam becomes a waterfall outlet for a stream meandering through a muddy marsh. The usual life-span of a dam and its reservoir ranges from 1 to 3 centuries. Data for 83 reservoirs in the United States show an annual water storage capacity loss varying from 0.07 percent (Schoharie at Prattsville, New York) to 5.0 percent (Bennington at Rago, Kansas). The average annual loss for the 83 reservoirs is 0.64 percent of storage capacity. If present silting rates continue, 35 of these reservoirs will be swamps in less than 200 years. Unless an effective method can be developed to deal with silt, or unless watershed erosion can be greatly reduced, hydroelectric power may have about the same life expectancy as our supply of fossil fuels.

Increasing demands for power encourage engineers to press for more dams. Many people, however, feel that our rivers should be something more than a series of pools that all too soon will become mud puddles. The Tennessee River drainage system is already punctuated with 33 dams, only one of which is less than 70 feet high. Many of these are flood-control dams. Building dams for flood control, although useful, is somewhat similar to taking aspirin for a toothache—it alleviates the symptoms but does not remove the cause. As pointed out in Chapter 4, the water uptake of the soil has been reduced by cultivation, and too much water runs off the land.

WATER POLLUTION

What constitutes polluted water is a matter of opinion. Many think of water pollution only in terms of drinking water safety. Those resigned to the status quo become concerned only when a situation becomes worse. As used in this text, water pollution occurs when any factor lessens the usefulness of a body of water.

Types of Water Pollution

Water pollution can be cataloged under several general types. The most common, and certainly one that has been with us as long as any other, is

Fig. 3-4. Reclamation of sewage water in Amarillo, Texas. Solids settle to the bottom of these settling tanks; water is then pumped to closed tanks where bacteria are added; and chlorine is added as water is pumped to using agency. (*Courtesy U.S. Department of the Interior, Bureau of Reclamation.*)

Fig. 3-5. Pollution of a mountain stream because of lack of proper sewage disposal. *(Courtesy U.S. Department of Agriculture, Soil Conservation Service.)*

oxygen-demanding waste

oxygen-demanding waste. Domestic sewage contributes the major portion, but animal waste, both natural from stockyards and feed pens, and commercial waste from food processing, contributes a sizable amount. Other businesses, tanneries and paper mills, for example, add to the load. The problem, discussed more fully later in this chapter, concerns the removal of dissolved oxygen from the water by processes that degrade organic matter. Oxygen-demanding waste must be controlled or aquatic animal life dies.

disease

Water-borne disease-causing organisms have plagued man for countless centuries. Today in the United States, the problem mostly concerns waters used for recreation, although wells sometimes are involved. Improper sewage treatment, septic tanks, and outdoor pit toilets too near shallow wells or draining into streams and lakes contribute to the spread of disease (Fig. 3-5).

eutrophication

A third category of pollution results from the heavy applications of fertilizer now being used in an effort to increase crop yields. Nitrogen and phosphorus stimulate the growth of algae and aquatic weeds. These nutrients are present in natural waters but not in sufficient quantity to cause a problem in most cases. When fertilizer that has leached into groundwater seeps into streams and lakes, eutrophication or excessive growth results. Sewage and industrial wastes also add to the problem. Secondary sewage treatment, rather than removal of these nutrients, creates more growth by converting organic matter to mineral form.

Other inorganic chemicals of a wide variety reach surface and underground water supplies from many sources. These include metal salts and acids from mining operations, oil fields, manufacturing, and agricultural practices and, to some extent, from natural leaching of rocks and soil. Most acids in surface waters come from operating or abandoned mines.

Synthetic organic chemicals that find their way into water include many that are toxic to fish, and in turn to other members of the food chain. These include some chemicals, many of which are harmful at low concentrations, that cannot be removed from wastewater by conventional waste treatment.

sediment

Sediment also constitutes a form of water pollution. The sand, silt, and clay that washes off the land and into waterways constitute a problem because of their volume. Rivers and harbors must be dredged to maintain shipping channels. Sand erodes pumping equipment and power turbines. Fish nests and oyster beds are covered by blankets of mud. The penetration of sunlight, essential for algae and other aquatic plants that oxygenate the water, is sharply reduced by sediment. Industrial and municipal water supply costs increase as a result of sediment loads.

Several sources contribute to the pollution of water with radioactive material. These include the mining and processing of radioactive ore; the radioactive material used in medicine and research, nuclear power plants, and other industries; and radioactive fallout from nuclear testing.

Raising the temperature of water is also a form of pollution. Tremendous volumes of water used for industrial cooling are returned to water sources at a higher temperature. This affects the capacity of the water to support life.

SEWAGE TREATMENT

Solids in Water

sewage treatment

A sewer system emptying into some kind of sewage treatment plant is the most common form of water pollution control in the United States. Sewer lines conduct wastewater and suspended solids from homes, businesses, and factories to a central collecting plant. Treatment at the collecting plant, which varies considerably from one locality to another, is intended to restore the water to a quality suitable for discharge into natural waterways or for reuse.

combined sewer system

Two general types of sewer systems have been built. The most common, because of lower costs, combines sewage collecting and storm water runoff (Fig. 3-6). Rainwater from roofs, yards, and streets flows into the same lines that carry sewage. The disadvantage lies in the fact that the sewage treatment plant cannot handle the volume of sewage combined with the extra water of a heavy rain. When such rains occur, a bypass valve is opened and the excess flow—that beyond the capacity of the treatment

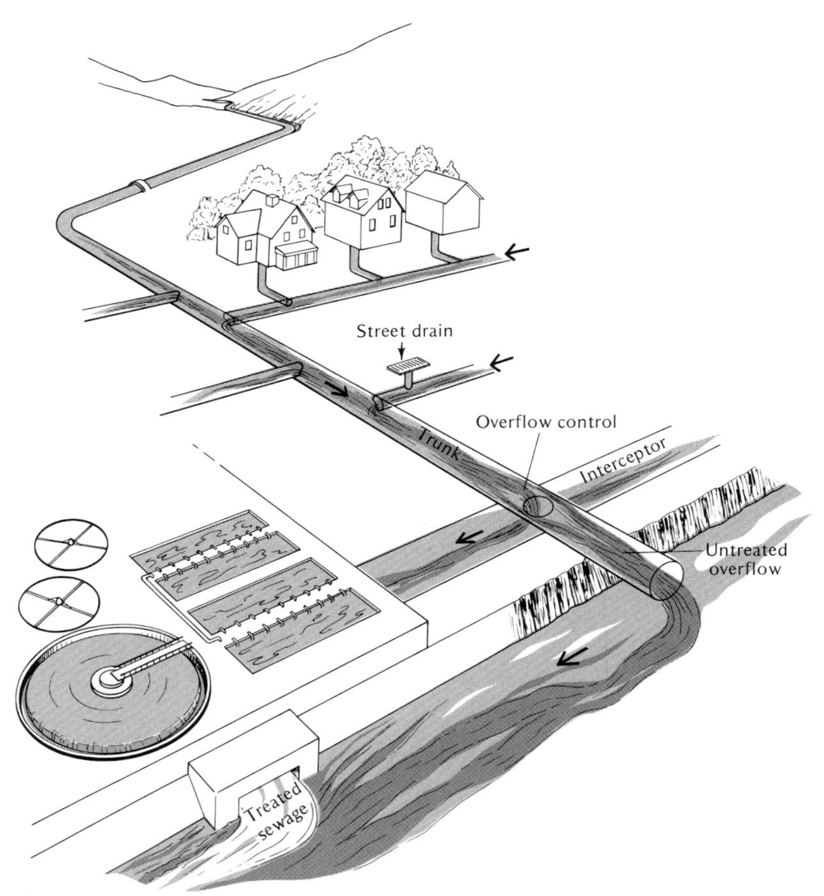

Fig. 3-6. Combined sewer system. (Redrawn with permission from *A Primer on Waste Water Treatment,* U.S. Department of the Interior, Federal Water Pollution Control Administration, 1969.)

plant—empties directly into the waterway. The amount of raw sewage that bypasses the treatment plant depends on the amount of rain that falls.

A second system uses parallel sewers to keep sewage and storm water runoff separate. Water draining from streets flows into a separate storm sewer.

The Function of a Sewage Treatment Plant

Basically, a sewage treatment plant speeds the processes that occur in nature as water regains purity. Our streams and lakes have always purified the water of the waste products of organisms that live in and near them. The volume of waste from such sources was low, however, and normal functioning of the nitrogen, carbon, phosphorus, and other cycles broke down the organic material. In these cycles the organic material was returned to the nonliving world for reuse by the next generation of organisms. As long as man's population was sparse, these cycles maintained relatively pure streams and lakes. In fact, one of the gems of American folklore which persisted beyond its safe application was the belief that a stream exposed to sunlight for a mile is purified. No allowance was made for the quantity of human and animal waste.

When people are crowded into cities, disposal of sewage becomes an overwhelming problem. According to Carpenter (1967), the volume of sewage in most large cities is close to 100 gallons a day per person. During the January 1970 strike of city employees in Cincinnati, Ohio, maintenance problems at one of the treatment plants resulted in the pouring of 260 million gallons of raw sewage into the Ohio River over a 4-day period.

Domestic sewage is roughly 1 percent (0.5 to 1.2 percent) solid material and contains little if any dissolved oxygen. Microorganisms per milliliter may number anywhere from 500,000 to 20 million. Most of them are regular inhabitants of the human intestinal tract and not harmful. However, pathogenic bacteria, particularly those of intestinal origin, do gain access to sewage. The ones most likely to survive long enough to cause harm are principally those that cause Asiatic cholera, typhoid, paratyphoid, and dysentery. Soil microorganisms are also abundant.

diseases in water

The relatively low amount of organic matter in sewage—1 percent—may seem insignificant, but very little food is needed for bacteria to produce a high-density population. In the laboratory nutrient media containing less than 1 percent organic matter may support a bacterial population numbering more than one billion cells per milliliter. All the nutrients necessary for rapid growth of many kinds of bacteria are present in sewage. Because the oxygen content of sewage is low, most of the natural breakdown of organic matter results from the activities of anaerobic species. This releases foul-smelling gases such as hydrogen sulfide and ammonia.

Formerly, the objectives of sewage treatment were to remove or decompose the organic matter that supports bacterial growth and to destroy or remove dangerous microorganisms. To these have now been added the reduction of those inorganic substances that act as nutrients for green plants, particularly algae. This concerns primarily phosphates and nitrates.

Primary Treatment

primary treatment

Primary treatment (Fig. 3-7), the oldest and most primitive procedure used in sewage treatment plants, relies on gravity and a few screens to take out the heavier material. Screens, placed at an angle to aid in cleaning, hold back the larger chunks. The debris thus caught is raked off by hand or by mechanical means. In some plants a shredding device replaces the screen, and the chopped material stays in the sewage water.

Screened or shredded sewage then passes through a grit chamber. The size of the chamber and the rate of flow are such that sand, cinders, and small stones settle to the bottom. This stage is not intended to remove or-

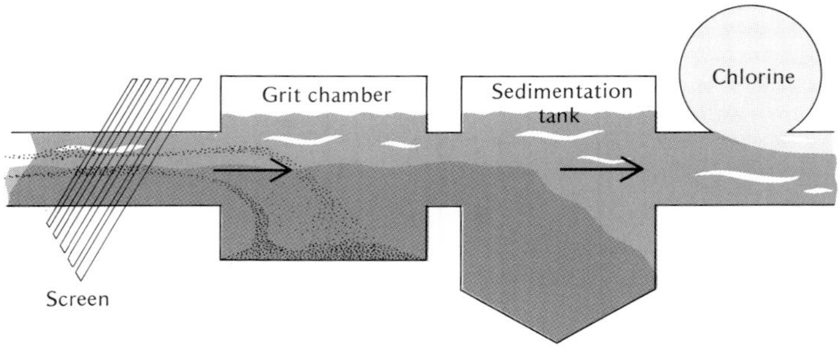

Fig. 3-7. Plan of a primary treatment plant. (Redrawn with permission from *A Primer on Waste Water Treatment,* U.S. Department of the Interior, Federal Water Pollution Control Administration, 1969.)

ganic matter. The grit chamber is most important in cities that use combined storm and sewage systems. Large quantities of grit and gravel washed off the streets and into the sewer system finally reach the treatment plant. Most of the material collected in the grit chamber serves as landfill.

The water with its load of suspended material then enters a sedimentation tank. Here a reduction in the velocity of sewage flow allows the suspended solids to settle to the bottom. More than one sedimentation tank is required. After a particular tank has been in operation for a few days or weeks, depending on the quantity of sewage and the tank size, flow is diverted to another tank. The settled material (sludge) is removed either by hand, or by mechanical means in most plants built since 1940.

The liquid leaving the sedimentation tank is treated with chlorine to kill bacteria and to reduce odor. This completes the primary treatment. The effluent is then discharged into some natural water—a stream, river, lake, or the ocean. Except for the chlorine, a sewage disposal plant using only primary treatment is little more than an improved septic tank. About 30 percent of United States municipalities carry their sewage operation no farther than primary treatment.

Secondary Treatment

secondary treatment

The liquid leaving the sedimentation tanks contains a variety of materials including colloidal suspensions of carbohydrates, fats, and proteins. Secondary treatment makes use of biological processes to remove up to 90 percent of the organic matter. Conditions are provided that promote the growth of the bacteria already in the sewage. The breakdown of organic matter occurs as the bacteria use these molecules for energy and as a materials source. The energy liberated when the chemical bonds are broken sustains the life processes of the bacteria.

Among the gases produced by bacterial action are hydrogen sulfide (H_2S), ammonia (NH_3), carbon dioxide (CO_2), and hydrogen (H_2). Most of these gases—carbon dioxide is an exception—are combustible and are used as a source of power in larger treatment plants.

The Trickling Filter System

trickling filter

One of the two most common methods of secondary treatment, known as the trickling filter system, utilizes a bed of stones through which sewage passes. Sewage water from a sedimentation tank is sprayed onto the stones from fixed pipes spaced throughout the bed or from rotating horizontal pipes. Figure 3-8 illustrates the latter. The spray action provides an even distribution of the material across the surface of the stone bed and adds oxygen. The bed may be 3 to 10 feet deep. Bacteria grow on the wet stones and break down sewage material that comes in contact with them. The liquid trickles through the stone bed slowly enough that the spaces between stones do not become waterlogged. Air thus provides the oxygen essential for a complete breakdown of the organic matter.

The Activated Sludge System

activated sludge

Another method of secondary treatment, the activated sludge system, is preferred by many chemical and sanitary engineers. This system, outlined in Fig. 3-9, increases the oxidation process by bringing air and bacteria-laden sludge in contact with the effluent from a sedimentation tank. During the several hours that the load remains in the aeration tank, bacteria break down organic matter. A final settling tank removes any undigestible solids. Part of the sludge from the final settling tank, because of its high bacterial population, can be returned to the aeration tank to provide the quantity of microorganisms essential for efficient operation.

The effluent from the final settling tank should contain no organic matter except bacteria. Relatively few pathogenic bacteria remain alive because of the aeration that has taken place. Most human pathogens are anaerobic and high oxygen concentrations are lethal for many. The addition of chlorine to the final discharge water with either a trickling filter or an activated sludge system controls bacteria. Liquid chlorine is converted to a gas and injected into the effluent 15 to 30 minutes before the water is released into a natural water course. Two to 5 parts per million of chlorine

Fig. 3-8. A trickling filter system for sewage treatment in Des Moines, Iowa. (*Courtesy U.S. Environmental Protection Agency.*)

Fig. 3-9. Plan for secondary treatment using activated sludge. (Redrawn with permission from *A Primer on Waste Water Treatment,* U.S. Department of the Interior, Federal Water Pollution Control Administration, 1969.)

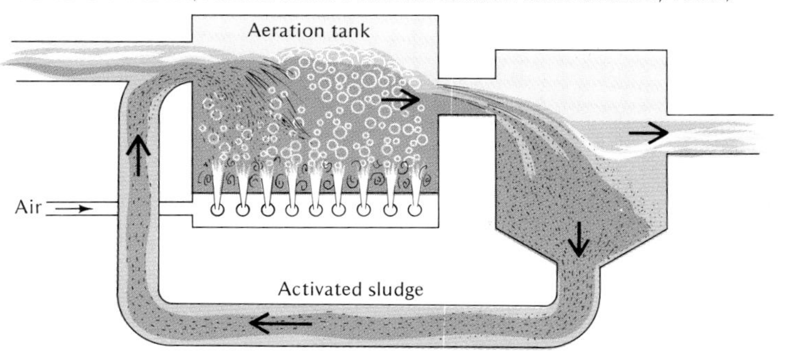

Fig. 3-10. Sewage lagoon system in New Hampshire for removal of pollutants in Cochico River. *(Courtesy U.S. Department of Agriculture, Soil Conservation Service.)*

properly administered can kill more than 99 percent of the harmful bacteria. It does, however, have a detrimental effect on aquatic life in the stream.

An activated sludge system has some advantages over a trickling filter system. The components of the former are smaller in size, and thus the treatment plant occupies less land area. Also, because it is enclosed, it is free of flies and odors. Higher operating expense is involved, however, and some industrial wastes reduce its effectiveness.

Lagoons

lagoons

As used for sewage treatment, a lagoon is simply a large pond 3 to 5 feet deep (Fig. 3-10). Sunlight, algae, bacteria, and oxygen interact to provide the equivalent of secondary treatment. The effluent from these oxidation ponds, as they are sometimes called, may be of better quality than that of traditional secondary treatment plants.

Lagoons can be used to supplement secondary treatment, as is being done at a water reclamation project at Santee, California. Following primary and secondary treatment, the town's wastewater remains in a lagoon for 30 days. After chlorination it is pumped onto land and allowed to filter through a series of lakes. The water of these lakes is suitable for swimming and other recreational uses.

EUTROPHICATION OF SURFACE WATERS

Productivity of Lakes

Areas that have undergone geological change within comparatively recent times contain most of the world's natural lakes. In North America the glaciated region around the Great Lakes is pitted with thousands of small lakes.

The large number of lakes in Florida is associated with a recent uplifting of that area. The western mountains contain more lakes than the older eastern mountains.

A useful classification of lakes can be based on productivity. Newly formed, deep, clear lakes contain relatively little nutrient material (minerals) and support limited plant and animal life. Lake Superior, Crater Lake in Oregon, Lake Tahoe in Nevada, and Lake Baikal in Russia are examples of *oligotrophic* ("few foods") lakes. Such conditions may persist for thousands or even millions of years. Lake Baikal, the deepest lake in the world, was formed before the Mesozoic era came to an end about 63 million years ago and is of special interest to biologists because it contains 405 species of animals, including 29 kinds of fish, that are found nowhere else in the world.

oligotrophic lakes

eutrophic lakes

Given sufficient time, lakes become *eutrophic* ("good foods"). From the moment a lake is formed, natural processes begin to fill it with sediment. Any incomplete decay of plants and animals dying in the water or falling in at the edge adds material to bottom sediments. Soil eroded by tributary streams and from the shores is carried into the lake. The lake shrinks; it finally becomes a grassy marsh and then a mixed bush and tree swamp; eventually a climax forest may grow where once a deep lake seemed eternal.

Evidence that man's activities are rapidly aging lakes can be found in many areas. Engineering activities on Russia's Volga River, for example, have caused shrinking of the Caspian Sea. This has been most noticeable since irrigation and power dam projects were started on the Caspian watershed in 1932. Since then the level of the Caspian Sea has fallen progressively, except during World War II when dam construction was halted temporarily.

Irrigation projects on the Amu Darya and Syr Darya, which empty into the Aral Sea (175 miles east of the Caspian Sea and formerly connected to it, covering about 24,500 square miles) began in 1940. These projects will eventually divert most if not all the flow of these rivers to irrigation. Estimates in 1956 indicated that the Aral Sea will be reduced to half its former size by 1981. Such a decrease would expose 13,000 square miles of new land. Thus man can bring about in less than a century what nature would have accomplished over thousands or even millions of years.

The Death of the Great Lakes

Great Lakes

In the heartland of North America lies the largest reservoir of fresh water in the world—the Great Lakes. They cover 95,200 square miles and contain 5457 cubic miles of water. Compared in size with other freshwater lakes of the world, they rank first (Lake Superior), fourth (Lake Huron), fifth (Lake Michigan), eleventh (Lake Erie), and thirteenth (Lake Ontario). These lakes have been of immeasurable value to Canada and the United States in providing a waterway to the continental interior. Duluth, at the western tip of Lake Superior, is 2000 miles by water from the mouth of the St. Lawrence River. The Great Lakes represent what was once assumed to be an inexhaustible supply of water for household and industrial use, for sport and commerical fishing, and for recreation. In spite of their enormous size,

Fig. 3-11. The Cuyahoga River pours its pollution (lighter shades) into the Cleveland, Ohio, harbor and then into Lake Erie where it mixes with cleaner water (darker shades). *(Courtesy U.S. Environmental Protection Agency.)*

however, ecologists warn us that these lakes show signs of rapid aging and early death (Fig. 3-11).

The death of a body of water occurs when it can no longer support fish and other oxygen-consuming aquatic animals—when the ecosystem is destroyed.

Generally, the exhaustion of dissolved oxygen occurs when excessive amounts of dead organic matter enter a waterway. The oxygen demand of the decomposer organisms that break down these wastes exceeds that produced by aquatic plants combined with that diffusing into the water from the air.

glaciers

The Great Lakes were formed by four glacial thrusts during the last 600,000 years of the Pleistocene epoch. Prior to glaciation river valleys covered the present Great Lakes area. The last glacier began retreating about 20,000 years ago and left the Great Lakes as they are today. Geologically speaking, the lakes are relatively young.

At present, Lake Erie is the most seriously polluted of the Great Lakes, although Lake Ontario, which receives the outflow from Lake Erie, is little better. Lake Erie does not have the water volume to handle the waste discharged from the metropolitan areas along its shores. Lake Erie's average depth is only 58 feet, whereas Lake Superior averages 487 feet.

Although Lake Erie seems doomed to become a long, muddy swamp in about 25,000 years, when the erosion of Niagara Falls eats a canyon back to

Fig. 3-12. Fish kill along an Illinois stream due either to depletion of oxygen supply or the presence of toxic pollutants. (*Courtesy U.S. Environmental Protection Agency.*)

the lake and drains it, it might be worth cleaning up for current use. On a much smaller scale, Lake Washington in Seattle, Washington, at one time about as polluted as Lake Erie, was "raised from the dead" by determined citizen action. Nitrates and phosphates from 10 sewage disposal plants created an algae problem and rendered the 18-mile lake unfit for recreation. The people around Lake Washington were willing to pay the price required to stop the dumping of sewage and industrial garbage into their lake. Bringing Lake Erie back would also require the cooperation of residents and businesses along the entire lakeshore and drainage basin. Corporations would have to take a new look at the relative value of dividends. Rivers so polluted that they catch fire (the Cuyahoga River near Cleveland, Ohio) should not be tolerated.

The cleaning of Lake Michigan may present a greater long-range problem than the cleaning of Lake Erie. Once pollution is controlled, Lake Erie eventually will flush itself, but Lake Michigan water moves much more slowly. If all pollution flowing into the lake were stopped now, some already in the lake would still remain for 100 years.

If the pollution of Lake Erie can be controlled and the lake made useful again, the same practices can prevent further deterioration of the other lakes. If corrective measures are not taken, all the Great Lakes, even Lake Superior, the largest, deepest, and cleanest of the five, will age rapidly because of the activities of man. The cities dependent on the lakes for drinking water would do well to remember the New York City water crisis of 1965 during which water use had to be curtailed drastically while the polluted Hudson River flowed uselessly by.

THERMAL POLLUTION

In fossil fuel and nuclear power plants, steam is the final source of energy that spins the turbines. After the steam has done its work, it is cooled to a liquid state in condensers and returned to the boiler. Because water has a high heat capacity and is readily available, it is the most suitable medium for cooling hot condensers. The successful operation of any steam plant, then, demands a continual supply of cold water which is heated and returned to the waterway.

As power demands increase, more water will be used for cooling. Electric power companies now use more than 111 billion gallons of water per day just for this purpose. This amounts to 80 percent of the water used in industry, exclusive of agriculture. Electric power companies, in the American tradition, sense an urgency for planning now to be able to meet future power demands. It is of more than passing interest to note, however, that these companies and the public in general tend to refer to these "demands" as if they were "needs." In view of the environmental crisis problems facing the human species, we might ask, somewhat parenthetically, if our power demands and our power needs are really synonymous.

thermal pollution

Thermal pollution results from the discharge of waste heat into natural waters. Sometimes the term is restricted to cases in which actual harm to living organisms occurs. However, any unnatural change in water temperature is more likely to be harmful than helpful. The organisms within ecosystems have adapted to their environments over long periods of time. Some may have the genetic flexibility to adjust to higher temperatures, but others are restricted to a narrow temperature range. Since damage may not be apparent immediately, or may go unnoticed in species assumed to be of little economic importance, thermal pollution, as used here refers to any significant increase in the temperature of natural waters as a result of the activities of man. In other specific cases thermal pollution also includes incidents of high water temperature not under the direct influence of man. The temporary surge of warm ocean currents into cold-water areas is an example.

Temperature Tolerance of Living Organisms

temperature tolerance

Living organisms have an optimum temperature or a limited temperature range at which they function best. For a particular organism this may be different at different stages in the life cycle. Many ecologists think that temperature exerts more control over living organisms than any other environmental factor. Cold-blooded animals—all animals except birds and mammals—can regulate their body temperature only within a few degrees of the environmental temperature. Conditioning fish to a higher temperature reduces the range in which they can survive (see Fig. 3-13). The gain in the upper limit of tolerance is always less than the loss in the lower limits.

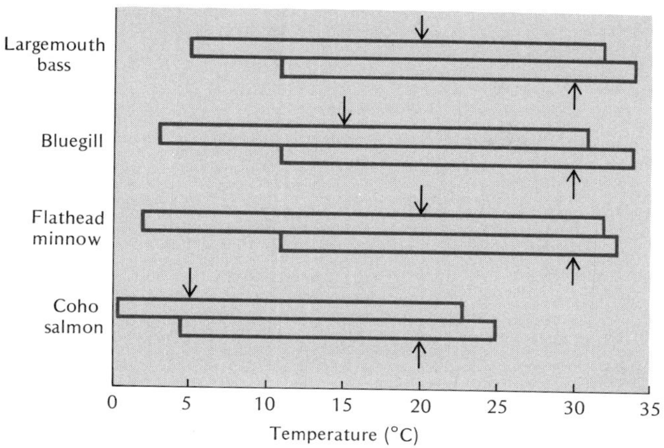

Fig. 3-13. Effect of acclimatization to different temperatures on the survival of four kinds of fish. Arrowhead indicates the temperature at which the fish were acclimatized. Limits of the bar indicate the LD_{50} temperature. (Based on W. S. Spector, *Handbook of Biological Data,* Philadelphia: Saunders, 1956.)

For example, raising the temperature 10°C (18°F) for flathead minnows raises their upper tolerance level only 1°. At the same time, they would suffer 50-percent mortality at a temperature 9° higher than before. This bears out the principle that high temperatures are less hospitable to life than low temperatures. This is particularly true in aquatic habitats because the amount of oxygen that dissolves in water decreases as the temperature increases. Also, an increase in temperature produces an increase in metabolic rate, thus increasing the demand for oxygen.

The speedup in metabolic rate caused by high temperatures shortens life in many aquatic animals. *Daphnia,* a water flea, has a life-span of 108 days at 46°F but only 29 days at 83°F (Fig. 3-14). Similarly, another water flea, *Moina,* lives 14 days at its optimum temperature of 55°F. At 91°F it lives only 5 days.

Temperature affects the reproductive cycles of many organisms. Above 72°F the banded sunfish fails to develop eggs. Cell division in carp eggs stops when the temperature increases to about 70°F. The possum shrimp, an estuary species, fails to lay eggs when the water temperature exceeds 45°F.

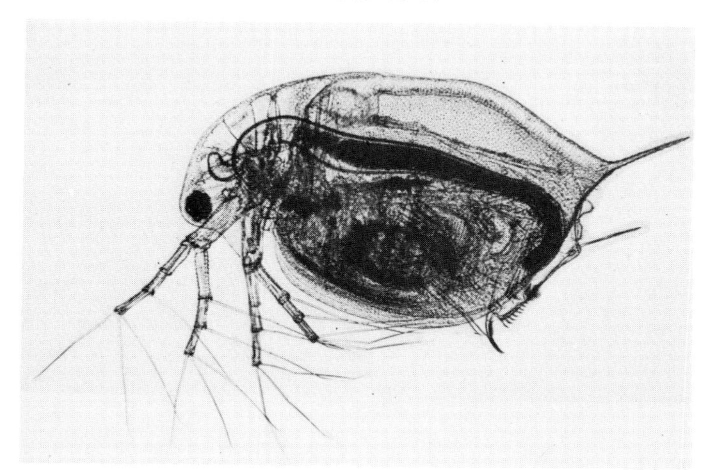

Fig. 3-14. *Daphnia,* water flea. Higher water temperature even though not hot by our standards, shortens the life-span of these minute animals. (*Courtesy Caroline Biological Supply Company.*)

Body size may also be influenced by temperature. In general, animals inhabiting cold water grow more slowly but attain greater final size than those in warmer water. This has been observed in molluscs and in several kinds of fish.

Sources of Thermal Pollution

sources

While many industries use great quantities of water for cooling, more than three-fourths of the water used for industrial purposes is used to cool the condensers of electric power plants. The 1968 annual use of water for industrial cooling in the United States amounted to 60 trillion gallons. While this has not caused serious trouble except locally, the prospect of a ninefold increase in power production is cause for alarm. Figure 3-15 shows the projected water requirement for steam-produced electric power up to the year 2020. This projection takes into account the increasing proportion of nuclear power plants which produce 60 percent more waste heat per kilowatt-hour than conventional fossil fuel plants. At the present growth rate, power demands may reach 2 million megawatts by the year 2000. Depending on the ratio of nuclear power to oil and coal-fired plants used, waste heat could amount to 20 million billion British thermal units. Since one British thermal unit raises the temperature of 1 pound of water 1°F, about one-third of the average daily freshwater runoff would circulate through power plants. During summer periods of limited water supply, all the daily water runoff in the United States would flow through power plants. The temperature of a river flowing at 3000 cubic feet per second would be raised 10°F by a 1000-megawatt power plant. Studies conducted by one power company indicate that the river temperature returns to normal by the time the water has flowed 3 miles from the plant. However, many variables (river volume fluctuations, width of the river, normal and abnormal air and water temperatures, and so on) influence how far downstream the thermal effects are observed.

In many streams during low summer water, the water temperature may reach 90°F. This is only 3° below what many feel is the maximum tempera-

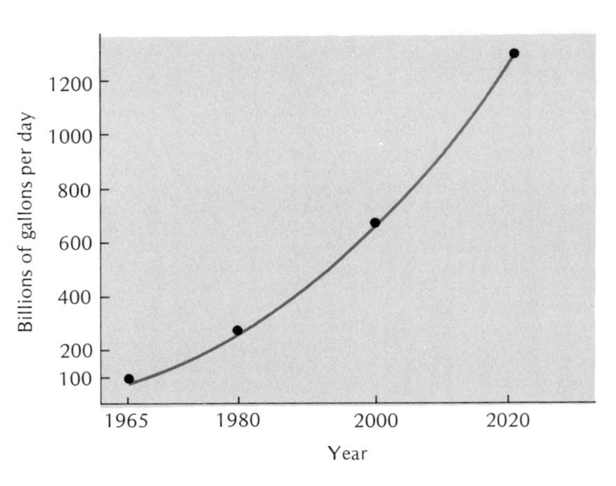

Fig. 3-15. Projected water requirements for steam electric power production in the United States, 1965–2020. 1965, 97.4; 1980, 265.0; 2000, 677.4; 2020, 1292.0 billion gallons per day. (Based on data from D. K. Todd, *The Water Encyclopedia,* Port Washington, New York: Water Information Center, 1970.)

ture that most fish in United States streams can tolerate. It has been suggested that a series of power plants on any United States river could well eliminate fish in most of the river during the warm months and interfere with migration patterns during the spring.

Solutions

wet towers

An alternative to returning hot water to the natural water source is the use of cooling towers. One type of tower, generally referred to as a wet tower, cools the condenser water through the process of evaporation. Hot water is sprayed onto a series of baffles in a silolike tower about 300 to 450 feet high and 300 feet in diameter. The hot water is cooled by evaporation as it falls through a draft of rising warm air. The water that collects at the bottom of the tower is about 20 to 28° cooler than when it entered the system. The cooled water may be reused in the plant cooling system, or it may be returned to the natural water source.

Large volumes of water are needed for cooling by this method. The coal-fueled plant built by the Tennessee Valley Authority on the Green River in Kentucky has three wet towers, each 320 feet in diameter at the base and 437 feet high. The combined capacity of these towers amounts to 846,000 gallons per minute. They reduce the water temperature 27.5°F.

Wet tower cooling by a 1000-megawatt power plant evaporates 20,000 to 25,000 gallons of water vapor per minute. This amounts to the equivalent of 1 inch of rainfall each day over 2 square miles. During cold weather

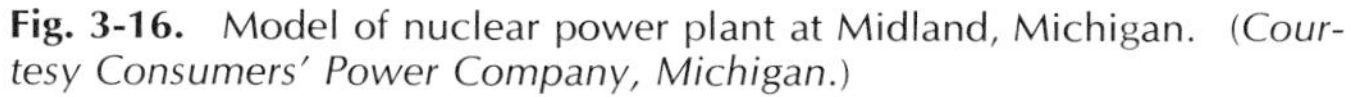

Fig. 3-16. Model of nuclear power plant at Midland, Michigan. (*Courtesy Consumers' Power Company, Michigan.*)

an ice-forming fog coats the roads and countryside downwind. Wet towers should not be used where salt water is the coolant because salt spray from the towers would coat the surrounding area. As much as 160 square miles could be affected.

dry towers

Dry towers operate similarly to the closed cooling system of an automobile. The tower contains a huge radiator. No water is lost or returned to natural sources. This type of system costs two and one-half to three times more to build and operate than a wet tower system.

artificial lakes

Artificial lakes may provide an answer to the cooling needs of power plants and other industries, particularly where several plants use the same water source. A 3-square-mile lake would provide all the cooling water needed for a 1000-megawatt power plant. Such a lake need be only a few feet deep at the discharge end and slope to a depth of 50 feet at the water intake. About 500,000 gallons per minute would be pumped from the lake for cooling steam condensers and then returned 20° warmer. At this pumping rate all the water in the lake would circulate through the plant every 15 days.

The lake is not a closed system, however. A water source must be available to replace water lost by evaporation and seepage. In some respects a cooling lake resembles a giant wet tower. Heat still has to be dissipated into the atmosphere, and evaporation plays a significant role. Whether this would produce excessive fog and icing in winter remains to be determined.

The National Environmental Policy Act, in effect since January 1970, requires exploration of the environmental impact before the construction of potentially hazardous installations such as nuclear power plants. A decision by the U.S. Court of Appeals for the District of Columbia issued July 23, 1971, sharply criticized the U.S. Atomic Energy Commission on four counts. Implied in the decision was the feeling that the intent of the National Environmental Policy Act had been subverted because of what the U.S. Atomic Energy Commission considered necessary to avoid a national power crisis. The result of the decision may be a requirement for cooling tower installation in nuclear power plants rather than discharge of heated water into rivers and lakes. This may apply to plants already in operation as well as to those under construction.

DISPOSAL BY SEA—EVERYBODY'S CESSPOOL

Much of man's pollution eventually finds its way into the sea. The 186 cubic miles of rainwater falling onto the oceans each day carry with it soil from Kansas, soot from Chicago, Illinois, sulfur from West Virginia, and hydrocarbons from automobile exhausts.

The oceans are large—they consist of over 350 million cubic miles of seawater containing more than 200,000 kinds of living things. It seems

Fig. 3-17. Acids being dumped into the ocean off the California coast. Approximately 3000 gallons are dumped per minute for 45 minutes from this barge. (*Courtesy California Regional Water Quality Control Board.*)

Carson

Ehrlich

difficult if not impossible to destroy this ecosystem. In her book, *The Sea around Us,* Rachel Carson (1951) states, "Man cannot control or change the ocean as, in his brief tenancy on earth, he has subdued and plundered the continents." That statement was made in 1951, however. Man has greatly increased his ability to exploit, contaminate, and destroy since Rachel Carson expressed this commonly held opinion. Many informed people who see what is happening to world oceans no longer ask, "Will the ocean survive?," but rather, "How long will the ocean sustain life?" Paul Ehrlich believes that if we continue pouring in wastes (Fig. 3-17) all life in the sea could end by 1979; other observers, less pessimistic, say it may last another 50 years. While these dire predictions are not likely to be fulfilled, they do point out how serious our problems are thought to be.

sick fish

Certain facts enter into any consideration of the destruction of a marine ecosystem. Productivity is not uniform either vertically or horizontally. Most production takes place in the upper few feet and over the relatively shallow waters of the continental shelves. About 80 percent of the marine fish catch is taken from continental shelf waters. And, more critical, 70 percent of useful fish and shellfish (clams, oysters, and so on) spend at least a part of their lives in coastal estuaries. Estuaries (river–ocean junctions) and coastal wet lands (tidal marshes) are seven times more productive in terms of the capture of solar energy than a wheatfield, and 20 times more productive than the open ocean. Yet we have allowed these lands to be drained and filled at an alarming rate of 900 square miles in the last 20 years. At the same time, our estuaries receive 30 billion gallons of sewage and industrial waste every day. We note that increasing numbers of ocean fish taken from the New York area have a disease that rots their tails and fins, that more deformed and diseased fish are being caught off southern California, and that catches of mackerel taken off central California contain

DDT

so much DDT that they must be impounded. The fishermen of Panacea, Florida, bring in only one-tenth as many crabs as they did 5 years ago, and the California sardine fishery is a thing of the past. In 1968 the 37 million tons of solid waste dumped into ocean waters off the United States coasts included garbage, trash, oil, dredging spoils, cleaning solutions, industrial acids, airplane parts, junked cars, spoiled food, and radioactive waste, to mention a few.

waste chemicals

A relatively new aspect of ocean pollution is seen in the increased dumping of barrels filled with chemical wastes. The shallow, heavily fished waters of the North Sea have yielded 80 of these barrels to the nets of fishermen in the last few years, and many others have washed ashore. Barrels being loaded at Rotterdam, the Netherlands, were supposed to be dropped north of the 65th parallel of north latitude, in depths no shallower than 2000 meters (beyond the 1000-fathom line) and at least 100 miles from shore. One of the substances found in these barrels, endosulfan, is toxic to fish in a concentration of 0.01 part per million. The drums also damage fishing nets. The number of such drums that have been dumped is estimated at tens of thousands.

Cousteau

Piccard

When Jacques Cousteau was diving in the Sargasso Sea in 1942, he could see at depths down to 300 feet. According to the magazine *Time*,[1] he reports that visibility is now restricted to the first 100 feet. Cousteau states further that large fish are very difficult to find in the Mediterranean. Macroscopic plant and animal life in the ocean has declined 30 to 50 percent in the last 20 years. Jacques Piccard, testifying before a United Nations symposium in Geneva, Switzerland, stated that if nothing is done the oceans will be dead in 30 years. If Piccard's estimate, that the yearly release of petroleum products onto the surface of the oceans is somewhere between 5 and 10 million tons, is close to correct, great sections of the ocean may now be loaded beyond their capacity. Toxic material, although biodegradable, may persist in the ocean depths for long periods of time.

Pesticides in the Ocean

DDT

Although the use of DDT has declined in the United States during the past 10 years, samples of phytoplankton collected in Monterey Bay, California, in 1969 contained three times more DDT than 1955 samples. This increase indicates that a considerable time lag may follow the reduction in domestic use of DDT before a similar reduction occurs in the oceans.

Sea slicks are surface films of organic matter floating on the water. These films retain chlorinated pesticides in concentrations up to 100,000 times those of the subsurface water (Seba and Corcoran, 1969). High biological activity occurs in the slicks. The tendency of DDT and other chlorinated hydrocarbons to concentrate in these surface films may lead to a much more rapid accumulation in marine food chains than if the distribution of the pesticides were homogeneous.

[1] "Dying Oceans, Poisoned Seas." *Time*, pp. 74–76, November 8, 1971.

The common crab of the California surf zone, *Emerita analoga,* can be used as an index of DDT pollution in coastal waters. When 25 crabs were placed for 24 hours in 25 liters of water containing 7.8 parts per trillion DDT, the average DDT concentration in the tissues of the crabs rose to 1016 parts per trillion. This proved to be 325 times the final seawater concentration and resulted from an active uptake process. Dead crabs absorbed only 2 percent as much as live crabs. The concentration was not a food chain phenomenon because the water had been filtered and no feeding took place.

Crabs that had recently spawned had a much lower DDT content than those of the same area that were carrying eggs. The concentration of DDT in the egg yolk indicates the phase of the life cycle most vulnerable to DDT—reproduction.

Three areas along the California coast had peak concentrations of DDT and related compounds. The crabs along the shore at the mouth of San Francisco Bay contained DDT concentrations slightly above 100 parts per billion. This is attributed to drainage from the agricultural regions of the Sacramento and San Joaquin Valleys. Slightly higher concentrations in the vicinity of Monterey Bay probably stem from the Salinas Valley. The Los Angeles County sewer carries effluent from the Montrose Chemical Corporation (a manufacturer of DDT) and empties into the Pacific Ocean at White's Point. The crabs collected in the vicinity had a DDT concentration 45 times that of the other populations. Sediments at a depth of 15 centimeters taken 3.7 miles from the sewer outfall had 310 parts per billion of DDT. Just how good are the chances for survival of offshore animals?

Oil on Troubled Waters

oil

Oil discharge regulations at the Federal level were published in the *Federal Register* (July 24, 1971). These regulations, applying to ocean waters within United States jurisdiction, prohibit any discharge of oil except from a normally functioning engine. These regulations cannot, however, prevent bilge flushing into the open seas. Also, laws can assure use of more safety devices but laws cannot prevent accidents.

the Torrey Canyon

In March 1967, the Torrey Canyon rammed the Seven Stones Rocks off the southwestern tip of England and spilled 35 million gallons of oil. On February 20, 1971, an oil tanker, the Esso Gettysburg, ran aground at the mouth of the New Haven, Connecticut, harbor and lost 386,000 gallons of heating oil. The resulting oil slick was ½ mile wide and 3½ miles long. Several days before the New Haven incident, two tankers collided in heavy fog under the Golden Gate Bridge. Between 500,000 and 1½ million gallons of oil were spilled.

Santa Barbara oil spill

The short- and long-term effects of the 1969 Santa Barbara oil blowout (Figs. 3-18 and 3-19) were investigated through a grant of $240,000 from the Western Oil and Gas Association to the Allen Hancock Foundation at

Fig. 3-18. Oil leak from a well off the Santa Barbara Coast. *(Courtesy U.S. Department of the Interior, National Park Service.)*

Fig. 3-19. Oil-covered beaches at Santa Barbara, California, 1969. *(Courtesy U.S. Department of the Interior, National Park Service.)*

the University of Southern California. The study lasted 18 months and 40 scientists participated. A review[2] of the findings reported that the spill apparently caused relatively little biological damage and that recovery was rapid. Oil may have killed some sea lion pups. This report by Humble Oil Company fails to list details on any species that were discouraged by the oil. Assuming the complete integrity of the research on the Santa Barbara spill, we may conclude that the type of petroleum product involved makes a big difference. Much of the Santa Barbara oil was heavy crude oil and settled to the bottom. The effect on marine life, then, did not approach the disaster of the West Falmouth, Massachusetts, spill in the fall of 1969. At that time a fuel barge crunched onto submerged rocks and lost 650 tons of fuel oil. This quantity is small compared with the Torrey Canyon wreck or the offshore oil rig failures along the Gulf and Pacific coasts.

Cape Cod oil spill

The West Falmouth accident, however, occurred near the Woods Hole Oceanographic Institution. In contrast to most other places where spills have occurred, this area had been studied extensively before the accident. An immediate and extensive kill of fish, worms, shellfish, and crustaceans took place. Ninety-five percent of the animals living within 10 feet of the surface died immediately, and a lingering death took many others.

Eighteen months later oil had covered 5500 acres of sea and tidal river bottom down to depths of 42 feet and was still spreading. No repopulation of the original disaster area had taken place during the 18 months.

[2] "Santa Barbara, Alive and Well." *The Humble Way*, 10(3):2–7, 1971.

Toxicity of Oil

toxicity of oil

The variety of toxic substances in crude oil renders all marine organisms susceptible to oil poisoning. Refined oil may be even worse because many of the toxic substances are more concentrated. The most toxic components of oil are low-boiling aromatic hydrocarbons such as benzene, xylene, and toluene. These dissolve in water more readily than saturated hydrocarbons and kill even in dilute concentrations. They are just as toxic to man as to marine forms. A low concentration of saturated hydrocarbons immobilizes many marine animals by acting as an anesthetic. In higher concentrations they cause cell damage, particularly in young animals. Olefinic hydrocarbons, found in refined petroleum products, are intermediate in toxicity between saturated hydrocarbons and aromatics.

Fig. 3-20. "Oiled" birds. *(Courtesy Michigan Department of Natural Resources.)*

Crude oils in general contain carcinogenic agents. Some of them, such as benzpyrene, are among the components that distill at high temperatures. A higher rate of skin cancer among refinery workers has been linked to prolonged contact with oil. In machine shops cutting tools on lathes, broachers, hand-screw machines, and reamers are cooled by a continuous flow of cutting oil. This oil, homogenized in water, may be sprayed onto the operator during part of the machine operation cycle. A worker who spends an 8-hour shift with a shirt sleeve or pants leg soaked with cutting oil may break out in a rash of blisters where his skin has been in contact with oil-soaked clothes.

The physical effect of oil pollution can be just as damaging. When seabird feathers become oil-clogged, they lose the air space that provides buoyancy and insulation. At low temperatures the birds freeze to death. Birds attempting to preen their feathers eat some of the oil. Oil is a severe hazard to diving birds such as puffins, anhingas, cormorants, and auks (Fig. 3-20).

Some Results of Oil Spills

auks

A rapid decline in the English Channel auk population has paralleled increased shipping. A 1908 census showed more than 100,000 auks along the English Channel; in 1967 only 200 were found. How much of the decline was due to oil and how much to DDT and to other causes is difficult to determine. Large oil spills, however, do claim the lives of thousands of birds. Carcasses of "oiled" birds, as many as 460 along 1 mile of shoreline, make up part of the drift that falling tides leave on the Newfoundland coast. Oil-caused bird kills reaching 10,000 have occurred on the Shetland Islands. These islands have nesting rookeries of shag, guillemot, gannet, fulmer, skua—in fact, of most northern European seabirds.

seals

Oil affects fur-bearing seals and otters much as it affects birds. Loss of insulation and buoyancy results in freezing and drowning. Oil spilling from a grounded oil tanker destroyed a sea otter colony on Kamchatka Island.

The threat of oil hangs over the Louisiana fur industry. Each year 3.5 million fur-bearing animals (muskrat, mink, raccoon, otter, nutria) are taken in the swamps of Louisiana. These animals spend most of their lives in or along water and many inhabit the coastal wetlands. Many of them could be victims of offshore oil well accidents.

sea otters

Some animals, given a reprieve from extinction from hunters, have made remarkable recoveries only to be threatened by oil spills. Nearly 1000 sea otters now live along the California coast. They belong to the southern race (*Enhydra lutris nereis*), which is distinguished from the northern race primarily by minor differences in the structure of the skull. The two once occurred over a wide stretch of the Pacific coast from Baja, California, northward to Alaska and westward across the Aleutian Islands to the Kurile Islands. The southern race occupied the coast from Washington southward.

The effect of an oil spill can differ considerably from place to place or from time to time at the same location. Elephant seal and sea lion pups on San Miguel Island might have suffered a high casualty rate had the Santa Barbara blowout occurred during the time they were nursing.

The bacteria that decompose oil use oxygen in large amounts. The breakdown of 1 gallon of oil may require the dissolved oxygen of 320,000 gallons of oxygen-saturated water. Where animals are already under the stress of toxins and low oxygen supply due to sewage pollution, an oil spill may push them across the thin line between marginal survival and extermination.

Current research at Esso Research and Engineering Laboratories involves efforts to find suitable organisms that can help clean up oil spills. Some soil bacteria appear to have this capacity and may have marine counterparts. Before the water used by oil refineries is returned to its natural water source, it can be treated to remove organic compounds derived from the oil. This is done by aeration of the effluent and then allowing microorganisms to break down the hydrocarbons through the action of enzymes.

Suggested Readings

Aron, W. I. "Ship Canals and Aquatic Ecosystems." *Science*, 174(4004):13–20, 1971.

Bazell, R. J. "Water Pollution: Conservationists Criticize New Permit Program." *Science*, 171(3968):266–268, 1971.

Burnett, R. "DDT Residues: Distribution of Concentration in *Emerita analoga* (Stimpson) along Coastal California." *Science*, 174(4009):606–608, 1971.

Carpenter, P. L. *Microbiology*. Philadelphia: Saunders, 1967.

Carson, R. *The Sea Around Us*. New York: Oxford University Press, 1951.

Cox, J. L. "DDT Residues in Marine Phytoplankton: Increase from 1955 to 1969." *Science*, 170(3953):71–73, 1970.

Editor. "Santa Barbara, Alive and Well." *The Humble Way*, 10(3):2–7, 1971.

Editor. "Dying Oceans, Poisoned Seas." *Time*, p. 74–76, November 8, 1971.

Greve, P. A. "Chemical Wastes in the Sea: New Forms of Marine Pollution." *Science*, 173(4001):1021–1022, 1971.

Hartley, E. R. "Recycled Water." *Science Digest,* 70(5):82–86, 1971.

Hoff, J. G. "Water Pollution—A Case History." *Science Teacher,* 38(1):31–33, 1971.

Huling, E. E. and T. C. Hollocher. "Groundwater Contamination by Road Salt: Steady-State Concentrations in East Central Massachusetts." *Science,* 176 (4032):288–290, 1972.

Jannasch, H. W., K. Eimhjellen, C. O. Wirsen, and A. Farmanfarmaian. "Microbial Degradation of Organic Matter in the Deep Sea." *Science,* 171(3972):672–675, 1971.

Kohl, D. H., G. B. Shearer, and B. Commoner. "Fertilizer Nitrogen: Contribution to Nitrate in Surface Water in a Corn Belt Watershed." *Science,* 174(4016): 1331–1334, 1971.

Krenkel, P. A. and F. L. Parker. *Biological Aspects of Thermal Pollution.* Nashville, Tennessee: Vanderbilt University Press, 1968.

Kuhn, D. J. "Comparison Test of Oxygen Content of Water Samples." *Science Teacher,* 37(9):58–59, 1970.

Mathews, D. "They Like Oil for Lunch." *The Humble Way,* 10(3):9–11, 1971.

Nelson, G. "We're Making a Cesspool of the Sea." *National Wildlife,* 8(5):14–16, 1970.

Powers, C. F. and A. Robertson. "The Aging Great Lakes." *Scientific American,* 215(5):94–100, 102, 104, 1966.

Ryther, J. H. "Nitrogen, Phosphorus, and Eutrophication in the Coastal Marine Environment." *Science,* 171(3975):1008–1013, 1971.

Schlee, S. "All the Fat and Sullage Fuddy Murriners." *Natural History,* 79(10): 18–29, 1970.

Seba, D. B. and E. F. Corcoran. "Surface Slicks as Concentrators of Pesticides in the Marine Environment." *Pesticide Monitoring Journal* 3(3):190–193, 1969.

Todd, D. K., Ed. *The Water Encyclopedia.* Port Washington, New York: Water Information Center, 1970.

Wagner, K. A. "Materiel Command Tropical Tests in Panama." *Air Service Command Report,* Patterson Field, Dayton, Ohio, 1944.

Wolman, M. G. "The Nation's Rivers." *Science,* 174(4012):905–917, 1971.

Depletion of the soil results in abandoned farmland. (*Courtesy Michigan Department of Natural Resources.*)

Soil

MAN'S DEPENDENCE ON THE SOIL

man and the soil

Throughout man's history, his progress has been closely tied to the soil. Fertility of the soil and the development of techniques to utilize this fertility spell the difference between bare subsistence and the production of a food surplus. Only when man produced more food than he needed for himself and his immediate family could some individuals be freed for the development of other arts. The growth of cities has always depended on the excess production of the farmer to feed those who did not till the land. The increased concentration of our population in cities places more stress on land which must now be worked by fewer people. At the same time farm land is being lost to erosion, housing developments, rural industrial sites and highways. A close look at soil will help us understand the magnitude of the problems arising from an increasing world population.

Some Basic Facts about the Land

1. Most food is produced from land crops. Foods that come from the sea may be strongly affected by what happens on land.
2. Exhausted, unproductive land can result only in hunger. A majority of the world population was underfed even early in this century.
3. Uncontrolled erosion is reducing the already limited amount of arable land in the world.

4 The United States has about 500 million acres of arable land which is being farmed or could be farmed with acceptable land-improvement practices. This is below the long-accepted criterion requiring 2½ acres per person to obtain an adequate diet. Increased yields may have given us a reprieve from the 2½-acre standard, but for how long?

5 More than 200 million acres of formerly arable land have been lost for cultivation or severely damaged by erosion.

6 Everyone has a vital stake in conservation of the soil in the interest of maintaining a reasonable standard of living.

the North Atlantic

Although the amount of food obtained from the ocean has increased in recent years, the ocean remains the ultimate recipient of the waste products of our civilization. The North Atlantic receives a disproportionate share of the world's pollution from the Rhone, the Seine, the Thames, the Rhine, the Hudson, the Delaware, the St. Lawrence, and countless smaller rivers. The threatened extinction of the pelican because of thin eggshells is really a problem involving contamination of the ocean with DDT. By the time man develops the technology to farm the ocean, the ocean may be too polluted to produce food safe for human consumption. In the foreseeable future most food must still come from the land.

Pearson and Harper (1945) calculated that 85 percent of the world's food comes directly from plants. Most of the rest comes from domesticated animals which have been fed land-grown plants. Thus man looks to the soil for his food.

soil products

The soil, through the plants grown on it, provides much more than just food. Since the invention of weaving, fiber-producing plants have provided raw materials for cloth. Manilla rope from hemp, linen from flax plants, sisal cordage and rope from several of the century plants (*Agave* species), and cotton of many varieties illustrate this plant dependence. Cellulose, the primary constituent of plants, continues to be the basis of most synthetic fibers. Wool and most other animal fibers trace back to the materials and energy that animals obtained from land-grown plants. Plants in ancient and in modern times have provided many additional items used by man. From plants we obtain shelter materials, furniture, gums, resins, dyes, oils, medicines, intoxicants, narcotics, alcohols, spices, perfumes, plastics, and other synthetics. All these depend on soil.

THE ORIGIN AND DEVELOPMENT OF SOIL

Soil is more than crushed rock. From a biologist's point of view, soil consists not only of particles derived from rocks (sand, silt, and clay) but also minerals, humus from decaying plants and animals, water, air, and microscopic life.

Soils differ considerably from one location to another. Parent rock from which soil is derived varies from one locality to another, and the processes of soil formation are not always the same. C. E. Kellogg of the U.S. Department of Agriculture places the number of different kinds of soil in the United States at 80,000 and in the world at 500,000 to 600,000. Soils can be moved considerable distances. The soils of much of the northern United States have been moved to their present locations by glaciers. Over much of the central United States, the soils are derived from loess—wind-blown particles. Water and gravity also play a part in moving soils.

Whatever its origin, soil is important because it supports, both physically and nutritionally, green plants, the producers in the various ecosystems. Where soil has been lost after timber has been destroyed or ranges grazed beyond their capacity, man's institutions have fallen. Darby (1956) describes the effects that the clearing of European woodlands had on the soil. Southern Europe along the Mediterranean formerly had an open forest of pines and evergreen oaks. This type of forest has had less regenerative success than forests in other parts of Europe. By the dawn of the Christian Era, most of the Mediterranean forest had been converted to scrublands or open, bare areas. In both, erosion was severe. Natural reforestation, which might have saved the soil, was hindered by the grazing of goats.

podsolization

Three processes of soil development, dependent on climate and vegetation, result in major differences in soil from one part of the world to another. The first of these, *podsolization,* occurs in the cold, rainy climates that support coniferous forests. These forests add organic matter to the soil very slowly. Needles and other litter accumulate on the surface, and decay occurs over a long period of time. The acidity of the litter, because of its gums and resins, inhibits the activities of many bacteria and fungi. Rain falling on this material becomes acidic and so dissolves minerals from the upper soil layer. These minerals may be carried deep into the soil by percolation of the water, or may be washed away. In either event the top layer (the A horizon) becomes little more than sand. The deposit of minerals, primarily iron and aluminum salts, in the subsoil (the B horizon) results in a dark-red heavy clay. The soil of the Canadian boreal forest illustrates an extreme degree of podsolization.

calcification

The second process of soil formation, characteristic of grassland areas and drier climates, is known as *calcification*. Grass vegetation decomposes more readily than the litter of coniferous forests and does not produce an acid soil. Water leaches very little of the soil minerals, partly because of lower rainfall and also because the soil is not highly acid. The minerals that leach most are calcium and other carbonates which tend to be redeposited in the subsoil, hence the term calcification is used to describe this process of soil development. Typical calcified soils, best developed in open grasslands with moderate rainfall, have a dark, deep topsoil rich in humus. These grade into desert soils which are also calcified to some extent. More sparse vegetation, and therefore less humus, coupled with lower annual rainfall and less leaching, results in a topsoil of high mineral content. The soil classified as pedocal covers all more-or-less calcified soils: grassland, scrub, and desert.

Tropical soils develop under conditions of high rainfall (as do podsolized soils) and high temperature (as do calcified soils). This combination, however, produces a soil different from the other two. Leaching, accelerated by copious rainfall and high temperature, removes most of the nutrients from the topsoil. Deep-penetrating root systems accomplish the recycling necessary to maintain growth in a lush tropical rainforest. According to C. E. Kellogg, a drip pan under a tropical rainforest would collect in 1 year about enough nutrients to account for the annual growth of the forest. Under such conditions, although the biomass is large, productivity is low. Slow turnover explains why slash-and-burn (see p. 83) plots must be left so long before they can be cropped a second time. In some cases they are left more than 20 years.

Fig. 4-1. Intense fires destroy not only vegetation and wildlife but also burn humus from the soil. This lowers soil quality and delays recovery. *(Courtesy Michigan Department of Natural Resources.)*

laterization

The development of tropical soils, particularly in rainforest areas, follows, then, a pattern known as *laterization,* the third of the three major processes of soil formation. A characteristic of laterized soil is the accumulation of iron and aluminum salts in the A horizon. Under extreme conditions, especially exposure to the baking action of direct sunlight when a forest is destroyed, topsoil becomes rocklike (laterite) and cannot be readily cultivated.

loams

The soils of most agricultural lands are broadly classified as loams. Such soils consist of coarse particles (sand and gravel) as well as fine particles (silt and clay). Soil structure depends on the amount of clay and humus present. Both these materials absorb water on their surfaces and become linked with larger soil particles to form crumblike aggregates. If both clay and humus are lacking, as in pure sand, the soil has no definite structure. Soils without larger particles and with too much clay are classed as heavy and may form clay pans nearly impervious to water. They are difficult for plant roots to penetrate.

water-holding capacity

The ability of the soil to hold water determines its value for agriculture. Water penetrates readily into light, structureless sands and gravels, but little water remains near the surface—the normal root zone. Heavy clay soils, however, hold the water that does penetrate and thus may be moist long after other soils have dried. The water in these heavy soils adheres so tightly to the clay particles, however, that most plant roots cannot absorb much of it.

In contrast to heavy clay, soils with both large and small particles provide the best water reservoir for root systems. In such soils minute channels of capillary size bring water from deep in the soil to the root zone as the upper soil water is depleted by evaporation and absorption by plants. This capillary water is essential for agricultural success.

Thus several factors contribute to the value of this vital resource, soil. When particle size, water-holding capacity, or oxygen, mineral, or humus content changes, the capacity of the soil to support agriculture may be drastically altered. Some requirements for specific crops are discussed in the following section.

Land Use Classification

The conservation of our soil depends upon intelligent management which can be achieved only through an understanding of the capability of the land. Features such as slope, actual and potential erosion, stoniness, salt in the soil, danger from flooding, level of the water table, capacity of the soil for holding water, general fertility, and intended use enter into the development of a land use plan. From this information a land capability plan can be drawn which maps an area for its optimum use. Eight classes are convenient to designate the potential of the land.

Land suitable for cultivation:

Class 1. Land that is free of erosion or nearly so; good productivity therefore makes it suitable for cultivation without special practices.

Class 2. Land requiring moderate protection from erosion, such as planting cover crops and contour farming; some draining may be necessary (small ditches); moderate to good productivity.

Class 3. Land requiring extensive erosion control, such as strip cropping, terracing and/or extensive draining; moderate to good productivity.

Land suitable only for limited cultivation:

Class 4. Land that should be kept in grass most of the time; might be cultivated 1 out of 6 years; erosion would be critical if cultivated regularly (too steep); moderate productivity makes it suitable for hay crops and pasture.

Land that should never be cultivated:

Class 5. Land that could be used for forests or grassland with moderate precautions to assure continued use.

Class 6. Land suitable for forestry or grazing but only with strict precautions to check erosion.

Class 7. Similar to Class 6, but extreme precautions must be enforced if land is to be used at all.

Class 8. Land that may have wildlife and recreation value but is not suitable for any kind of cultivation, grazing, or even tree farming. Generally, such land is too wet, salty, sandy, steep, or rough to be used.

The U.S. Soil Conservation Service assists farmers in the preparation of land capability maps for their farms. These maps, generally marked off on aerial photographs, enable farmers to put each plot of their ground to its best use.

CROP REQUIREMENTS—SOIL AND CLIMATE

pH

In many cases crop plants have specific soil and climate requirements. Alfalfa grows well even in acid soil if calcium is abundant. Azaleas grow best in acid soil, but their high requirement for iron may be the critical factor rather than a low pH as such. Iron is more readily available to plants in acid soils. Most crops tolerate a range of acidity varying from

Fig. 4-2. Slash-and-burn agriculture in central Cuba. A root crop is grown with the corn. Rock outcrop prevents mechanized cultivation.

pH 5.8 to 7 (neutral). Corn, small grains, many legumes, forage grasses, and many vegetables grow well under slightly acid soil conditions.

Basically, soil requirements for the production of crops include several conditions. Soil must be such that the tools used for cultivation can operate. Figure 4-2 shows a corn and taro planting in the province of Santa Clara in central Cuba. In such rocky soils mechanized cultivation is not possible. Even using a horse to pull a plow would be difficult. Hand labor and low yields result only in subsistence agriculture. Also, the soil must be resistant to destructive erosion and must store enough moisture to meet the requirements of the crop within the pattern of rainfall or irrigation. At the same time waterlogged soils can be detrimental to many crops. Sufficient oxygen must be present in the soil to permit development of an adequate root system. The availability of sufficient nutrients to insure a good yield and, at the same time, the absence of harmful concentrations of soluble chemicals, particularly salt, help determine the selection of crops for an area and even whether or not the tract should be cropped at all.

Soil characteristics affect plants at all stages of growth. Seed germination, depth of roots, woodiness of the stem, the general vigor of the plant, its final size, and even susceptibility to parasites, drought, and cold may vary for the same species in different soils.

General farm practices in the United States over the last 80 years have tended to upgrade poor soils and downgrade good ones. The use of organic fertilizer, crop rotation, and drainage can increase the fertility of relatively poor land. However, farmers who worked the rich prairie soils saw no need to maintain the high initial fertility.

A few examples illustrate additional differences in soil and climate requirements and indicate why many crops thrive only in certain areas.

Wheat

wheat

Many people depend on wheat as their principal food because this crop thrives under a variety of climate and soil conditions. Wheat grows well, for example, in typical short-grass prairie soils. These occur in the South African veldt, a grassland with scattered shrubs, in southeastern Europe and across central Asia below the evergreen Taiga. In North America semiarid grassland extends from Canada (Edmonton, Alberta) southward almost to Mexico City, a span of 33 degrees latitude (about 2400 statute miles). Wheat yields well on windswept prairies too dry and too cold for rice or corn (Fig. 4-3). Wheat needs moderate moisture and cool weather for early growth, followed by bright summers which turn dry as harvest approaches. A bit of the folklore of the Midwest is embodied in a saying formerly heard: "Cool and dry for wheat and rye."

Corn

corn

Corn is a major crop wherever the continental climate produces a humid warm summer and a humid severe-temperate winter. The coldest month generally has an average temperature below freezing and the warmest is usually above 50°F. About half of the world's annual corn crop is grown in the United States. Few crops have attained the degree of sophistication now found in hybrid corn (Fig. 4-4). Modern corn is totally a creation of man and unable to survive in the wild.

Fig. 4-3. Wheat is one of the basic foods consumed by man. Its ability to grow under conditions too dry for corn and rice has contributed materially to man's achieving his present population.

Fig. 4-4. The use of hybrid seed has greatly increased corn yields.

Rice

rice

Rice is the principal cereal crop of humid subtropical climates. Rainfall of 30 to 65 inches annually may be concentrated in the summer. Summer temperatures average 75 to 80°F and winter temperatures 40 to 50°F. Rainfall and temperature must be sufficient to mature the crops in late summer or early fall. For successful rice growing the soil must hold water for long periods.

Cotton

cotton

Cotton does best where the climate provides a span of 200 frost-free days. The average temperature during that time should be above 60°F. Although cotton is grown under a variety of conditions, the best results are obtained on light soils.

Potato

potatoes

The potato, which originated in the tropics of South America but at a high elevation, does not yield well where average temperatures exceed 70°F. Well-drained sandy loams and soils with high organic content provide the best areas for potato growth. The potato is affected by photoperiod. Long days promote stem elongation; tubers develop best during short days. Short days with low light intensity increase foliage attack by blight, however. Varieties susceptible to blight when grown in Florida during the winter season may be relatively immune when grown in Maine during the summer.

Manioc

manioc

Manioc, the source of tapioca and a staple diet item for many primitive people, grows in tropical lowlands. It yields reasonably well with little attention and on poor soils. This may account for its widespread cultivation in areas not yet touched by modern agriculture.

MAN AGAINST THE LAND

The tendency to view the destruction of the land from agricultural practices as something new or at least a phenomenon of the nineteenth and twentieth centuries overlooks similar happenings in the past. A review of the origins of agriculture and the results of various land use practices provides a basis for evaluating present trends.

The Origin of Agricultural Systems

Two approaches to farming, mixed planting and monoculture, originated early in the agricultural revolution and persist today. Each is suited to specific areas and each has its own goals, but the environmental consequences of the two differ appreciably.

Shifting Agriculture

tropical agriculture

Slash-burn-abandon designates a farming method common in the tropics. A few acres are cleared by hand and the fallen trees and other debris burned during the dry season. The mineral storehouse in the tropics is in the dense canopy of vegetation overhead. The burning of trees deposits these nutrients in the soil. In extreme cases only one crop is produced and then the farmer moves to a new area. More often a clearing can be farmed for 3 to 5 years.

Planting is begun as soon as the soil is softened at the start of the rainy season. Holes are punched in the ground and a seed or other reproductive plant part is dropped in. Hand weeding, if done at all, constitutes the only cultivation practice.

The shifting agriculture of tropical areas is an efficient use of the land and is productive as long as population pressures are low. Planting and harvesting may be continuous the year around because a variety of species is grown together and planting can be done at almost any time of the year. The *milpas* of Mexico are planted with corn, squash, beans, and other annuals; the *conucos* of more tropical America are generally planted with root crops such as taro and manioc, and cotton and bananas. The seeming disorder of these mixed plantings actually represents an efficient use of moisture and light. Erosion is reduced because the diverse vegetation protects the soil from the onslaught of falling rain. Abandonment after a time allows deep-rooted returning trees to bring fresh nutrients to the surface.

Seed Sowing—The Rise of Monoculture

Evidence of repeated cultivation of a single crop at the same site, and preparation of the seedbed prior to sowing, has been found at early neolithic sites in the East. As a result of this practice, then as today, the soil lies exposed to erosion for long periods. In early stages of agriculture, this method was used primarily for growing cereal grains. The crops matured together and were all harvested at the same time. In contrast to the irregularly shaped milpa clearing adapted to hillsides, the seed sowers chose fields on flatlands. Long, narrow fields were preferred because they reduce the number of times draft animals are turned around while cultivating.

monoculture

Planting a single crop (monoculture) in dense stands was a more effective use of the better agricultural lands and supported a larger population. Land might lie fallow but was not abandoned.

Herding animals have long, if not always, been associated with the

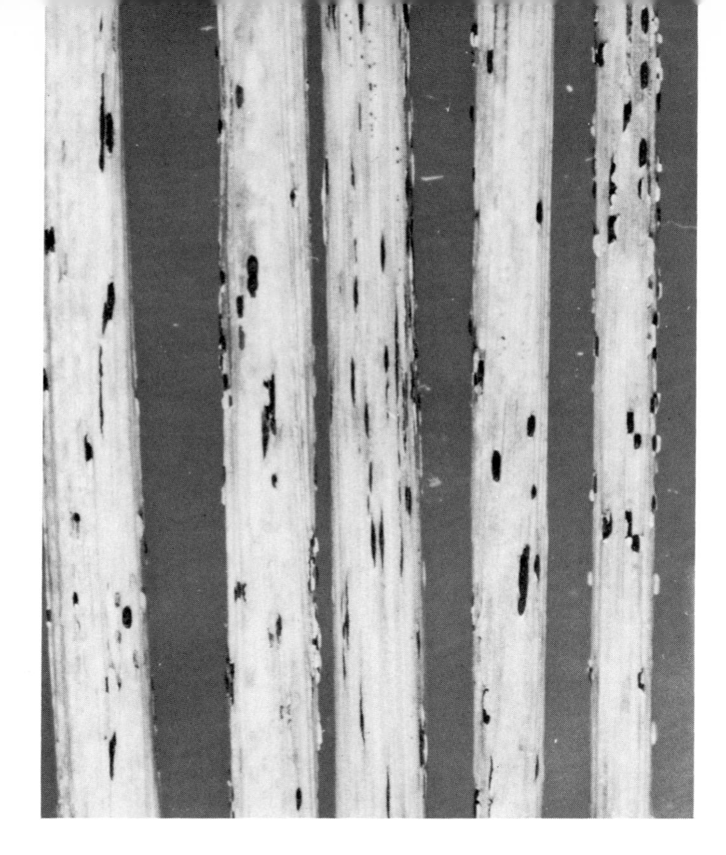

Fig. 4-5. Rust pustules on wheat stems. Monoculture favors the rapid spread of disease.

monoculture form of agriculture. Land less suited to cultivation was used for grazing, and the cropped fields were grazed after harvest.

nomads

From the Near East the seed-sowing agriculture spread in three directions. One thrust was onto the steppes of Eurasia where it lost most of the sowing aspects and became primarily pastoral. This produced a nomad culture. The original domestication of all grazing animals, except reindeer and caribou, was accomplished by sedentary agriculturists in an area between India and the Mediterranean.

rye and oats

In northern Europe, Celtic, Germanic, and Slavic people drifted westward predominately as cattle and horse raisers. This migration led them into cooler areas with shorter summers and higher humidity. The climate here was better suited to rye and oats, the weeds of former wheat and barley fields. Winnowing, a threshing process in which a mixture of grain and chaff is tossed into the air and the wind blows the chaff farther than the seeds, failed to separate the "weeds" from the intended crop. Thus rye and oats may have gradually become domesticated as the original crops failed.

hay and pasture grasses

Northwest and central Europe appear to be the original home of our principal hay and pasture grasses and clovers. With burning, openings were made in forests, which allowed the development of native grasses and legumes. Here more than elsewhere farmers were concerned with the production of crops for livestock food rather than just grazing. Provision had to be made to carry the animals through the winter. House and barn were a single structure. Prosperity was measured in terms of the number of livestock, not acres of tillable land. Such a practice conserved soil nutrients and resulted in little soil erosion. By returning manure to the fields, man established a closed ecological system.

In northwest and central Europe, clay soils, rich in nutrients but deficient in drainage, were widespread in the lowlands. The modern plow may have originated here and certainly had its major development here. The plow was beneficial in aerating the soil under such conditions, but it also exposed the land to erosion.

the Mediterranean area

Sowing and herding cultures also spread westward along both sides of the Mediterranean. Since there was little climatic change from their Near East origin, wheat and barley continued as the principal grains. Sheep and goats were of greater importance than cows and horses. The hazard of drought was present continuously. The lands on both sides of the Mediterranean became worn and eroded from persistent use. Limestone near the surface may have been a factor. Shallow soil was readily eroded when plowed. Droughts aggravated the erosion by leaving plowed hillsides bare and the soil without protection during the rainy season. The exact causes of the development of deserts along the Mediterranean coast are not fully known. These areas underwent the same kind of deterioration as that now taking place in the western United States as overgrazing magnifies the effect of any drought. More and more valuable species of forage plants disappear, and less palatable shrubs replace them.

Soil and the Iron Plow

ard

From its origin in the Fertile Crescent (and other places), agriculture had expanded to the approaches to western Europe prior to 3000 B.C., and farmers colonized Britain about 2000 B.C. The light plow (the ard), suited to cultivation of the loose, sandy soils of the Mediterranean region could not handle the heavy clays underlying the mixed oak forests. The development of an iron shear, with a moldboard to turn the furrow slice, reached southern England in pre-Roman times. The plow has been with us since, and its use has been an assumed necessary technique in agriculture. However, some side effects of the plow merit investigation (Fig. 4-6).

Fig. 4-6. Plowing exposes the soil to wind and water erosion by destroying plant cover. *(Courtesy U.S. Department of the Interior, Bureau of Land Management.)*

Fall versus spring as a time to plow and to apply fertilizer must be considered in relation to the economy of the soil, effect on production, and water pollution from runoff. The increase in fall plowing, in Indiana and Illinois, for example, results from the trend toward increased farm acreage and from earlier planting schedules. Fall plowing, which leaves the soil surface without protection even by the stubble of the harvested crop, exposes the land to wind and water erosion. Wind erosion is particularly severe on well-drained silt-loam soils. As regards water erosion, not as much difference can be found between that resulting from fall plowing and that from spring plowing. In the corn belt the heaviest rains fall in April, May, and early June, after spring plowing. Well-drained, medium-textured Illinois soil on a 2-percent slope loses 4.4 tons of soil per acre with fall plowing and 4.0 tons with spring plowing over a 1-year period.

effects of plowing

Iowa was the first state to attempt to legislate soil erosion control. If erosion standards are not met, a farmer can be required to improve his farming practices to reduce erosion.

The question of the best time to apply fertilizer parallels the plowing problem. Purdue University agronomists recommend fall only if the fertilizer can be turned under soon after application. Nitrate fertilizer should not be applied to waterlogged soils. Such soils are deficient in oxygen and favor denitrifying action by bacteria. These bacteria change valuable nitrate (NO_3) fertilizer into the nitrite (NO_2) form when bacterial action removes oxygen from the nitrate molecule. The rate of nitrate loss can be reduced by applying fertilizer after the soil temperature drops below 50°F at a depth of 4 inches.

chisel plow

One answer to the problem of erosion has been found in chisel plowing or in no plowing at all. The chisel plow cultivates without turning over soil and tends to work surface organic matter into the ground rather than bring it to the surface. On the Illinois 2-percent slope mentioned above, water erosion after chisel plowing carries away only 0.9 ton of soil per acre compared with the 4 tons per acre lost after spring plowing of the traditional type. Zero tillage increases the availability of water during drought and hot weather, but uniform stands and weed control are more difficult. In experiments at Mount Morris, Illinois, the following advantages were reported for farming without plowing.

1. Less pollution of streams and lakes with fertilizer, herbicides, and insecticides because less water runs off the land.
2. More acres can be planted by one person. Corn acreage was boosted from 60 to 250 acres on one farm without increasing manpower.
3. Some reduction in overhead costs.
4. Reduction in soil erosion. With contour, no-tillage planting on a 6-degree slope in a silt-loam soil, erosion was reduced to 3 tons of soil per acre per year. This is an improvement over "normal" losses of 8 tons per acre annually with conventional plowing on the contours and 24 tons per acre farming up and down the slope. Each ton of topsoil lost carries with it about $2.75 worth of fertilizer.

The effect of cultivation, particularly plowing, on the structure of the

Seabrook Farms

soil was revealed by studies made at Seabrook Farms in New Jersey. The Seabrook Company raises and processes large quantities of vegetables. The wastewater from washing, fluming, and other operations carried too much organic matter to permit dumping into nearby streams. (Fish would have died as bacteria used all available oxygen in degrading the organic matter.) In hopes of safely disposing of the wastes and obtaining benefits from the nutrients in the water, the wastewater was sprayed onto one of Seabrook's cultivated fields but the infiltration rate was very slow. An inch of wastewater made the field soupy. The spray was then turned onto a woodland 600 feet away which was of the same soil type. In the initial test 5 inches of water per hour for 10 hours were absorbed by the forest soil and there seemed to be no limit to the amount it could utilize. During the subsequent 15 years, 10 million gallons of wastewater per day were disposed of in the forest soils. This is the equivalent of 600 inches of rainfall per year. Vegetation flourished.

Agriculture in the New World

In the New World sugarcane became the major crop on tropical plantations. Although close planting protected the soil from severe erosion, fertility was reduced unless the waste was returned to the soil. The British colonies were the most conservative in soil management. The waste from cane harvests was fed to cattle and the manure spread on the cane fields.

row crops

Soils suffered, however, where row crops were planted. Tobacco, indigo, cotton, and even coffee plantings increased soil erosion. This stemmed from clean cultivation, which left nothing on the ground, and the use of steeper slopes and thinner soils. Tobacco cultivation produced the greatest damage, particularly in the South. Virginia, Maryland, and North Carolina were ravaged first. Ohio and eastern Missouri repeated these mistakes in the midnineteenth century.

The abandonment of cultivated land provided opportunity for stockmen. The early stages of succession afford good grazing. Grasses and legumes replace the weeds typical of the first year or two after the termination of cultivation. Throughout Spanish America stock ranches appeared on former Indian agricultural lands as the native population declined.

When the Spaniards turned from trading and looting to successful prospecting, the areas around mines were denuded of trees for fuel and building. Grass was depleted by overgrazing of pack and other work animals. Even today, many of the old mine sites of North and South America are surrounded by broad areas of impoverished plants on eroded soils.

The surges of migration that began in the late eighteenth century and lasted until World War I stemmed from the impact of the eighteenth century industrial revolution and also from a less noticed agricultural revolution in Europe. The latter was marked by the spread of potato growing, the development of beets and turnips as field crops, and the recognition of the importance of rotation of clover in the scheme of planting. Improved methods of cultivation and better breeds of livestock were also factors. When this system was transported to American soils, corn was added.

cotton

In 1800 Mexican upland cotton was introduced into the southern United States. The invention of the cotton gin made possible a rapid advance of the cotton frontier. Cotton planting moved from Georgia across to Texas during the first half of the century, sweeping the Indians away in the process. This southern parallel of the westward drive of tobacco had essentially the same disastrous effect on the land. Cotton farming, because of row cultivation and clean culture, greatly increased erosion. Southern upland soils lost organic top layers. Extensive gullies began before the Civil War. One gully that started in 1870 now cuts across all of Stewart County, Georgia, and into two adjoining counties. It has permanently ruined more than 100,000 acres of the better land of that area.

gullies

The year 1825 marked the opening of the Erie Canal. This provided a market in the East for grain and meat from the West. Agriculture pushed into Kansas and Nebraska by the middle of the century, and plowing of the prairies began. Many of the prairie counties reached maximum population in less than a generation. The droughts of the early 1890s caused a failure of corn; wheat then became the dominant crop.

We paid a high price for turning the grasslands of the prairies to the cultivation of wheat. The white pine stands of the Great Lakes region were destroyed to build barns and houses on the farms and in the towns that sprang up. The cut-over lands suffered catastrophic burning. At one time a fire in Michigan's lower peninsula burned from one lakeshore to the other.

Farming then became speculative rather than being managed as a sound investment on a long-term basis. Wheat was cropped heavily with minimum overhead for maximum short-term gain—no rotation and no organic fertilizers were used. In reality this became shifting agriculture. After exhausting the soil, farmers sold their land and moved. What happened in the United States was repeated on other great world grasslands and at about the same time. Similar destruction occurred in southern Russia, South Africa, Australia, and the Argentine pampas. The momentum of the industrial revolution was sustained by plowing the great nontropical grasslands of the world.

Erosion

Because erosion has such direct effects on the lives of so many people, and in fact affects most living things, even marine plants and animals (DDT in California coastal waters), this subject is treated in more detail (Fig. 4-7).

erosion

A prominent factor in erosion is the capacity of moving water to carry solids with it. The South Atlantic–Gulf region has 5100 miles of river subject to significant erosion but loses only about 350 acres of land surface per year. In contrast, the lower Mississippi River loses 4705 acres from only 1044 miles of river. The greatest losses occur along rivers that drain the central plains states. The combined losses of the lower Mississippi, Mis-

Fig. 4-7. An example of badly eroded land in the Tennessee Valley 35 years ago. Improved farm practices restored these areas and prevented the reoccurrence of damage. (*Courtesy Tennessee Valley Authority.*)

river system losses

souri, and the Arkansas–White–Red River systems amount to 17,005 acres per year. This is an average of approximately 3 acres per mile annually for the river mileage with significant erosion. Erosion losses along ocean beaches are slight by comparison. From Maine to Mobile, Alabama, about 0.1 acre per mile is lost each year. Tropical storms and northeasters account for most of these losses. Occasionally, local losses of greater magnitude occur during intense storms. Harrison and Wagner (1964), in a report of work on beach level fluctuation, recorded the destruction of a main sand dune along part of the Virginia shoreline. During one 2-day storm (March 7–8, 1962), the dune protecting a row of houses at Virginia Beach suffered a minimum vertical loss of 5½ feet of sand. The seaward edge of the dune was lowered more than 12 feet. An indication of similar loss along the Florida coast is shown in Fig. 4-8.

The quantity of suspended solids carried by a river correlates (other than with velocity) more with the type of agriculture than with water volume. Figure 4-9 shows the yearly sediment load for five rivers in terms of the amount of soil removed from each square mile of the drainage area. The Amazon, although it carries by far the largest volume of water, removes the least soil per square mile from its watershed. This contrasts sharply with Asian rivers flowing through densely populated agricultural areas. The heavy sediment load in the Ching Ho, a tributary of the Huang Ho (Yellow River) of China, is picked up in regions of intensive subsistence agriculture where wheat and millet are the main crops.

sediment loads

The Mississippi River drains approximately 1,144,500 square miles, and its discharge rate at Red River Landing, Louisiana, averages close to 569,500 cubic feet of water per second. This water, enriched with the soil and the minerals of the land it drains, carries a load of 385.7 million tons of topsoil into the Gulf of Mexico each year. This amounts to the erosion of 87.6 tons of soluble material and 249.4 tons of solids per acre per year for

Fig. 4-8. Evidence of shore erosion along the Florida Gulf coast. The junction of root system and tree trunk indicates the former soil level at which this slash pine tree grew.

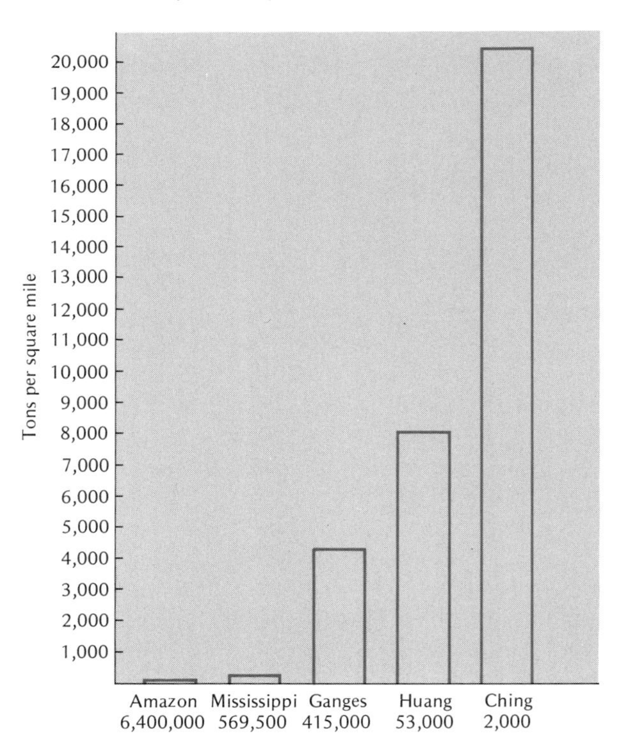

Fig. 4-9. Yearly sediment load of some major rivers in tons per square mile of drainage area. The number below the name of the river is the river volume at the mouth expressed in cubic feet of flow per second. (Data compiled from D. K. Todd, *The Water Encyclopedia*, Port Washington, N.Y.: Water Information Center, 1970.)

Some Rainfall Records[1]

Some of the world's rainfall records illustrate the potentials for erosion disaster that lie in this essential phase of the hydrological cycle. Cilaos, Réunion (an island in the Indian Ocean), had 73.6 inches of rain in 24 hours on March 15–16, 1952. Holt, Missouri, received 12 inches of rain in 42 minutes on June 22, 1947. The title for the wettest place on earth might well be claimed by Cherrapunji, India, which holds the records for the world's greatest rainfall in a 1-year period (1042 inches, August 1860 through July 1861), for 1 month (366 inches in July 1861), and for a 5-day period (150 inches in August 1841). The largest "bucket" of water ever dumped from the clouds at one time fell on July 4, 1956, when Unionville, Maryland, received 1.2 inches in 1 minute. In contrast, Bagdad, California, had no rain for 767 days (October 1912 through November 1914)—the record drought for the United States. Bagdad is wet compared with Wadi Halfa, Sudan, however, where no rain was recorded over a 19-year span. The unequal distribution of rainfall in the world and unpredictable weather behavior, in conjunction with the present practice of growing most food on the land, greatly increases the risk involved in allowing the world population to double as it will in 35 years or less at present rates.

[1] Based on data in D. K. Todd, *The Water Encyclopedia,* Port Washington, N.Y.: Water Information Center, p. 54, 1970.

the entire drainage area. When in flood stage, the muddy Mississippi dumps enough topsoil into the Gulf of Mexico each hour to cover 2700 acres of land to a depth of 7 inches. Under forest conditions a 7-inch layer of humus-rich soil represents a state of equilibrium. Soil formation equals soil loss. However, soil formation replacing that lost by erosion proceeds slowly. For all practical considerations, soil is a nonrenewable resource (Fig. 4-10). The 7 inches of soil over an acre weigh about 1000 tons. Soil

Fig. 4-10. Flood erosion near Lawrence, Washington, undercut this barn and destroyed valuable farmland. (*Courtesy U.S. Department of Agriculture, Soil Conservation Service.*)

cannot be hauled back economically even though redeposited only a short distance away. Under favorable climatic conditions and good forest, grass, or other vegetation cover, 1 inch of topsoil can build from the subsoil in 200 to 1000 years. The 3 billion tons of soil washed off American croplands each year are enough to load a freight train long enough to encircle the earth's equator 18 times. We can ill afford this loss.

effect of slope and crop rotation

As the pressure for increased food production mounts, more of the same crops may be grown in the same fields year after year, thus less crop rotation (corn one year, wheat the next, and clover the third). On a 3.6-percent slope (a drop of 3.6 feet in 100 feet) having a loam soil (a mixture of sand, clay, silt, and organic matter) in the area around Columbia, Missouri, soil loss with crop rotation was 2.7 tons per acre per year. With continuous corn production, the loss amounted to 19.7 tons per acre each year. The practice of leaving land fallow 1 or 2 years but cultivating a fine dust mulch on the surface to conserve soil moisture results in even greater susceptibility to erosion. Erosion on bare land on the 3.6-percent Missouri slope was 41 tons per acre per year. Failure to practice good soil conservation in order to feed increasing numbers of people who are unwilling or unable to limit their own reproduction rate offers no solution to the basic problem of overpopulation.

Gullies

Sometimes erosion removes the soil so evenly over large surfaces that it goes almost unnoticed. The farmer eventually sees that his ground has become lighter in color, or unproductive spots begin to show. Perhaps the plow touches rocks which are now nearer the surface than before. The humus layer of topsoil is being lost as more water falls than soaks in. This is sheet erosion, the least conspicuous but at the same time the most extensive and therefore the most destructive erosion in America.

sheet erosion

When sheet erosion progresses to the formation of microgullies, rill erosion takes place. This is more apparent in the countless grooves that run down the slopes, but it too is often ignored. The next plowing erases the rills. Rill erosion, about as serious as sheet erosion, occurs most frequently on freshly plowed soils in which the silt content is high and the slope more than 4 percent.

rill erosion

Gullies form with the combination of soft soil, heavy rainfall, and fast runoff. If water continues to flow through the same rills, the incisions may be cut deeply enough to be classified as gullies. Generally, the destruction is called a gully if the channel measures at least 1½ feet wide and 1 foot deep. Woodland erosion is described as a gully if the channel exposes main lateral roots. These gullies often start where runoff water funnels into a wooded section from a cultivated plot of higher elevation. Severely burned woodland is very susceptible to gully formation.

gullies

Unlike rills, gullies cannot be erased by cultivation. Most gullies cannot be crossed by farm machinery and some have gutted the land to depths of over 100 feet. Generally, two kinds are recognized: those with V-shaped

The Terminology of Soil Erosion

The term *gully* is used in the east-central, south-central, and southeastern states. Listed below are some of the terms used in other areas for this same landscape feature.

Arroyo: used in the Southwest, east of the Pacific slope.
Barranca: used in the Pacific Southwest.
Coulee: used in the Northwest.
Ditch: used in the Northeast and in the north-central states.
Draw: used in the West.
Wash: used in the Southwest to describe a wide, shallow gully.

bottoms due to sloping sides, and those with wide, flat bottoms and vertical sides. The former are more common; the latter develop where lower soil levels are soft and easily undercut. Undercutting results in large chunks of land breaking away and produces a U-shaped basin. Because the soft lower strata erode easily, U-shaped gullies are more difficult to control and have been known to advance uphill more than 100 feet during a single hard rain. Normally, gullies carry water only during and immediately after rains or as a result of snow melt.

control

Gully control is often a matter of recognizing the beginning stages, for this is by far the easiest time to stop the damage (Figs. 4-11 and 4-12). In general, the entire gully should be planted with some type of vegetation cover. In cases in which this is not practical, at least the head of the gully, the uphill end, must be controlled. The gully grows more by cutting uphill

Fig. 4-11. Erosion of the unprotected strip between rows of citrus trees in Tulare County, California. This destruction results from weed-free cultivation. *(Courtesy U.S. Department of Agriculture, Soil Conservation Service.)*

Fig. 4-12. The use of cover crops prevents erosion of the strip between rows of citrus trees in Tulare County, California. Wimmera 62 ryegrass is shown 1 year after seeding at 9 pounds per acre. *(Courtesy U.S. Department of Agriculture, Soil Conservation Service.)*

than by widening of the sides. Development of a vegetation cover may be hampered by violent runoff before the plants can become established. During this critical period, "porous" dams (made of brush, rolls of wire, rocks, trash) may slow the water and prevent damage. Diversion channels to shunt the water away from freshly planted areas sometimes help if they do not start new gullies.

Sodding may be employed to control gullies (Table 4-1, Fig. 4-13) although it is rather expensive. A shallow gully can be turned into a grassy flume with a high resistance to further erosion. Where the gully is steep and the quantity of available sod limited, sod strips at right angles to the fall line provide bases from which the grass can spread.

Fig. 4-13. Permanent grass-legume cover secured by heavy applications of lime and fertilizers is the most striking feature of the Tennessee Valley Authority extension test demonstration farm program. Erosion is prevented, fertility restored, labor more efficiently used, and profits from livestock, and at times from seed, frequently exceed those from cash crops. *(Courtesy Tennessee Valley Authority.)*

TABLE 4-1
Some Plants Used in Gully Control

Trees and shrubs	*Legumes and vines*	*Grasses*
Black locust	Alfalfa	Bermuda
Eucalyptus	Blackberry	Bluegrass
Honey locust	Dewberry	Brome
Pine	Kudzu	Centipede
Poplar	*Lespedeza*	Fescue
Tamarisk	Raspberry	Johnson
Walnut	Sweet clover	Kikuyu
Wild plum	Virginia creeper	Orchard
Willow		Red top
		Sudan
		Wheat grasses

Temporary dams can be used successfully to slow water and thus reduce its earth-moving capacity. The dam must be set deeply enough and built far enough into the sides of the gully so that it will not be undermined or washed out by end channels. Several low check dams provide more satisfactory control than a single large dam. The spillway in the dam should provide for any anticipated crest of water without overflowing the dam.

Livestock can defeat all efforts directed toward gully control. In fact, gullies often start in paths worn bare by sheep and cows. These animals tend to follow one another in single file when going to and from water holes. This creates a bare path which erodes more readily than adjacent grass sods. Many of the gullies on the western prairies started as bison trails. Prehistoric man in North America used these ditches to kill or disable bison by stampeding herds across them. Fencing livestock out of gullies may be the first step toward arresting erosion and eventually returning the areas to pasture.

Wind Erosion

dust bowl

The U.S. Soil Erosion Service was established in the fall of 1933, a few months before the need for soil conservation was so vividly demonstrated. The following May (1934), dust clouds rose over the parched fields of Texas, Oklahoma, Kansas, and parts of New Mexico and Colorado. The topsoil from the Great Plains, carried on high wind currents, floated eastward, darkening the sun and filtering onto windowsills and sidewalks as far as the Atlantic seaboard. Kansas soil muddied the middle Atlantic. This was wind erosion, a product of intensive agriculture plagued by drought. The storm began May 12, 1934, and dropped about 320 pounds of dust per acre over a wide belt of the United States. While this may not appear to be high, it amounts to the removal of 200 to 300 million tons of the most fertile part of the Great Plains topsoils (Fig. 4-14).

Fig. 4-14. Blowing soil has buried a hay rake on an abandoned farm in Texas County, Oklahoma (1937). *(Courtesy U.S. Department of Agriculture Office of Information.)*

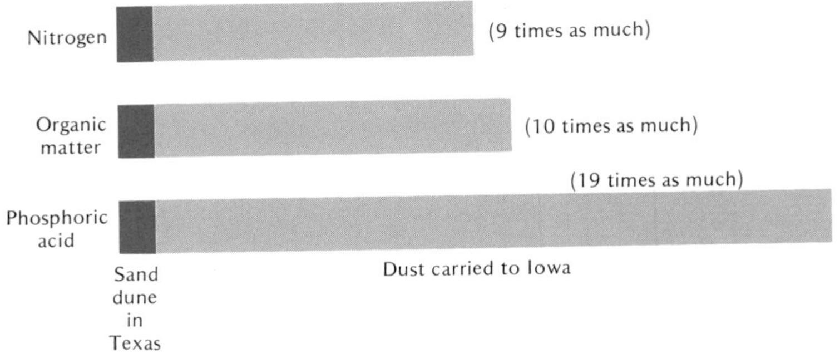

Fig. 4-15. Comparison of nitrogen, organic matter, and phosphoric acid in a sand dune formed at the origin of a dust storm with the relative quantities of these components in dust carried 500 miles. The quantity remaining in the sand dune, represented by the darker portion of the column, is considered to represent one unit.

The disaster of a dust storm stems from the kinds of materials that are moved. A storm that originated in the Texas–Oklahoma Panhandle early in 1937 traveled northeastward into Canada. An analysis of the dust that this storm dropped onto the snows in Iowa, when compared with what was left behind in a small sand dune at Dalhart, Texas, formed by the same storm, indicates which soil components moved farthest. Figure 4-15 illustrates the relative quantities of nitrogen, phosphoric acid, and humus in the material left behind and in that carried to Iowa. This shows that the dust contained 9 times as much nitrogen, 10 times as much organic matter, and 19 times as much phosphoric acid (a plant nutrient) as remained in the sand dune. Unplowed soils in the vicinity of the origin of this dust storm contained 79 percent sand; the dunes nearby, formed from cultivated soils, were 92 percent sand.

Suggested Readings

Bennett, H. H. *Elements of Soil Conservation.* New York: McGraw-Hill, 1955.

Budd, T. "Soil Loss Is Toll for Fall Plowing." *Prairie Farmer,* p. 16, November 6, 1971.

Clapper, L. E. "Washington Report." *National Wildlife,* 7(2):16–17, 1969.

Darby, H. C. "The Clearing of the Woodland in Europe." *In* W. L. Thomas, Ed., *Man's Role in Changing the Face of the Earth.* Chicago: University of Chicago Press, 1956.

Dasman, R. F. *Environmental Conservation.* New York: John Wiley, 1968.

Harlan, J. R. "Agricultural Origins: Centers and Noncenters." *Science,* 174(4008): 468–474, 1971.

Harrison, W. and K. A. Wagner. "Beach Changes at Virginia Beach, Virginia." *Army Corps of Engineers Miscellaneous Paper,* 6–64:1–25, 1964.

Hartley, W. and E. Hartley. "Can Land be Developed without Wrecking Nature?" *Science Digest,* 71(1):73–78, 1972.

Kovarik, T. J. "Raped Earth of Strip-Miners—Can It Be Healed?." *Science Digest,* 71(4):64–69, 1972.

Lemon, E., D. W. Stewart, and R. W. Shawcroft. "The Sun's Work in a Cornfield." *Science,* 174(4007):371–378, 1971.

Pearson, F. A. and F. A. Harper. *The World's Hunger.* Ithaca, N.Y.: Cornell University Press, 1945.

Thomas, W. L., Ed. *Man's Role in Changing the Face of the Earth.* Chicago: University of Chicago Press, 1956.

Wilsie, C. P. *Crop Adaptation and Distribution.* San Francisco: Freeman, 1962.

Consumers Power Company Big Rock nuclear power plant. *(Courtesy Consumers Power Company, Michigan.)*

Atomic Radiation

energy spectrum

A broad spectrum of radiant energy pervades the universe. In terms of frequency (1 Hertz = 1 cycle per second) and wavelength, electric power is slowest (60 Hertz) and longest (5000 kilometers). Radio broadcast spans a range of ½ million to 20 million Hertz and varies in wavelength from 15 to 600 meters. At the other end of the spectrum, nuclear reactors that use accelerators achieve frequencies of 3×10^{20} Hertz or more and wavelengths as short as one-trillionth of a millimeter. Between these extremes, and at progressively shorter wavelengths, lie the electromagnetic radiations of television and frequency modulated radio, microwaves, invisible heat rays, visible light, ultraviolet light, and X rays.

The shorter the wavelength, the greater the energy, and therefore the greater the potential danger to living cells. This chapter discusses the biological effects of the high-energy radiation of nuclear reactions. An understanding of the role radiation plays in cell physiology requires development of some of the concepts of atomic structure, followed by sufficient additional information to enable the student to understand the effect of high-energy radiation on man and other organisms.

subatomic units

However continuous and solid objects appear, all matter is made up of very small units called atoms. Atoms in turn are composed of three fundamental particles: electrons, protons, and neutrons. Because disintegrating atoms release about 23 different kinds of particles, atoms were once thought to contain many kinds of subatomic units. These particles, however, represent the different forms of energy that are released when atoms disintegrate. They are not part of the original atomic structure. These particles are to an atom as sawdust is to a log. A log contains no sawdust as such; it does contain material that can be broken down into sawdust as the log is cut into boards. The discussion that follows centers on the "sawdust" of the atom—the energy that is released as fundamental particles disintegrate and atoms are transmuted (changed) into other kinds of atoms.

NUCLEAR STRUCTURE

protons
neutrons

The structure of an atom resembles a minute solar system in which electrons whirl in orbit about a central nucleus. Positively charged protons and neutrons which have no charge form the bulk of the nucleus (Fig. 5-1). The neutron was first thought to be a different kind of matter, but investigations now indicate that it consists of an electron and a proton. A neutron may break down into an electron and proton when it is removed from the nucleus of an atom. Its half-life outside a nucleus is only 13 minutes, although it can remain stable within a nucleus for billions of years.

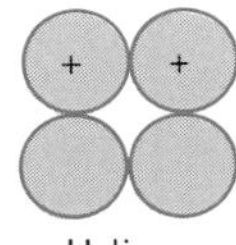

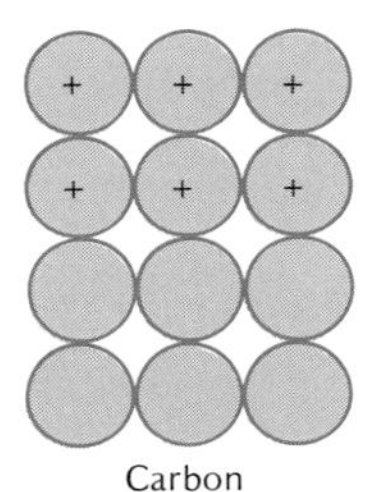

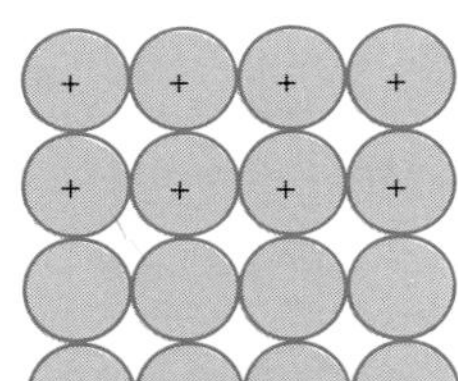

Fig. 5-1. Balance of protons (+) and neutrons in three elements.

Elements differ in the number of protons contained in their atoms. All atoms that contain 1 proton are hydrogen; those with 2 are helium; oxygen atoms contain 8 protons; uranium atoms have 92. The *atomic number* of any element states the number of protons that the atoms of that element carry. One element can be changed into another by losing or gaining protons. This process is called transformation.

proton-neutron ratio

The ratio of protons to neutrons in atomic nuclei varies. Smaller, lighter atoms tend to have these particles on a one-to-one basis. Thus helium has two protons and two neutrons; carbon has six of each and oxygen eight of each (Fig. 5-1). As the atomic number increases, so does the proportion of neutrons to protons. Some examples, starting with the lightest and proceeding to progressively heavier elements, illustrate this change. The proton-to-neutron ratio for hydrogen is 1:1, for calcium 20:20 (1:1), but for zinc 30:35, strontium 38:50, gold 79:98, and radium 88:138. This consistent shift toward more neutrons in proportion to protons with increased atomic weight is shown in Fig. 5-2.

mass number

Different atoms of the same element may vary in the number of neutrons they contain. These variant forms of an element are known as isotopes. For example, one isotope of hydrogen contains no neutrons, another has one, and a third has two. The total number of protons and neutrons in any isotope of an element is its *mass number*. The isotopes of hydrogen have mass numbers of 1, 2, or 3 (Fig. 5-3). Oxygen, the most abundant element on earth, occurs in three isotopic forms. The most common form has a mass number of 16. The other two have 9 and 10 neutrons and thus mass numbers of 17 and 18, respectively. All three isotopes have eight protons in the nucleus. Figure 5-4 illustrates three isotopes of lithium, a light metal. The number following the name of an element specifies a particular isotope of that element (carbon-12, carbon-14, strontium-90, uranium-235).

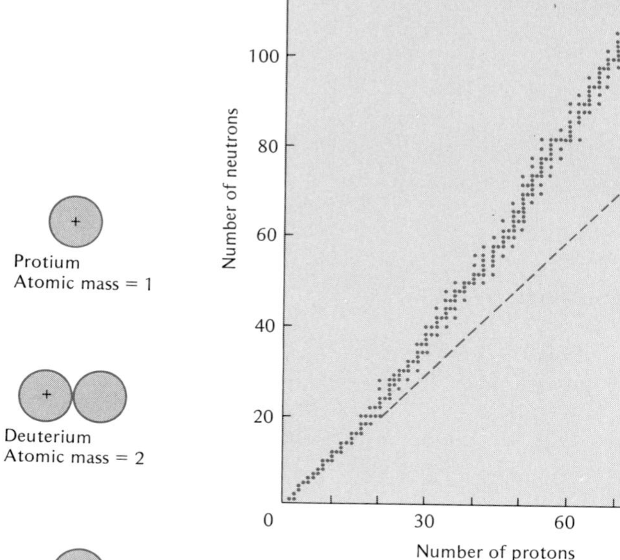

Fig. 5-3. Isotopes of hydrogen. + indicates a proton.

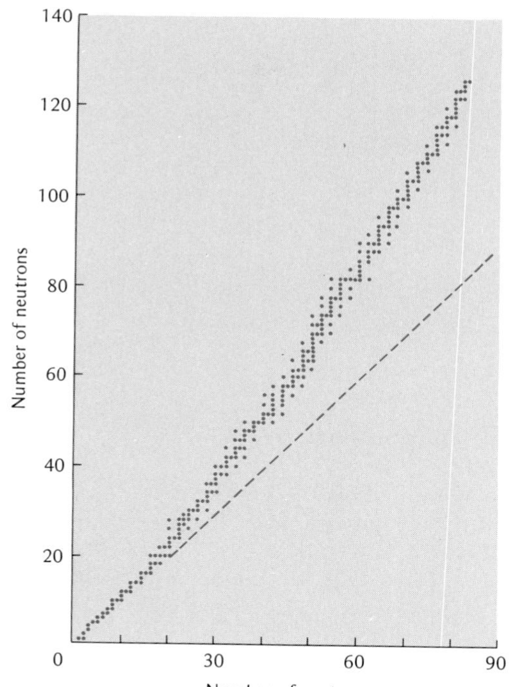

Fig. 5-2. The more protons in a nucleus, the more neutrons per proton are needed to keep the nucleus stable. Each point represents a known nucleus containing a specific number of protons and neutrons. The straight line corresponds to the positions the nuclei would be in if they contained equal numbers of protons and neutrons.

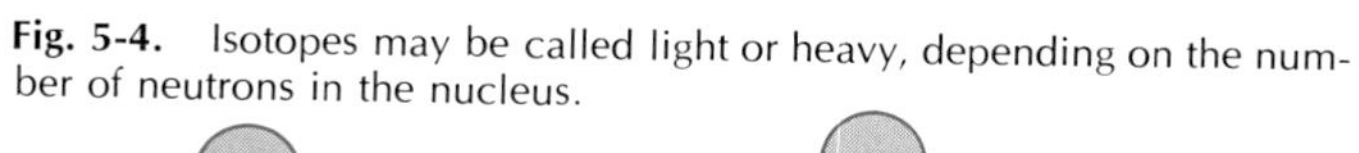

Fig. 5-4. Isotopes may be called light or heavy, depending on the number of neutrons in the nucleus.

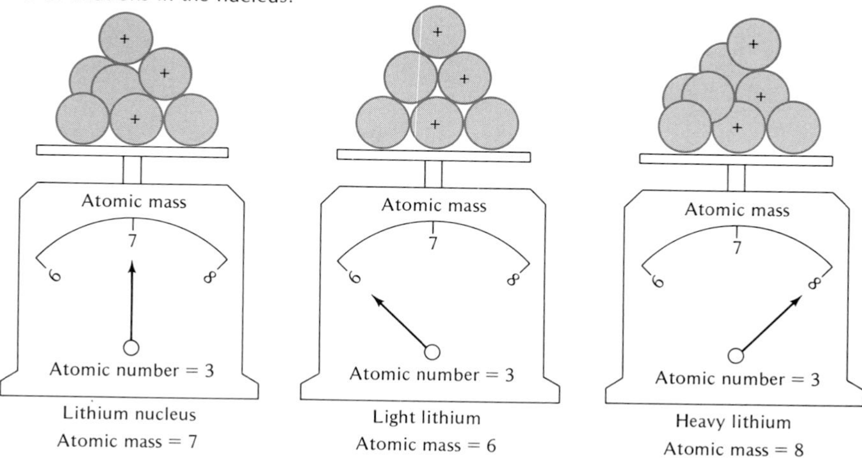

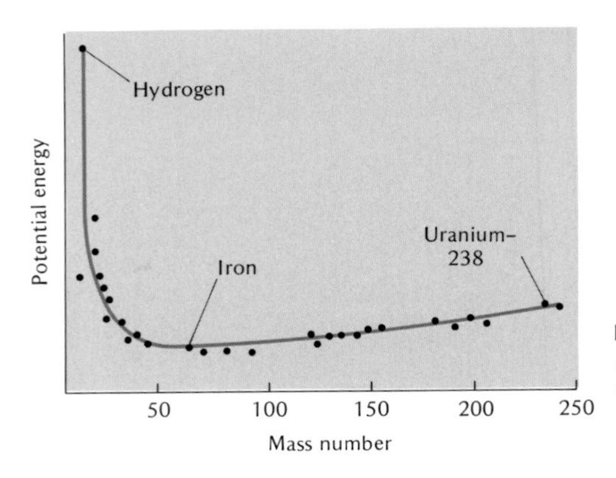

Fig. 5-5. Potential energy of nuclei. Atoms at either end of the energy curve tend to roll downhill toward iron, liberating energy as they transmute to other elements.

Nuclear Transformations and Energy

The energy within atoms powers the universe. Sunlight reaching earth originated as radiation resulting from nuclear fusion reactions within the sun. When such reactions occur, the end product weighs less than the sum of the reacting materials. This loss in weight results from the transformation of matter into energy. If all the atoms in 1 pound of matter were changed into energy, the energy released would equal that of 10 million pounds of TNT. Nuclear transformations, however, convert only part of the atom into energy.

fission and fusion

Nuclear energy can be released by either of two processes: heavy atoms can split and lightweight atoms can fuse. The fission of uranium-236 to form tellurium-137 and zirconium-97 and release two neutrons is an example of the former. The fusion of two hydrogen atoms to form helium illustrates the latter. As implied in Fig. 5-5, elements whose nuclei fall on the right side of iron release energy by fission; those on the left side by fusion. Atoms at either end of the energy curve in Fig. 5-5 tend to "roll downhill" toward iron. The steeper curve on the left side indicates the much greater energy release by fusion than by fission reactions. The energy released by a hydrogen bomb far exceeds that of a uranium bomb.

Kinds of Radiation

As radioactive (unstable) nuclei decay and change to other kinds of atoms at lower energy levels, energy is released in the form of radiation. This radiation may be in an electromagnetic form, or it may consist of subatomic particles which leave the atom at high velocity (Fig. 5-6).

alpha particles

Alpha particles have the structure of a helium atom stripped of its electrons. They consist of two protons and two neutrons. These particles emerge from a disintegrating nucleus and travel at speeds up to 10,000 miles per second. Because an alpha particle carries two positive charges, charged particles near its pathway cause deflection. Alpha particles are absorbed or deflected by a piece of paper or by the outer layer of human

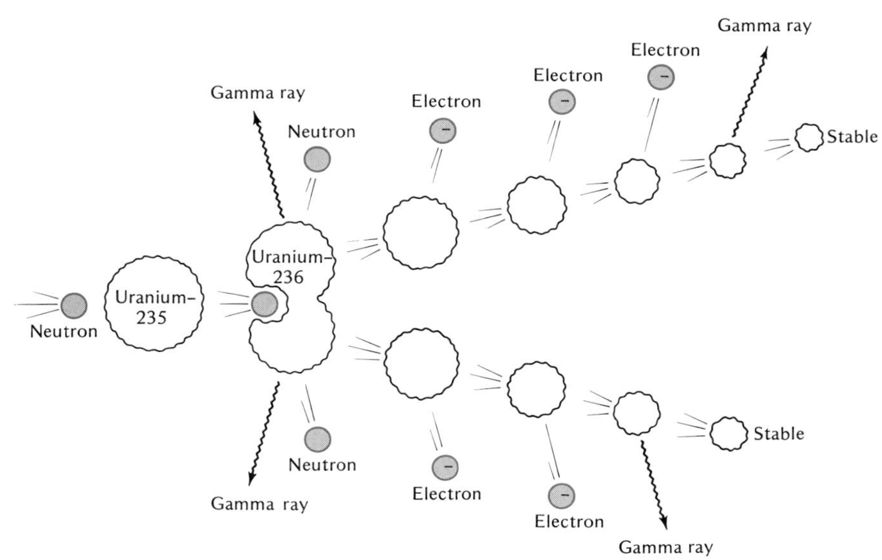

Fig. 5-6. The chain decay following the splitting of a uranium atom. The fragments give off a short series of emissions in several stages until a stable form is reached.

skin. Damage can occur, however, if a person swallows or inhales material that emits alpha particles (Fig. 5-7).

beta particles

Beta particles are high-energy electrons which escape the atomic nucleus. This occurs when a neutron breaks apart and becomes a proton. The electron is emitted and a proton is left behind. The increase in the number of protons changes the atom into a different kind of element. Beta particles vary in energy but are known to penetrate ⅛ inch of aluminum or ¼ inch of wood.

Single protons may be emitted by radioactive isotopes and are intermediate in penetrating power between alpha and beta particles. Neutrons, however, are much more penetrating because they carry no charge, thus they are not deflected by charged particles. Neutrons travel in a straight line until they collide with the nucleus of an atom. Atoms are largely open space, and as a result, a neutron may travel considerable distance

Fig. 5-7. The penetrating capabilities of three kinds of radiation.

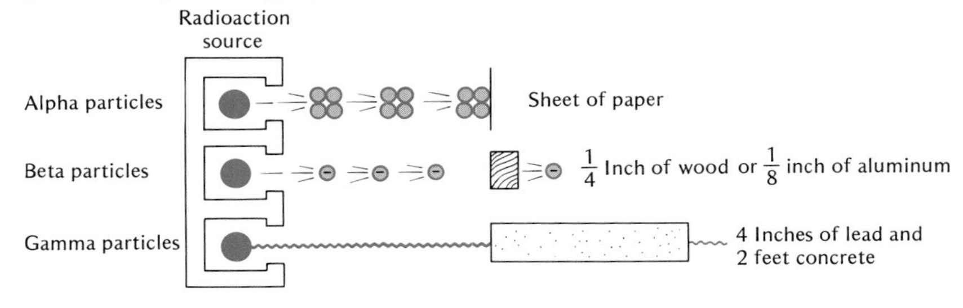

before striking the nucleus of an atom. As a result, deep body tissues receive about the same radiation as skin surfaces.

Neutron bombardment can produce radioactive isotopes from stable atoms. The detonation of an atomic bomb releases neutrons. These particles produce a variety of radioactive isotopes from the material in the vicinity of the explosion. Such radioactive material becomes part of the fallout which may harm living organisms great distances from the site of the explosion.

gamma rays

Gamma rays are short-wavelength, high-energy radiation released by radioactive isotopes. These rays, unlike alpha and beta radiation, are not made up of particles but are similar to visible light. Penetrating power is highest for the shorter wavelengths. These rays may pass through 4 inches of lead or 2 feet of concrete. The longer-wavelength gamma rays merge into the area of X rays.

cosmic radiation

Cosmic radiation, in the form of carbon, oxygen, nitrogen, and other atomic nuclei stripped of their electrons, reaches us from outer space. This primary cosmic radiation has limited penetration into the earth's atmosphere. The total insulating effect of the atmosphere at sea level is the equivalent of 3 feet of lead. The collision of these particles with atmospheric nuclei, however, produces a shower of particulate and electromagnetic radiation. Extensive cosmic showers have been recorded in salt mines 1968 feet below the earth's surface. Were not the intensity of such radiation very low, life on the surface of the earth would be impossible.

As far as is known, none of man's sense receptors respond to high-energy radiation. He may see or feel the results of exposure to radiation but not the radiation itself. Some animals, for example, small planarian flatworms (Fig. 5-8), appear to detect gamma rays. If an evenly dispersed population of planarians is exposed to cobalt-60 (an emitter of gamma rays) in such a way that part of the area is shielded by a lead plate, the worms crawl under the shield. If the eyespots of these worms are removed, they do not seek the shielding of the lead plate. Such experiments must be carried out in total darkness to avoid the same response to visible light.

Fig. 5-8. *Planaria,* freshwater flatworms. Some tests suggest that these animals can avoid radiation by moving under a lead shelter. (*Courtesy Carolina Biological Supply Company.*)

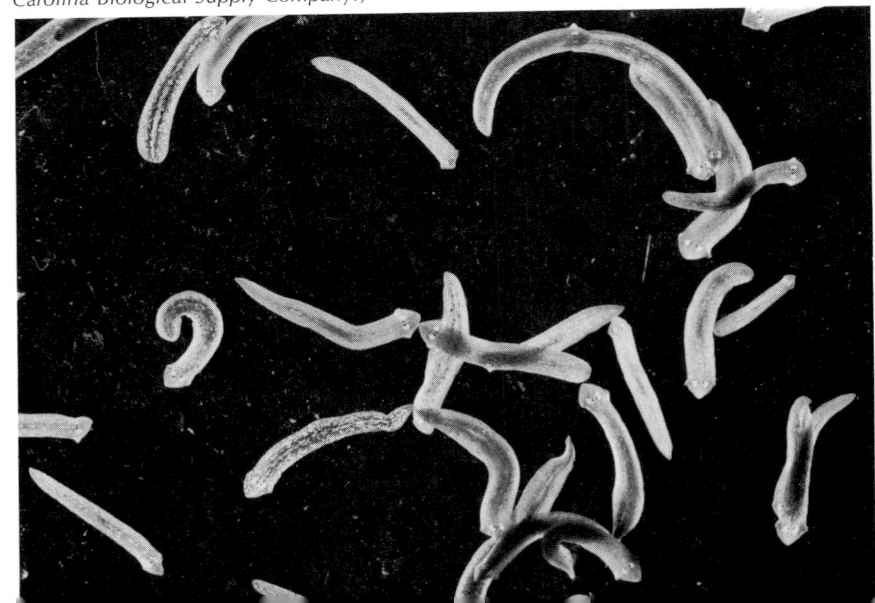

Half-Life

half-life

Scientists use the concept *half-life* to describe the decay or disintegration rate of radioactive isotopes. Radioactive isotopes undergo decay at a fixed rate independent of variations in pressure and temperature. Bismuth-210 has a half-life of 5 days. Over this period of time, 1 pound of bismuth-210 is reduced to 8 ounces. After 5 more days, 4 ounces remain. At the end of 15 days, the sample contains only 2 ounces of bismuth-210. In other words, half-life is the length of time it takes for one-half of any given amount of an istope to decay. Mathematically, no matter how many half-lives occur some radiation is always present.

Half-lives differ for different elements and for different isotopes of the same element. The half-life of polonium-212 is less than one-millionth of a second; that of potassium-40 is 1.3 billion years. In contrast to the latter, a man-made isotope of potassium, potassium-42, has a half-life of 12.4 hours. The two additional neutrons in potassium-42 make a great difference.

SOME USES OF RADIOACTIVITY

All branches of science make use of radioisotopes. Plants and animals metabolize radioisotopes in essentially the same way as the nonradioactive or stable forms of the same material. After an animal has been given small quantities of radioactive iodine, a radiation counter (Geiger counter) shows high radioactivity in the region of the thyroid gland in the neck. Some, but much less, activity is recorded in other parts of the body. Iodine concentrates in the thyroid gland. In some cases of cancer of the thyroid, pieces of this gland break off and are carried to other parts of the body.

iodine-131

Radioactive iodine (^{131}I) concentrates in these fragments, thus the migrating cancerous tissue can be located by use of a Geiger counter. Such tissue can be removed by surgery.

Rapidly dividing cells are more susceptible to destruction by radiation than nondividing cells. Chromosomes are broken by the radiation during mitosis or meiosis. Since cells of cancerous tissues undergo rapid divisions, radiation can be used as a tool in the treatment of cancer. Cancer

cancer therapy

therapy based on radiation subjects the region of the cancerous tissue to high-energy beta and gamma radiation. Hopefully, the cancerous cells will be destroyed without excessive damage to the normal tissues.

Radioactive carbon (^{14}C) ranks among the most useful isotopes in biological investigations. Carbon-14, absorbed by green plants in the uptake of carbon dioxide, played a vital role in the discovery of the com-

carbon-14

plex steps in the processes of photosynthesis. The use of carbon-14 and other radioactive isotopes has helped to solve problems related to the diffusion of materials through membranes, human blood flow, and many other aspects of metabolism.

phosphorus-32

Radioisotopes are very useful in the study of food chains and production rates in ecosystems. A measured amount of phosphorus-32 added to the water in a pond can be detected first in the algae and then in the herbivores. Finally, the larger fish and the birds that feed on the herbivores will also contain some radioactive phosphorus. Because plants remove phosphorus-32 from the water as readily as nonradioactive phosphorus, the radioactivity of the water at specified time intervals can be used to calculate the total removal of phosphorus by the plants. This gives a measure of food production in the pond.

In addition to carbon-14, neutron bombardment of atmospheric nitrogen produces tritium (^3H), a radioisotope of hydrogen. Tritium has a short half-life (12.3 years) and thus can be used to check the age of more recent events. The advertised age of alcoholic beverages can be verified by the use of tritium.

iron-59

Exposure of iron to neutrons in a nuclear reactor produces radioactive iron (iron-59). This isotope is used extensively to test for wear of metal parts. The wearing of a radioactive piston ring, for example, can be detected in the lubricating oil of an engine. This system has several advantages. The oil can be sampled while the engine is running; loss of metal as slight as one-millionth of an ounce can be detected; and the testing is rapid, economical, and simple.

carbon-14 dating

Extensive use has been made of radioactivity in determining the age of rocks, plant and animal remains, and cultural artifacts of biological origin. Plants utilizing carbon dioxide take in some radioactive carbon. As long as the plant is alive, the ratio of carbon-14 to carbon-12 in the plant tissue is the same as in the air or water in the environment. When the plant dies, the radioactive carbon in the plant continues to decay but no more is added. Carbon-14 has a half-life of 5730 years. Plant remains that are 5730 years old have half as much carbon-14 in proportion to carbon-12 as that found in living plants. Carbon-14 has been used to determine the age of materials such as wood from Egyptian tombs, ancient scrolls from the Dead Sea area, charcoal from the campsites of prehistoric man, and ancient bones. Wood thought to have been smashed by the last glacier over Wisconsin was determined by carbon-14 to be about 11,000 years old. This technique has been applied to deep-flowing ocean currents originating in the Arctic. In 1500 years the water had moved halfway to the equator. Much of our accurate knowledge of prehistoric man comes from radioactive isotope dating.

Some radiocarbon data need interpretation. Plant material and turtle shells from Montezuma's Well (north of Phoenix, Arizona) were sent to laboratories for carbon-14 dating. The results indicated an age of 14,000 to 16,000 years, although the plants and animals were alive when collected. In this case the date was not that of the plants and turtles but rather the time when the water in which they were living had fallen onto the earth. This water has been underground about 15,000 years and no new radioactive carbon from the air had been added during this time.

Because its half-life is less than 6000 years, the amount of carbon-14 remaining after 40,000 years is too small to be measured accurately. For

thorium and uranium

materials older than 30,000 to 40,000 years, age can better be determined by radioisotopes with longer half-lives. Thorium and uranium, each with a half-life of several billion years, can be used in solving problems such as the age of the earth's crust. On the basis of the transformation of uranium to lead, the oldest rocks found on earth go back about 3.8 to 4 billion years. This was a time, however, when the earth had cooled sufficiently to form rocks and raise mountains. It does not represent the beginning of the earth. Work with meteorites, which may have formed at about the same time as the earth's crust, indicates the age of the solar system is about 4.5 to 4.6 billion years. These calculations were based on potassium-to-argon and rubidium-to-strontium transformations.

IONIZATION AND RADIATION MEASUREMENT

ionization

Fast-moving, high-energy particles abruptly disturb the organization and electrical balance of atoms lying in their paths. When such particles strike an orbital electron, the impact drives the electron out of its orbit (Fig. 5-9). This disrupts the electrical balance and leaves the atom with a positive charge. The freed electron attaches to a neutral atom, giving it

Fig. 5-9. When an ionizing particle strikes an orbital electron, the impact drives the electron out of its orbit. The free electron becomes attached to a neutral atom, giving it an excess negative charge; it thus becomes a negative ion. The atom that lost the electron is overbalanced by one positive charge and therefore is a positive ion.

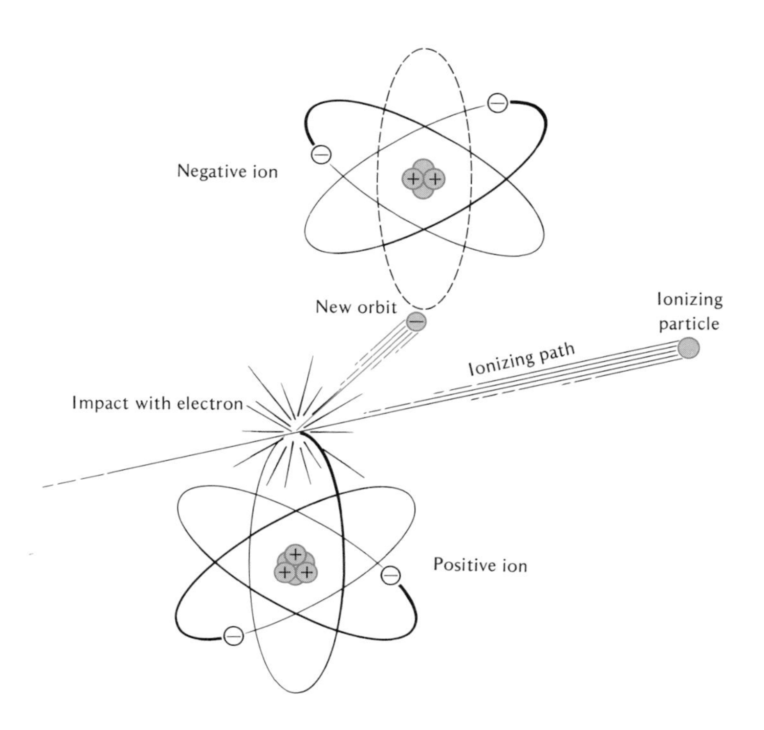

a negative charge. This process whereby neutral atoms become negatively or positively charged is called *ionization;* the charged atoms produced are called *ions.*

Secondary ionization may occur also. Uncharged particles, usually high-velocity neutrons, striking the nucleus of an atom cause it to emit a charged particle (proton). The proton, intercepting orbital electrons in its path, causes ionization on impact. Secondary ionization takes place when gamma rays interact with orbital electrons. The atom absorbs the energy of the radiation and then reestablishes equilibrium by throwing off an electron. The liberated electron may dislodge other electrons from their orbits, and they in turn dislodge others.

Ionization is the process by which radiation causes biological damage. The traditional unit for measuring radiation, the roentgen (named for Wilhelm Roentgen, the discoverer of X rays) is a measure of ionizations in the air. One roentgen is the amount of X or gamma radiation that produces 1.6×10^{12} ionizations per cubic centimeter of air at sea level. From the biologist's point of view, a more practical unit of measure is the rem (roentgen equivalent, man). The rem is used to measure ionizations or the absorbed dose of radiation in tissue, but for our purpose it can be considered about equal to a roentgen.

Background Radiation

background radiation

The intensity of the radiation that constantly surrounds us depends upon several variable factors. The earth's magnetic field deflects lower-energy cosmic rays toward the poles. Radiation from this source is greater in higher latitudes. The underlying rock strata also play a part. Granite, for example, contains more radioactive uranium, potassium, and thorium than is found in other kinds of rock. Even the kind of building material makes a difference. Brick, for example, contains more radioactive materials than wood.

The radiation exposure over the open sea at the equator averages about 53 millirems per year (1 millirem equals 0.001 rem). Denver, Colorado, founded on granite 1 mile deep, has a background count of 170 millirems per year. On 20,000-foot granite peaks at 55° North latitude, the background radiation reaches an impressive 560 millirems per year.

To this natural radiation is now added the ionizing radiation produced by man through the use of X-ray machines, television sets, and the radiation of fissionable waste materials from nuclear reactors and bomb testing. The production of nuclear energy results in a wide variety of fission products nearly all of which are radioactive. Most of these "artificial" radioisotopes have half-lives of less than 30 years but will remain hazardous for about 600 years. In terms of geological time, this addition to natural radioactivity is probably insignificant. In term's of man's lifespan, however, the effects may well be far-reaching and dangerous. If the ancient Egyptians had used nuclear energy at the magnitude we propose for the near future, we would still be affected by some of their radioactive wastes.

Dose Rate Effects

The Federal Radiation Council has established safety standards for persons who work with radioactive isotopes and for the general public. Their guidelines allow 5 rems per generation (30 years). This amounts to about 170 millirems per year above the background radiation level as the maximum permissible gonadal dosage, exclusive of radiation used for medical treatment. This is about double the average background level.

radiation damage

Certain body tissues are more susceptible to radiation damage than others. The spleen, lymph nodes, thymus in children, liver in the fetus, reproductive cells, cells of the digestive tract, all embryonic cells, and the blood-forming tissue of bone marrow are particularly vulnerable because they contain large numbers of actively dividing cells. The symptoms of exposure to high-level radiation include nausea, diarrhea, and vomiting due to the death of cells lining the alimentary canal. Cataracts may develop as a result of damage to the conjunctiva (the epithelial covering of the eye), and hair follicles may be destroyed with a resulting permanent loss of hair. These first symptoms of radiation sickness develop at exposures of 100 to 300 rems. Temporary recovery usually occurs but is often followed by a reduction in the number of red and white blood cells, hemorrhages below the surface of the skin, and ulcerations of the digestive tract and the mouth. The phase characterized by a low white blood cell count holds the greatest danger. These cells protect the body against disease. A person who might have survived the effects of radiation may die from bacterial or viral infection. Those who do survive an exposure of 100 to 300 rems suffer a reduction in fertility and an increased tendency to develop leukemia.

symptoms

Exposure to levels between 25 and 100 rems produces no obvious symptoms of damage except a reduction in the number of white blood cells. Below 25 rems no measurable change in irradiated persons can be detected, but delayed degenerative effects may occur.

lethal dose

All persons exposed to 1000 rems die because of extensive ionization within the cells. A liver cell, for example, undergoes approximately two million ionizations from an exposure of this intensity. Since each liver cell has about two billion protein molecules, only 1 in every 1000 is ionized. This is enough to destroy the cell, however.

Death from any factor, such as radiation, chemical poisons, temperature, or loss of blood, may depend on the magnitude of that factor. For statistical purposes the dosage or other appropriate quantity that kills half of a test population is a convenient reference point known as the LD_{50} (lethal dose for 50 percent). The radiation LD_{50} for human beings is 500 rems.

The United States—A Second Hiroshima

A point made by Muller (1955) concerning the long-range effects of low-level radiation seems well worth serious consideration. Each of the 160,000 survivors of the Hiroshima atom bomb blast is thought to have received about 100 rems of radiation. Although difficult to estimate precisely, a resulting 80,000 harmful gene mutations may well be the heritage of their descendants.

Based only on highly penetrating gamma radiation, the Atomic Energy Commission estimated that the total testing of nuclear devices by the United States and Russia subjects each American to an exposure of 1/10 rem. However, 100 rems produce the same genetic damage whether received quickly by one person or accumulated by many persons over a long period of time. If we multiply exposure by population, the 160,000 Hiroshima survivors received a total of 16 million "man-rems." If a similar calculation is applied to our own 200 million plus population, we are receiving an exposure of more than 20 million man-r as a result of nuclear tests. Thus the genetic damage passed to future generations in America as a direct consequence of testing may exceed that produced at Hiroshima.

In the interest of objective analysis, these mutant genes do constitute a small fraction, less than 1 percent, of the mutations inherited from past generations. Their effect alone cannot be expected to undermine the genetic constitution of the population as a whole. In fact, the careless use of unshielded X-ray equipment in all probability causes far greater damage each year than all nuclear testing to date.

Radiation Damage to Cells

Radiation results in two types of biological damage: somatic (damage to body cells) and genetic (changes that are transmitted to the next generation). In somatic and in reproductive cells, ionization breaks chemical bonds in the molecules. The new bonds that are established can change the molecular configuration. Figure 5-10 illustrates a possibility. These changes may not be compatible with cellular processes. Damaged DNA cannot program the proper genetic information required to sustain the life of the cells. Cancer of the lungs, skin, bone marrow, and other tissues can be caused by radiation. Then, too, a latent period of 5 to 20 years may occur after exposure before cancer appears.

Irradiation of somatic cells is a real hazard, but radiation of reproductive cells holds greater potential danger for the human race. Even a slight alteration of sperm or egg can lead to spontaneous abortions or deformed offspring, or produce genetic effects not immediately apparent. Random accidents to genes (mutations) may be passed unchanged to future generations. Between 5 and 12 percent of all natural mutations are thought to be caused by background radiation.

mutations

Most mutations are harmful, and yet improvement in the adaptation of a species starts with the rare gene mutations that are beneficial. The

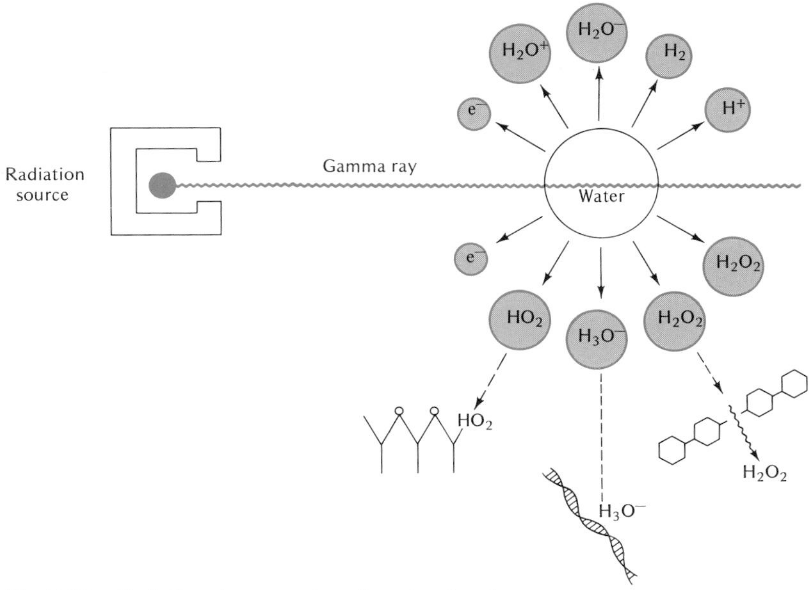

Fig. 5-10. Radiation damage. Ions from irradiated water cause structural changes in proteins, DNA, and other organic molecules. The changes may not be compatible with cellular processes.

Fig. 5-11. Death of trees due to chronic exposure (20 hours per day) to gamma radiation from a 9500-curie source of cesium-137 for about 6 months. *(Courtesy Brookhaven National Laboratory.)*

evolution of types better fitted to their ecological niches depends primarily on new gene combinations brought about by hybridization between individuals that vary in genetic makeup. These genetic differences arise through gene mutations that are random—not directed toward a planned change in a specific gene. This is somewhat similar to taking a television set that is already working well and indiscriminately changing the resistance of a circuit or switching two of the wires. So it is with plants and animals. Species present today are the result of adaptation through the pressure of natural selection over long periods of time—in some cases millions of years. Only a remote probability exists that a random mutation could further improve the effect of a gene that is already adequate. Exceptions do occur and these provide the grist for the evolutionary mill, but most mutations are detrimental and many are deadly. Mutations cause much of the illness and premature death among people.

The Essence of Heredity

Genetic information, chemically coded in the DNA molecules of reproductive cells, determines what a fertilized egg can become. The underlying principles governing the transmission of hereditary material from parent to offspring include the following generalizations. (Exceptions of limited application are ignored.)

1. The unit of inheritance is the gene, a complex molecule which controls the activities of the cell.
2. A reproductive cell (sperm or egg) contains one complete set of genes. Sexual reproduction (the union of sperm and egg) provides the offspring with two sets of genes, one inherited from each parent.
3. Genes act in pairs. The members of a pair—one received on a chromosome from each parent—may be identical, or each may promote a different expression of the same general characteristic. The cause of the albino condition in many animals including humans illustrates this. We use the symbol *A* to designate an allele (one member of a pair of genes) that promotes the development of normal skin and hair pigment; *a* symbolizes the allele that prevents pigment formation. Three combinations of this pair of alleles occur: *AA, Aa,* and *aa.*
4. In many cases a pair of different alleles (*Aa*) results in the development of one condition or the other rather than an intermediate. One allele is said to be *dominant* over the other. In the case of albinism, a person with the genotype *Aa* has the same pigment development as one with *AA*. Only the inheritance of the allele *a* from both parents produces an albino. When the alleles of a pair are alike, as in *AA* and *aa,* the individual is said to be *homozygous* for that pair. A genotype such as *Aa* is called heterozygous.
5. In other cases, such as sickle-cell anemia, the heterozygous person is intermediate between the extremes. This disease affects the shape and lowers the oxygen-carrying capacity of the red

blood cells. The gene symbols used by most geneticists for the normal allele is Hb^A. The allele that causes the sickle shape of the red blood cells is designated Hb^S. A person with no sickle-cell problem has the genotype Hb^AHb^A. Severe sickle-cell anemia results when an offspring inherits the Hb^S allele from both parents. Generally, the Hb^SHb^S genotype results in an early death. The heterozygous individual (Hb^AHb^S) has some sickle-shaped red blood cells and suffers from mild anemia. His condition is called sickle-cell trait. He should avoid high altitudes and other conditions involving low oxygen level.

6. Offspring are not necessarily similar to their parents. A marriage between normally pigmented persons can result in some albino children. This requires, however, that both parents are heterozygous for albinism (*Aa*). A characteristic that develops only in homozygous genotypes is said to be recessive. Albinism is recessive to normal pigmentation. In the same sense the allele *a* is recessive to *A*. The appearance of a recessive trait can occur only when the recessive allele is transmitted to the offspring by both parents. (The knowledge of this biological fact could reduce tension in some families.)
7. The consequences of the mating of known genotypes are mathematically predictable when large numbers of offspring are produced. The probability that the offspring from an $Aa \times Aa$ marriage will be albino is 1/4 for each child. Each gamete carries one member of each pair of alleles. The sperm will be *A* or *a*, and the egg it fertilizes likewise will be *A* or *a*. Thus the union of sperm and egg could be $A + A$, $A + a$, $a + A$, or $a + a$. Only the last combination, one out of four, results in an albino. This 25-percent chance for albinism applies separately to each conception. Such a couple might have all normal children, all albinos, or any combination. With 1000 such marriages, however, we would anticipate close to 250 albino children among the firstborn.
8. The frequency of a gene—how many people in a population carry it—is more important than dominance. Blue eye color is recessive, yet in some areas of northern Europe most of the people have blue eyes. The allele for albinism is much less frequent. In the United States one person in every 71 is a carrier (*Aa*). Thus we can expect one albino child in every 20,000 births ($1/71 \times 1/71 \times 1/4$).

Part of the problem arises from the long delay that may occur between the initial radiation damage to a gene in a reproductive cell and its visible effect on some individual in a future generation. This is possible because most radiation-induced mutations that can be passed on are recessive (see page 112, "The Essence of Heredity"). In this hidden condition a mutant gene may be carried by many of the direct descendants of the person in whom the mutation occurred. Recessive genes function in the control of structures or processes only when inherited from both parents. When two individuals, both of whom carry a recessive mutant, marry, each offspring has a 25-percent chance (one in four) of receiving the mutant gene from both parents (a 50-percent chance if each parent is considered separately).

Radiation and Aging

aging

The relationship between the aging of persons and their exposure to high-energy radiation has not been critically established. Radiation shortens the life-span of mice by about 1 week per rem, but it is difficult to translate these data in terms of man. One hypothesis suggests that the constant bombardment by background radiation gradually destroys the genetic code of somatic cells and leads to a cessation of their functions. Also, cells damaged by ionization may be treated as foreign and attacked by the body's own defense mechanisms. Radiation decreases the rate of cell division, and this could lead to organ failure.

NUCLEAR POWER AND REACTOR WASTES

electric power plant

One of the spectacular advances in the peaceful application of nuclear technology is the use of atomic energy in the production of electrical power. Figure 5-12 diagrams the basic parts of a nuclear power plant, a system that uses a fuel mixture of uranium-235 and -238. These radioactive isotopes are packed into stainless steel or zirconium alloy rods several feet long and about 1 inch in diameter. Groups of these rods are assembled far enough apart to prevent an atomic explosion but close enough to allow a sustained chain reaction. The rate of chain reaction is regulated by control rods made of cadmium. Water passing over the fuel elements changes to steam which drives the turbine generators producing electric power. The problems of thermal pollution from the cooling system were discussed in Chapter 3.

The spent fuel elements removed from the reactor core are extremely dangerous. Because of the high level of radioactivity, they are stored at the reactor site for about 150 days to allow the short-lived isotopes to

Fig. 5-12. Diagram of a nuclear power plant. (*Courtesy Consumers Power Company, Michigan.*)

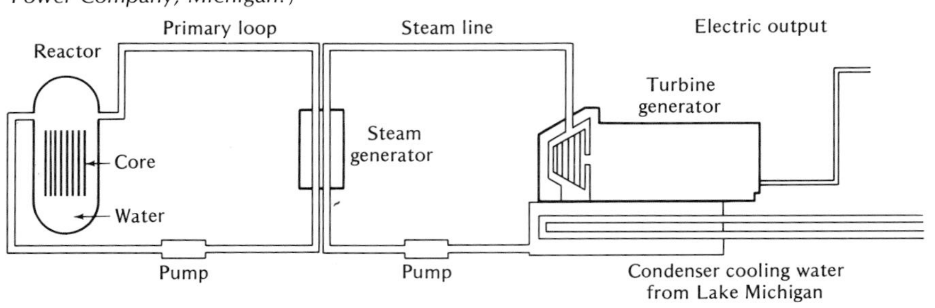

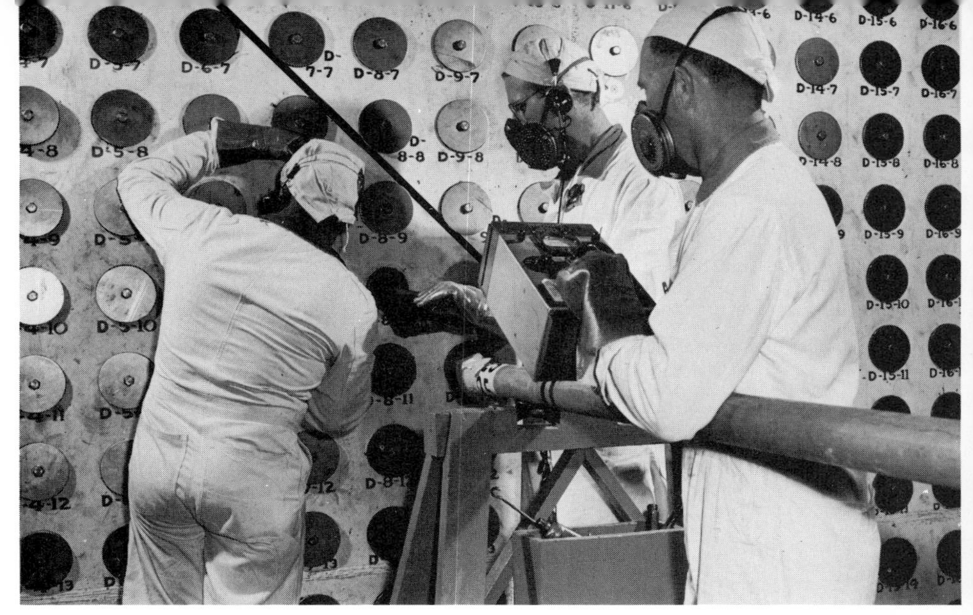

Fig. 5-13. Removal of uranium from a graphite research reactor after it has chain reacted for many months or years. (*Courtesy Brookhaven National Laboratory.*)

reprocessing

decay to a less dangerous level. Then they are moved to reprocessing plants in large lead and steel tanks weighing up to 70 tons (Figs. 5-13 and 5-14).

Only a small portion of the reactor fuel is actually burned out. A reprocessing plant separates and concentrates those materials with enough energy left to be useful, and they are returned to the reactor plant. The remainder, still very dangerous, is the waste that presents a disposal problem. After the reprocessing treatment most of the waste's radioactive isotopes are held in liquid form. This material, which contains, among other isotopes, strontium-90 and cesium-137 (hazardous lives of at least 600 years) is stored at the reprocessing plant for periods up to 10 years. Then it is condensed by drying to a solid state and shipped to a permanent storage site.

Almost ½ ton of dangerous wastes is produced for every 1000 megawatts of electric power generated. The millions of gallons of high-energy waste now in storage will remain hot enough to boil water for the next 1000 years and will outlast the containers in which they are stored (Fig. 5-15).

disposal

Basically, three methods can be used for radioactive waste disposal. The *dilute-and-disperse* method, used for low-level waste such as that found in the cooling water, involves diluting with more water and returning to the stream or source of the cooling water. With the use of wet towers, the radioactive wastes discharged from the high stacks are diluted to a "permissible" level by the large quantities of air flowing through the stacks. However, even low-level radiation release increases the hazard to man because of the potential of the organisms in food chains for concentrating radioactive materials.

The *delay-and-decay* system is used to treat most radioactive gaseous wastes of short-lived isotopes. The gases are compressed and stored until they have reached a less active level. Intermediate-level liquid wastes are

Fig. 5-14. Emptying highly radioactive waste from a lead container into a disposal vault with 17-inch-thick reinforced concrete walls, for ultimate disposal by burial. *(Courtesy Brookhaven National Laboratory.)*

discharged directly into the ground on the assumption that slow movement through the soil will allow time for the isotopes to decay to a less hazardous level.

The *concentrate-and-contain* approach is reserved for high-level-activity liquid wastes produced at fuel reprocessing plants. The wastes from these plants will eventually be dried to solids and stored (Fig. 5-16), but permanent storage methods are yet to be adopted.

salt mines

Abandoned salt mines appear to be the best disposal sites provided they are in areas not subject to earthquakes and are without flowing water. Salt beds 200 feet thick are required. Also, the bed should be at least 500 feet below the earth's surface to avoid exposure by some future glaciation, but not more than 2000 feet because of the cost of access.

Although ½ million square miles of the United States lie over salt deposits, only three areas are suitable for storage of radioactive waste. The salt mines under Detroit, Michigan, might be used. An area in western New York meets the requirements, and so does a vast area running from central Kansas southwestward into parts of Oklahoma, Texas, and New Mexico.

RADIOACTIVE MATERIALS IN FOOD CHAINS

The greatest danger from radioactive materials in the environment is that they are picked up by producer organisms and concentrated as they pass through a food chain. Two radioisotopes in particular, strontium-90 and cesium-137 (half-lives about 30 years), pose a serious problem.

strontium-90

Strontium functions in plant and animal metabolism similarly to calcium. Plants absorb strontium through their roots and use it in place of calcium in the formation of the pectate layer that cements cells together.

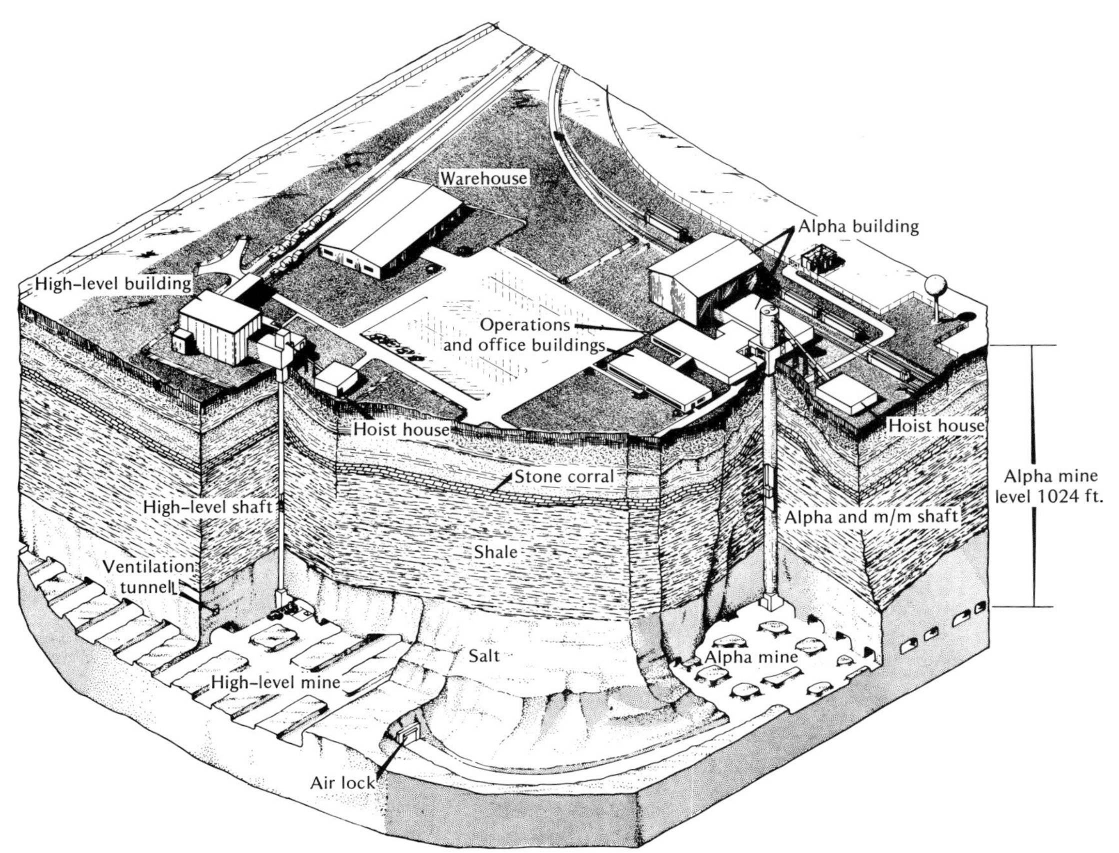

Fig. 5-15. Artist's concept of a federal repository for radioactive wastes near Lyons, Kansas. Commercial wastes from nuclear power plants are stored 1000 feet underground in the bedded salt formation. (*Courtesy Oak Ridge National Laboratory.*)

Plants appear to make no distinction between calcium and strontium; they use one as readily as the other. Thus cereal crops provide a link that carries radioactive strontium from the air to man by way of the soil.

Milk, a common source of the calcium needs of man, may contain strontium from the grass eaten by cows. But strontium is not as readily absorbed into the cow's bloodstream as is calcium. With equal amounts of calcium and strontium in the food, a cow absorbs 10 units of calcium for every unit of strontium. This filtering effect benefits those populations whose diet contains considerable milk. The strontium load in the bones of persons in Western countries is one-sixth that of the Japanese. The latter have abundant rice and little milk.

Radioactive strontium-90 concentrates in bone tissue because the body uses it in the same way as calcium. Radiation from strontium that has been incorporated into bone tissue seriously affects blood cell formation in the bone marrow. Fortunately, the amount of radioactive strontium in the soil at the present time is small, but any radiation is potentially harmful and adds to the overall body load.

cesium-137

Plants absorb cesium primarily through the leaves as a result of fallout brought down by rain. Translocation in the plant results in concentration in certain tissues, for example, potato tubers. The human body uses cesium much as it does potassium. Thus most radioactive cesium finds its way into muscle tissues. In the case of meat-eating animals, strontium may be left behind in discarded bones, but cesium becomes concentrated as it goes through a food chain. This is well illustrated in the Arctic. Slow-growing lichens (*Cladonia,* reindeer moss) accumulate cesium-137 at a level well above that found in other plants in the same area. Caribou and recently introduced reindeer feed extensively on lichens. Cesium-137 doubles in concentration from lichens to caribou and doubles again from caribou to man.

Following the 1956 hydrogen bomb test, Japanese fishermen found alarmingly high radioactivity in Pacific Ocean tuna. Geiger counter readings for some fish ran as high as 4500 counts per minute; the normal background count, although variable, is usually between 12 and 20 counts per minute. This high level of radiation was a matter of deep concern to the Japanese because of the quantity of fish in their diet. A Japanese investigation found that the ocean water over a wide area northwest of the bombsite at Bikini registered 100 counts per minute. The exceptionally high counts found in the tuna resulted from the concentration of radioactive materials as they passed through the food chain. Tuna are the final predators at the top of the food pyramid. More than 457 tons of tuna having radiation above 100 counts per minute were destroyed.

fusion reaction

The fusion of hydrogen to form heavier elements, as described earlier in this chapter, has been suggested as a "clean" way to produce power since the end product, helium, is not radioactive. Fusion methods produce an abundance of tritium (hydrogen-3) during the operating cycle, however, and as much as 13 percent may be lost from the reactor. Tritium losses from nuclear fusion could be greater than that from present fission reactors and fuel reprocessing plants combined. Clean nuclear fusion depends in part on better methods of tritium containment (Fig. 5-16).

Fig. 5-16. Tanks for temporary storage of liquid wastes near Richland, Washington. The tanks hold up to 1 million gallons each. (*Courtesy Battelle-Northwest.*)

How we dispose of our atomic wastes today will be of grave concern to mankind for a long time. Many of the high-energy wastes from nuclear plants have hazardous lives of about 600 years but others, such as plutonium-239 (half-life 24,360 years), have hazardous lives that could extend beyond the existence of the human race. The buildup of long-lived radioisotopes, for all practical purposes, will be irreversible within the generations of modern man.

Suggested Readings

"Anti-Radiation Treatment. Work by Russians in Armenia." *Science Digest,* 70(2):16, 1971.

Atomic Radiation, A Report by the RCA Service Company, U.S.A.F. Contract No. AF. 33(616)-3665, 1957.

Clegg, E. J. *The Study of Man.* New York: American Elsevier, 1968.

Dobzhansky, T. "Man's Evolutionary Future." *Science Teacher,* 39(1):17–20, 1972.

Fowler, T. K. and R. F. Post. "Progress toward Fusion Power." *Scientific American,* 215(6):21–31, 1966.

Henry, H. F. "Radioactivity—Safer Than You Think." *Science Teacher,* 37(8): 29–32, 1970.

Hogerton, J. F. "The Arrival of Nuclear Power." *Scientific American,* 218(2): 21–31, 1968.

Isaac, G. L., R. E. F. Leakey, and A. K. Behrensmeyer. "Archeological Traces of Early Hominid Activities East of Lake Rudolf, Kenya." *Science,* 173(4002): 1129–1134, 1971.

Lynch, T. F. and K. A. R. Kennedy. "Early Human Cultural and Skeletal Remains from Guitarrero Cave, Northern Peru." *Science,* 169(3952):1307–1309, 1970.

Muller, H. J. "The Genetic Damage Produced by Radiation." *Bulletin of the Atomic Scientists,* 11(6):210–212, 230, 1955.

Nelson, B. "Mobile TB X-ray Units: An Obsolete Technology Lingers." *Science,* 174(4014):1114–1115, 1971.

Novick, R. E. "Urban Anthropology: The Emerging Science of Modern Man." *Science Teacher,* 39(1):21–25, 1972.

Pryor, W. A. "Free Radicals in Biological Systems." *Scientific American,* 223(2): 70–83, 1970.

Radioisotopes in Science and Industry. Special Report of the Atomic Energy Commission, January 1960.

Rump, T. R. "A Third Generation of Breeder Reactors." *Scientific American,* 216(5):25–33, 1967.

Schaefer, H. J. "Radiation Exposure in Air Travel." *Science,* 173(3999):780–783, 1971.

Seif, M. "Fusion Power: Progress and Problems." *Science,* 173(3999):802–803, 1971.

Shapiro, H. L. "The Strange, Unfinished Saga of Peking Man." *Natural History,* 80(9):8–18, 74–83, 1971.

Shephard, J. "The Nuclear Threat inside America." *Look,* 34(25):21–27, 1970.

Williams, G., III. "Radioactive Accidents." *Science Digest,* 70(2):10–14, 1971.

Young, L. B., Ed. *Evolution of Man,* New York: Oxford University Press, 1970.

The cheetah, similar to most of the large cats, is an endangered species. The fur trade, primarily for women's coats, contributed heavily to its extermination in many areas. Fortunately, many companies have turned to synthetic products and no longer use the pelts of endangered species. (*Courtesy Lion Country Safari, Laguna Hills, California.*)

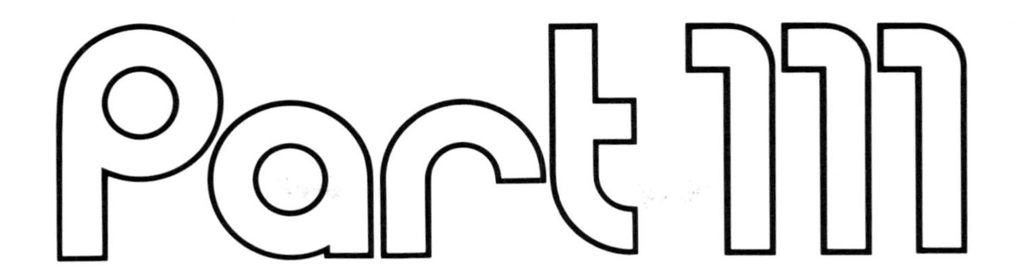

STRESSES: ECOLOGICAL AND PERSONAL

Three of the pests that plague man: flea (upper left), tick (lower left), and tapeworm (right). *(Courtesy Carolina Biological Supply Company.)*

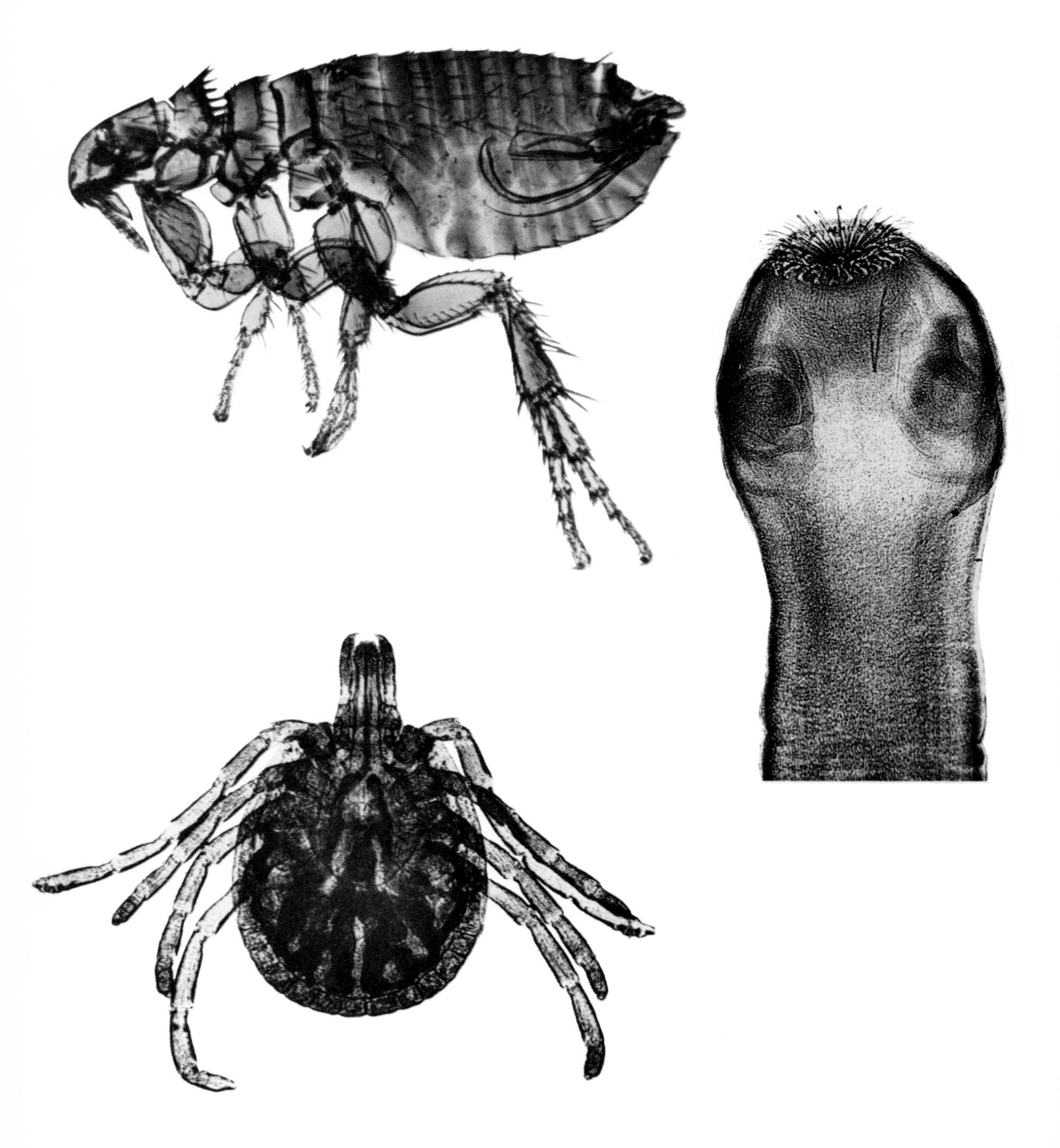

6

Disease

THE ECOLOGICAL BASIS OF DISEASE

biosphere

The biosphere, which extends over the entire surface of the earth and into the ocean depths, is made up of countless thousands of species of plants and animals, each in some way dependent on the others. This intricately complex relationship also embraces all those factors that are extrinsic to the organisms and that may in any way impinge upon them. As noted in previous chapters, this includes such factors as temperature, rainfall, humidity, topography and, of greatest importance, light because it is through the power of radiant energy that the biosphere is maintained. The energy of solar radiation enters the biosphere only through photosynthetic organisms: green and purple bacteria, algae, and a vast population of higher plants. Such organisms are of necessity confined to the portion of the biosphere that receives solar radiation by day, but the collected energy is passed through an intricate network of living things until it finds its way through the entire life spectrum and is dissipated at last in the form of heat. The real test of survival of any individual or species lies in its ability to capture, at some place in this system, enough energy to maintain it-self. Fierce competition for the available energy has resulted in the evolution of an enormous array of life-styles. Each of the thousands of species has cut out for itself a particular niche in which it can utilize a portion of the energy flowing through the ecosystem. Under the pressure of natural selection, organisms have come to occupy every conceivable place where energy is found. In other words, wherever energy is available, whether it be in the form of sunlight falling on a green leaf or in the body fluids of a host's digestive tract, some organism takes advantage of it. Without this competition there could be no selective pressures and no possibility for improving species. Consequently, life depends upon predation, not only as a means of capturing energy but also to drive evolutionary development throughout time. Cruel as it may seem, this process of capturing energy by "eating each other up" is the mainspring of the evolutionary process (Fig. 6-1).

No organism on earth can escape this system, but each in time falls prey to other creatures and yields an allotment of energy to the ecosystem. No animal in nature really lives to a ripe old age. Only man can approach this luxury, but even he finally succumbs to disease. Survival of a species is insured only if some of its members can live long enough to reproduce in sufficient numbers to maintain the population. It is through the process of death that organisms open the way for development of other living things by recycling essential materials and making possible the transfer of energy. Death then, as well as competition, is a highly important function of the biosphere. Life as we know it could not exist apart from death.

death

The relentless struggle for survival has resulted in the stratification of organisms into different levels of organization—the individual, the popula-

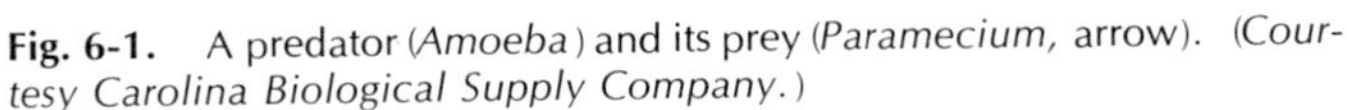

Fig. 6-1. A predator (*Amoeba*) and its prey (*Paramecium,* arrow). (*Courtesy Carolina Biological Supply Company.*)

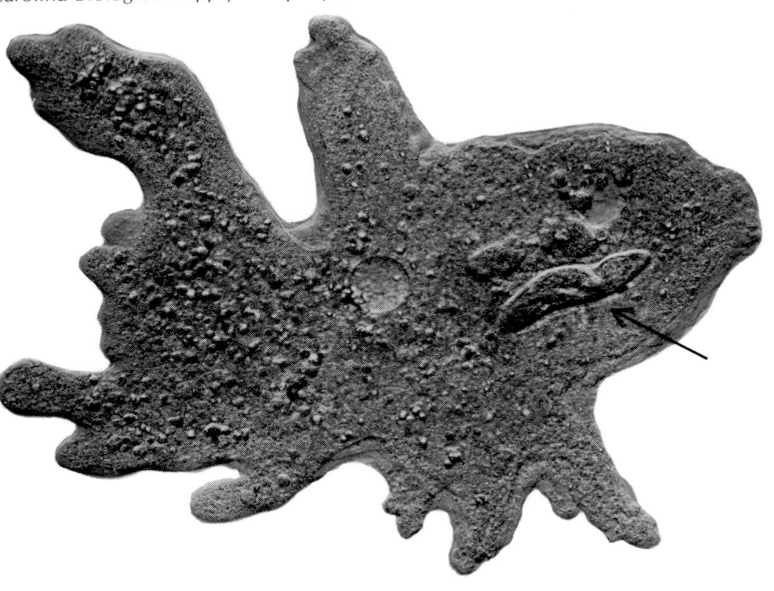

tion, and the community. It is customary to designate each sequence of organisms through which energy flows as a food chain. There are so many possible combinations of greatly intertwined food chains in most real communities that they are thought of as being grouped together in the form of a food web. Every food chain or web is made up of producers (green plants), consumers of various types, and finally decomposers, all of which are important in one way or another in maintaining the dynamic equilibrium and evolutionary progress of the ecosystem.

Within the various strata of the community are a great many kinds of interspecific interactions growing out of individual energy needs. These interactions fall under the general classification "symbiosis" which in a broad sense means living together.

symbiosis

There are different kinds of symbiotic relationships, the first of which is commensalism, a relationship in which one species benefits while the other is neither benefited nor harmed. A second is mutualism, in which both species benefit. The third is parasitism, in which one species benefits but the other is harmed.

All heterotrophic organisms, with the exception of saprophytes, feed on other living things. We generally think of this feeding in terms of large animals eating smaller ones, but competition demands that all feeding possibilities be explored and often very small, even one-cell forms exploit even the largest organisms to obtain their supply of energy. This infection of living things by small organisms is one form of *disease*. Disease plays an essential part in the ecosystem and has been an important factor in the evolution of species, including man. While disease is generally looked upon as detrimental to the individual, it is not necessarily detrimental to the host species. Just as wolves improve deer herds by removing the weak, disease tends to remove the genetically unfit from naturally occurring populations.

It can be shown that organisms living in the complete absence of microbes do not respond as vigorously as do those who live among the hordes of microscopic life in the ecosystem. Competition, having begun with the first living things, seems to be necessary for the well-being of the individual.

definition

Not all maladies we call disease are caused by organisms. Some are congenital defects resulting in the malfunction of some tissue or organ in somewhat the same way that infecting organisms upset the physiological equilibrium of organs and tissues. Disease can then be defined as *an alteration of living cells in such a way as to jeopardize their survival in the environment.* The environment is important in any concept of disease since a phenomenon may appear to be disease under one set of circumstances but not under another. For example, when we encounter eight or nine million red blood cells per cubic millimeter in the blood of people living at the extremely high altitudes of the Andes, we do not consider it pathological since the higher concentration represents an adaptation to low oxygen pressure. The same phenomenon at sea level would probably constitute a pathological situation. Likewise, a mentally ill person might be more readily accepted in a primitive society than at a social club.

A normal healthy person is surrounded by and even infected with a great many microorganisms with which he maintains a dynamic balance. What we call disease then, is a maladjustment to the environment, which occurs in connection with three factors: (1) a stimulus from the environment such as stress or infecting organisms; (2) a response from the host, which depends on its genetic inheritance and nutrition; and (3) a particular arrangement of thoughts and traits which we call culture.

stress

Many pathogens may be present in a healthy organism, and stress is often all that is needed to activate them. Apparently healthy pigs living peaceably on a farm may be carriers of latent salmonella infection. The stress of a trip to the slaughterhouse may activate the bacteria and make transmission of the disease possible. During a cholera epidemic in 1965, victims often tested negative for the bacilli until they were placed under stress before being checked. Under prolonged stress these latent pathogens increased their numbers and were released into the stools. In animals, including man, there is accumulating evidence that infections are spread from one area to another and among species by apparently healthy carriers. When carriers undergo stress, they too may develop the disease. Some complex relationships between man and the bacteria he normally harbors are described on page 130, "The Resident Bacteria: Friend and Foe."

The response of the host to environmental stimuli depends to a large degree on its genetic makeup. Although there are many instances in which a disease organism may infect more than one species, they tend to be limited to one. It is impossible to infect humans with certain bacteria found in animals, and diseases such as typhoid cannot be transmitted to animals.

leaf blight

Corn leaf blight has been of great concern to farmers and to the commodities market. It has important implications for plant breeding, plant pathology, and the dependability of our food production system. An uneasy equilibrium exists among many biotypes or strains of pathogens. The frequency of occurrence of various disease organisms depends on the genotype of the host crop and on the virulence and reproductive rates of individual components, both host and pathogen.

The development of a new high-yield crop variety may have unexpected results in the spread of disease. In regions where many varieties of a crop have been planted before, differences among them in susceptibility to various diseases have provided a margin of safety. Climatic conditions and other factors may have been just right for a particular disease, but plots of the suitable host were interspersed with resistant types. An epidemic was prevented, although an occasional field of the crop sustained extensive damage. When a new improved variety becomes available, all the acreage may be planted with this one type. Any disease to which the new variety is susceptible can then sweep through an entire agricultural region.

Various aspects of the ecology of animals predispose their involvement in disease cycles. Geographical distribution, altitude, affinities for specific habitats, territorial distribution within a habitat, and periodicity of activities all have a bearing on host–parasite predisposition to disease cycles. Diets and feeding habits determine the likelihood of infection, and reproductive characteristics determine whether or not a host species is a suitable reservoir. Behavior patterns such as nest site selection may predispose the involvement of the host with parasites and pathogens.

culture

The third factor that plays a role in shaping the pattern of disease is culture, the sum total of the techniques and concepts used by a population to control its environment. A cultural trait, of survival value or not, may have been developed in former times under different environmental conditions and passed down through the centuries. A public health worker finds it useful to consider whether or not the cultural traits of the population with which he is working tend to promote or reduce survival.

hookworm

Several decades ago it was observed that among certain Chinese villages about half the inhabitants were devastatingly infected with hookworm while others in the village were free of this disease. An investigation into their cultural practices revealed that the infected villagers were rice growers while the rest of the inhabitants raised silkworms. The rice growers spent their working days wading in rice paddies which had been fertilized with night soil (human excreta). This explained the cause of their disease since the water-borne larvae enter through the skin.

An equally interesting culture-disease relationship is found in North Vietnam, a country made up of two main types of terrain: a fertile delta and a less fertile hill region. Rice is grown in the delta, and the hilly regions are devoted mostly to lumber but have some upland rice farming. The delta rice growers built their houses on the ground with the stables on one side of the house and the kitchen on the other. Meals are brought into the house after being cooked outside. The people living in the hills, however, build their houses on poles and their living rooms are located about 10 feet above the ground. They cook their meals in the living room which is often filled with smoke. The animals are kept under the house.

malaria

The mosquito malaria vector, *Anopheles minimus,* is found only in the hill country, but its flight ceiling seldom exceeds 10 feet above the ground. Thus in the evening it seldom encounters humans but feeds on farm animals under the houses. The fumes that spread through the living room from the activity of people and animals aid in driving the mosquitoes away should any happen to reach this level.

Fig. 6-2. Mosquito egg raft. (*Courtesy Carolina Biological Supply Company.*)

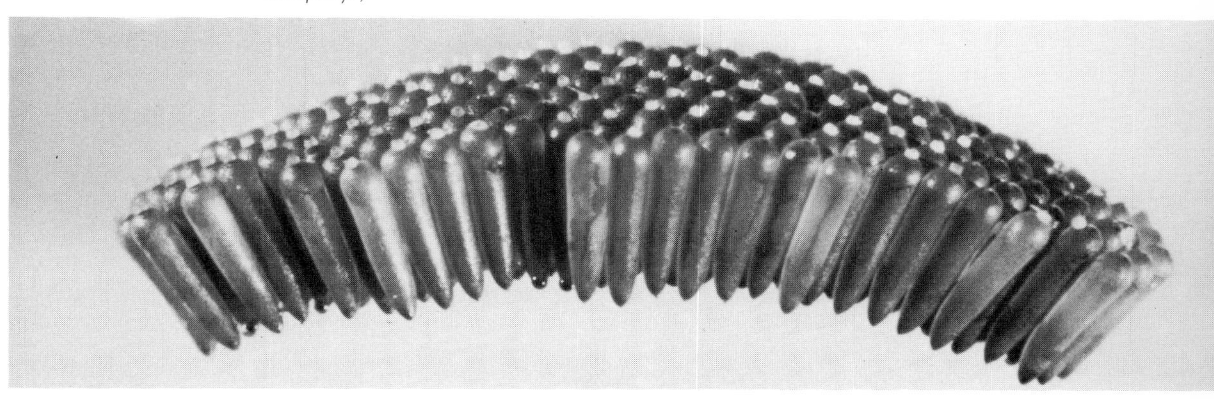

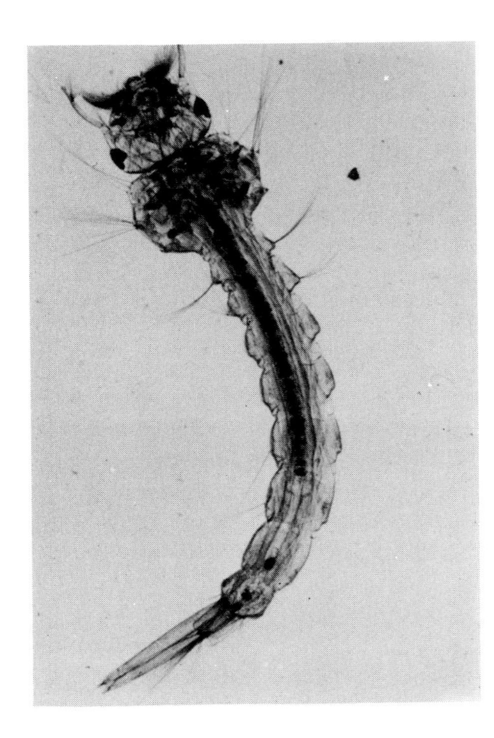

Fig. 6-3. Mosquito larva. (*Courtesy Carolina Biological Supply Company.*)

When the pressures of overpopulation among the delta people forced them to move into the less populated hill region, they took their delta culture with them. They built homes on the ground and did their cooking outside to insure smoke-free houses. This cultural practice was an open invitation to the *Anopheles* mosquito which prefers human blood to that of other animals, and the new inhabitants of the hills became infected with malaria.

Similar examples can be given to show the close interrelationship between culture and disease patterns in any part of the world. Disease is not a simple matter of one cause or one effect but rather is a biological expression of maladjustment which cannot be understood apart from an ecological approach.

diet

The maintenance of the delicate equilibrium that we call good health in an individual results not only from his genetic makeup but also from an adequate diet. The effect of the nutritional status on disease is best summed up in the historic relationship between famine and pestilence. Diet-disease experiments on man and other animals have clearly shown that an inadequate intake of food reduces resistance to disease, and infection becomes more frequent and severe (Scrimshaw, N. S., 1963). Only a few diseases, such as malaria, can be alleviated by a deficient diet. Malaria parasites do not thrive in human tissues when hemoglobin levels are low because of a grossly inadequate iron intake, but active malaria can result from even a low parasite count. Because nutritional deficiency greatly adds to the seriousness of disease, the cause of death among starving populations is often given as a gastroenteric or respiratory disease when in fact these ailments may only have administered a merciful last stroke.

KINDS OF DISEASES

Throughout most of history, man has thought that evil spirits caused illness. It was not until after the early 1400s that scientists began to suspect that at least some diseases were caused by invisible particles which early researchers called "living seeds of disease." They believed that these particles in the blood simply developed out of nothing, and in the 1500s doctors suggested that "germs" could be passed from one animal to another and in this way spread the disease.

germ theory

Koch

Pasteur

Bacteria and other microbes were first seen with primitive microscopes in the 1600s, but the *germ theory of disease* was not established until late in the nineteenth century. The relationship between germs and disease was first shown by Robert Koch, a German physician, and Louis Pasteur, a French chemist, both of whom worked with *anthrax,* a deadly infectious disease of both man and other animals. Koch showed that animals injected with anthrax microbes developed the disease, and Pasteur developed a vaccine that slowed the growth of the pathogens within the body.

Today almost everyone knows that microorganisms cause disease by attacking living tissue. Some multiply so rapidly that the tissues or even the host organisms die. Infectious diseases are caused by many kinds of organisms such as viruses, bacteria, fungi, protozoa, and multicellular parasites.

Virus Diseases

virus

Viral diseases are caused by extremely small infectious particles which exhibit only one property of life, that of reproduction. They reproduce only when they are within a host cell. All the hundreds of kinds of viruses are pathogenic, and each causes a particular disease in some living organism. They can be crystallized and retain their ability to invade defenseless cells for many years. Once within the host cell, viruses may multiply rapidly and produce large populations in a short time. For example, the genetic material injected into a bacterial cell by a single virus particle can result in the production of 100 to 200 new virus particles within 20 minutes. These are liberated and attack other bacterial cells within a few seconds, and each is as dangerous as the first.

Viruses cause a great many diseases, including smallpox, chickenpox, mumps, measles, influenza, many varieties of the common cold, polio, sleeping sickness, viral pneumonia, and hepatitis. Some kinds of viruses cause cancer.

Bacterial Diseases

bacteria

Bacteria, larger and more complex than viruses, are actually unicellular plants that are unable to manufacture their own food but are capable of utilizing most types of organic matter, living and dead, as an energy source.

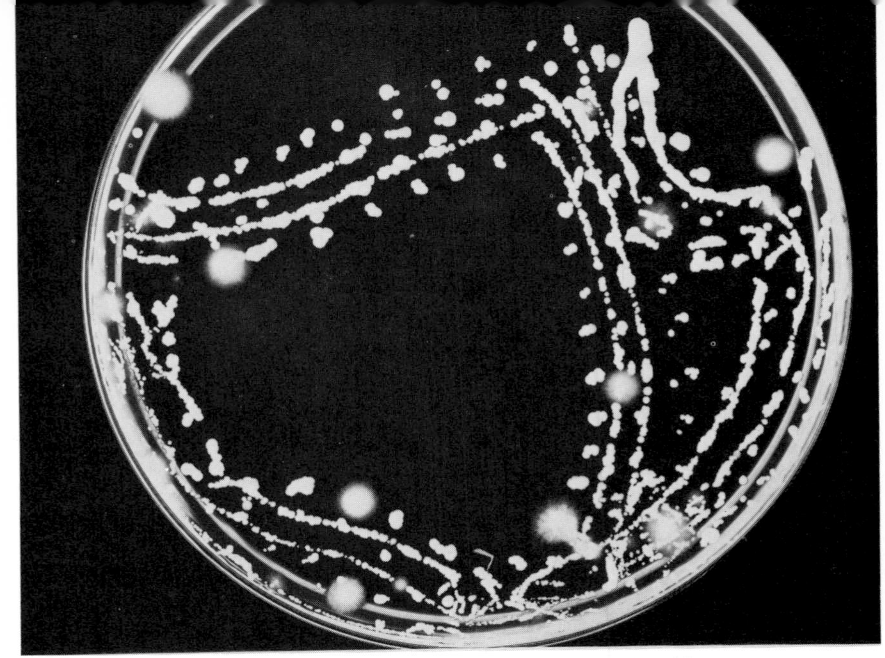

Fig. 6-4. Development of bacterial colonies (smooth outline) and mold colonies (fuzzy outline) where a fly walked on a nutrient agar plate. The two heavy parallel lines (center of track) are from the fore and hind pairs of legs, and the two outside rows are from the middle pair of legs.

Scientists do not know exactly how many kinds there are, and bacteria mutate rapidly to produce new strains. They are so numerous that a single gram of soil may harbor as many as 100 million bacteria of many different kinds. Most bacteria do not cause disease and are a very necessary part of the environment. They bring about the decay of all kinds of organic matter. They are used in the manufacture of many products including beverages, cheese, and other dairy products.

Various kinds of bacteria live on and in the bodies of plants and animals. Many species can be found occupying specific regions of the human skin, nose, mouth, and throat and in the lungs, stomach, and intestines where normally they do no harm.

The Resident Bacteria: Friend and Foe[1]

The human body is an ecosystem within itself. In addition to the parasites that man may support, several microorganisms regularly inhabit specific parts of the human body. As long as they remain in these "natural" areas, they are tolerated and may even be helpful. Difficulty arises when the microorganisms invade other parts of the body.

Coliform bacteria (*Escherichia coli* and *Enterobacter aerogenes*) living in the intestine of man contribute valuable nutrients such as vitamin K,

[1] Contributed by Walter Hoeksema, Ferris State College, Big Rapids, Michigan.

riboflavin, and pantothenic acid. Coliforms also prevent the growth of pathogens by antagonistic action. Some of the materials produced by the coliform bacteria inhibit the growth of disease-causing bacteria. Coliforms may have a stimulatory effect on antibody production. This provides added protection against pathogens that have common antigens.

Coliforms as a group, however, cause more infections of the urinary tract than any other group of microorganisms. Thus when they "escape" from the intestine, where they are beneficial, coliforms are definitely man's foe.

Streptococcus salivarius (viridans group) colonizes the human upper respiratory tract soon after birth. These bacteria probably provide no direct benefits to man but neither are they harmful. If, however, *Streptococcus salvarius* enters the bloodstream (which might occur after a tooth extraction), subacute bacterial endocarditis may result. This serious illness results from infection of the endocardium (a thin membrane lining the cavities of the heart). Bacterial endocarditis is more likely to occur if the heart has been previously damaged by rheumatic fever or congenital heart disease.

Many potentially pathogenic bacteria, such as *Staphylococcus aureus,* may be harbored in the nasopharynx with no apparent detrimental effects. These resident bacteria are tolerated by some persons but not by others. A healthy carrier may transmit such bacteria to individuals who cannot tolerate them, however, and disease results. The potential danger an undetected "staph" carrier represents can be envisioned by considering the consequences of a surgeon, who is such a carrier, working in an opened abdominal cavity. The carrier may also harm himself if the resident staphylococci gain entrance to the bloodstream and reproduce in tissues where they are not tolerated.

Bacteria can pass from one person to another, causing a great many diseases including whooping cough, scarlet fever, diphtheria, and tuberculosis. Other bacteria from the soil may cause tetanus (lockjaw).

Protozoan Diseases

protozoa

Unicellular animals, or protozoa, are found in great numbers almost everywhere in nature. Some are harmful to higher forms of life, including man. One of the more important protozoan diseases is malaria which is caused by three species of the genus *Plasmodium*. Other protozoa cause sleeping sickness, amebic dysentery, and many other diseases.

Diseases Caused by Multicellular Parasites

flatworms

Two kinds of worms cause disease. Parasitic flatworms (*Platyhelminthes*) must spend part of their life cycle in the bodies of humans or other animals. Probably the best known flatworm is the tapeworm, a parasite of the di-

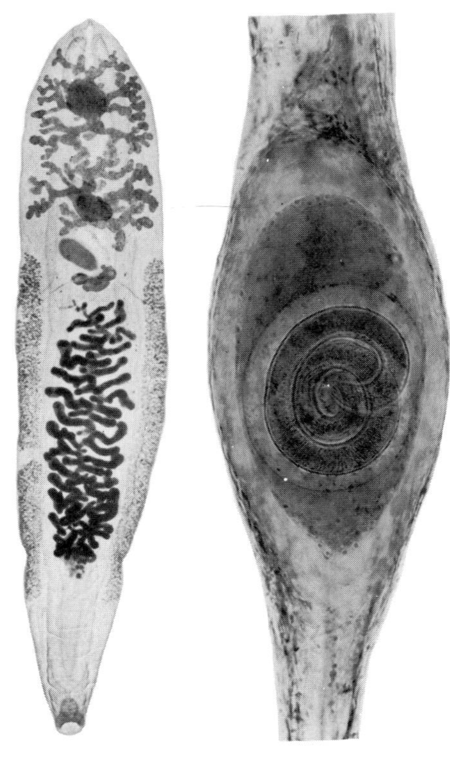

Fig. 6-5. Left. Chinese liver fluke. (*Courtesy Carolina Biological Supply Company.*)

Fig. 6-6. Right. Encapsulated larva of porkworm in muscle tissue. (*Courtesy Carolina Biological Supply Company.*)

gestive tract of humans and many animals. The other group of worms causing disease is made up of roundworms, *Nematodes,* which cause trichinosis, a disease in which both the digestive tract and muscles may be infected during different parts of the parasite's life cycle. Many other bizarre diseases are caused by members of this phylum. Figures 6-5 and 6-6 show two of the organisms which cause serious problems.

THE SPREAD OF DISEASE

Man is an agent of dispersal of his own diseases, and the spread of pathogenic organisms is both a biological and a cultural phenomenon. Contagious diseases such as measles, smallpox, and the like spread easily through the mobile populations of today, but it is difficult to visualize how they could have been maintained in the small, dispersed populations of Paleolithic hunting and gathering peoples. Diseases caused by pathogens that have intermediate hosts, such as yellow fever and malaria, move more slowly because of their specific environmental requirements. Unlike contagious diseases, however, they are more apt to persist in endemic form in scattered populations, and it is possible that relationships of this kind evolved with man. Figure 6-7 illustrates conditions which contribute to the spread of disease.

Fig. 6-7. A rural drainage and health problem. *(Courtesy U.S. Department of Agriculture, Soil Conservation Service.)*

VENEREAL DISEASES

The term *venereal disease* refers to several serious diseases. The name is derived from Venus, the goddess of love, because these diseases are most often contracted through sexual intercourse. The most common venereal diseases, *syphilis* and *gonorrhea,* could be stamped out if all infected persons sought treatment.

Syphilis

syphilis

Syphilis, an acutely infectious disease, is the most dangerous of the venereal diseases and invades every system of the body. If left untreated, it can disable or cause death. The syphilitic condition is caused by *Treponema pallidum,* a spirochete. This spiral-shaped microorganism is extremely sensitive to drying, heat, antiseptics, and soap. Because moisture is essential to its survival, it thrives in such areas as the genitals, the mouth, and the anus. The disease is transmitted only through intimate physical or sexual contact. It is almost impossible to contract the disease from toilet seats, towels, cups, or other inanimate objects.

Symptoms

Syphilis does not make a person very ill in its early stages but is highly contagious at this time. The primary stage usually develops about 3 weeks

chancre

after infection, but it can appear as late as 90 days. A *chancre sore* appears either on the genital organs, in the anus, or in some females in the cervix. The spirochete usually enters the body through microscopic breaks in the skin or mucous membranes. The sore develops at a site where white blood cells, attacking the infecting organisms, pack the infected area so tightly that tissue cells may be killed. Chancre sores that appear in the mouth usually result from kissing a person who has secondary sores. They are not itchy or painful and may actually go unnoticed. They are likely to disappear within a few days whether or not the disease is treated. This disappearance of the sores often leaves the infected person under the delusion that he does not have a venereal disease. This can be a tragic error in judgment. The disease is highly contagious during the primary stage.

second stage

If the disease is not treated, it progresses to the second stage. This phase may also go unnoticed or be misinterpreted because the symptoms (such as sore throat, skin rash, mouth sores, enlarged lymph nodes, swollen joints, pain in the bones, fever, headache, and loss of hair in patches) often imitate those of other common diseases. These symptoms tend to come and go for a period of about 4 years, after which they may disappear completely. The disappearance of the symptoms, however, is no indication that the disease has been cured, and the person is still highly infectious.

hidden stage

The disease then enters the *latent* (hidden) stage; it goes "underground," leaving no visible symptoms. An infectious relapse may occur, however, during the early part of this phase, with the reappearance of some of the earlier symptoms. This phase usually lasts 5 years or more, and as the disease progresses relapses no longer occur and the infected individual gradually loses the ability to infect others.

final stage

During the fourth and final stage, syphilis may settle in any body system, but the nervous and circulatory systems are the most likely to be affected. When syphilis attacks the circulatory system, the elastic fibers of the aorta are destroyed and aneurysms (enlarged areas) develop. The bicuspid valve between the left ventricle and the aorta may be destroyed. This results in enlargement of the heart which now works harder because of inefficiency in pumping. This condition is further aggravated when inflamed coronary arteries supply less blood to the heart muscles. Obviously, any one of these conditions has serious consequences and may drastically reduce life expectancy.

Syphilitic attack on the nervous system can take many forms. The meninges, or protective membranes of the brain, may be attacked, causing paralysis in varying degrees. If the brain and spinal cord are directly attacked, paralysis of the legs, a malady called *tabes dorsalis,* may develop in combination with other difficulties such as impotence and urinary ailments. Should the syphilis organism concentrate in the brain, insanity and general paralysis may follow. These symptoms can develop rapidly or slowly over a period of several years. Regardless of the rate at which the disease progresses, the results are the same: extreme mental and physical deterioration and finally death.

Diagnosis

diagnosis

Syphilis is very difficult to diagnose. The spirochete that causes the disease does not stain easily, and microscopic evidence is difficult to establish. Also, the causal organism cannot be grown in culture media in the laboratory. During the primary phase blood samples appear normal, that is, they test negatively. To make an accurate diagnosis, knowledge of the social activities of the individual is often of utmost importance to the doctor. He should know if there is a history of the disease in the patient's family and whether or not the patient has had sexual relations with someone likely to be infected. Fluid may be taken from primary sores when they are still present and examined by special microscopic techniques (dark-field) to determine the presence of the spirochetes.

Blood tests designed to confirm the disease become more effective during the second stage, but even then one negative test does not assure that the person is not infected. Additional tests should be made a few weeks following treatment to determine whether or not any disease organisms remain. Microscopic examination of glandular fluids can also be used to determine the presence of the disease.

During the latent stage repeated blood tests must be made to confirm the disease, and infection can remain undetected if the patient has been taking antibiotics to treat some other illness. More than one test is always necessary.

During the fourth phase the disease may be identified through blood tests, by examination of cerebrospinal fluid, or through the examination of any tumorous tissue that may be present.

Because syphilis can be contracted congenitally (from birth), all pregnant women are tested for syphilis. The disease can be cured if the mother is treated early in pregnancy and the baby will be born normal. If treatment is not given however, the baby will be born either diseased or dead. The tissues (the placenta) between the baby and the mother do not stop the movement of the spirochetes. The effect the disease has on the fetus depends on the extent of the mother's infection. Many states do not issue marriage licenses until a medical test has been made and the applicants are determined to be free of venereal disease. The Wassermann and Kahn blood tests, widely used to detect syphilis, can be effective when symptoms have disappeared.

Treatment

treatment

A German doctor, Paul Ehrlich, was the first to discover a drug that kills the syphilis organism, but treatment was difficult and slow. It can be treated with compounds of arsenic, bismuth, and mercury. Today, however, all stages of the disease are treated by injections of penicillin and often a single dose is sufficient, especially if it is given before the appearance of the first symptoms. At least two injections are advised. Treatment should not be terminated until at least two negative blood tests verify that the disease has been cured. It is important to bear in mind that the early stages of the disease can be completely cured. The organism can be killed during the later stages also, but by then permanent tissue damage may have occurred.

The incidence of syphilis is rising throughout the United States today, partly as a result of liberal attitudes concerning sex. It could be completely eradicated from society if all individuals who contract the disease could be reached with medical assistance. Free clinics for treatment of venereal diseases are available in most areas.

Gonorrhea

gonorrhea

Gonorrhea usually attacks the mucous membranes of the sex organs and eyes. It is caused by a bacterium, *Gonococcus,* which spreads immediately below the surface linings. It is similar to syphilis in that it is sensitive to drying and weak antiseptics and is spread through intimate sexual contact in adults. The disease is not transmitted from a mother to an unborn baby, but infants often contract the disease when they are born. The eyes of infants and the genitals of girls are very susceptible to infection. The eyes of newborn babies are treated to insure against the disease and possible blindness. The incubation period of gonorrhea (the time from infection until the disease can be detected) ranges from 2 days to about 3 weeks. The first symptoms are not obvious in women and may go unnoticed. The urethra and cervical canal are affected first. Some discomfort may be experienced when urinating and there may be a slight vaginal discharge. Complications begin to arise, however, if the woman is not treated for the disease. Glands of the genital area may swell and become very painful; the infection may then spread up the urethra to the bladder where it causes a disease known as *cystitis*. This makes urination painful and frequent. *Proctitis,* or inflammation of the rectum, commonly occurs during this stage. The most serious effect of gonorrhea, however, is *salpingitis,* inflammation of the fallopian tubes. In some cases women have no early symptoms of the disease until they become violently ill with abdominal pains accompanied by vomiting and fever. Similar attacks may recur for several months during which menstruation may be erratic in terms of both length of the cycle and the amount of flow. The later stages are characterized by twisted and scarred fallopian tubes, and sterilization may result. When gonococci organisms reach the bloodstream, they cause a form of arthritis.

Gonorrhea in males is characterized by a burning sensation during urination and a discharge of yellow pus from the urethra. This highly infectious discharge may be carelessly transferred to the eyes. If the disease is untreated, *Gonoccocus* invades the bladder, as in women, causing cystitis, and urination becomes painful and frequent. As the disease progresses, other areas of the urogenital tract may become infected. Infection of the sperm-carrying ducts, the epididymis, results in scar tissue which blocks the passage of sperm, rendering the individual sterile. During this stage, painful swelling of the testis occurs.

diagnosis

Diagnosis of gonorrhea can be made through laboratory cultures of the discharge or by direct microscopic examination of the discharge. Blood tests for gonorrhea are almost useless, but the syphilis blood test is often given during gonorrhea examinations since an individual can have both diseases at the same time.

Treatment

treatment

Gonorrhea is commonly cured with penicillin, although it has also been successfully treated with sulfa drugs. Sulfa drugs have also been taken as a preventive measure. Usually one or two injections of penicillin cure the disease, but certain strains have developed a resistance to the drug and can be successfully treated only with heavier dosages. Vietnam Rose, for example, is a strain of gonorrhea that developed a tolerance to antibiotics as a result of prostitutes taking low dosages to protect themselves from the disease. Tolerance is built up as successive generations of the bacteria adapt to the antibiotic through the process of natural selection.

Other types of venereal diseases include *chancroid, Vincent's infection,* and *venereal lymphogranuloma,* an infection that attacks the lymph nodes of the groin.

Venereal disease is a serious health problem that is becoming worse each year. Anyone needing treatment should go to a doctor just as he would for any other illness. Only by giving proper treatment to everyone with the disease will it be possible to eradicate this ancient scourge of mankind.

Suggested Readings

Hammond, A. L. "Aspirin: New Perspective on Everyman's Medicine." *Science,* 174(4004):48, 1971.

London, W. T. "Infectious Goiter." *Science,* 171(3974):928, 1971.

Loomis, W. F. "Rickets." *Scientific American,* 223(6):76–91, 1970.

Pichirallo, J. "Black Lung: Dispute about Diagnosis of Miners' Ailment." *Science,* 174(4005):132–134, 1971.

Scrimshaw, N. S. "Food." *Scientific American,* 209(3):72–80, 1963.

Snider, A. J. "Breakthrough for Bleeders—If They Can Afford It." *Science Digest,* 70(2):55–56, 1971.

Americans, along with many other nations, have developed a drug-oriented society. Drugs come in many varieties and are packaged in many sizes to meet the budget needs of the moment.

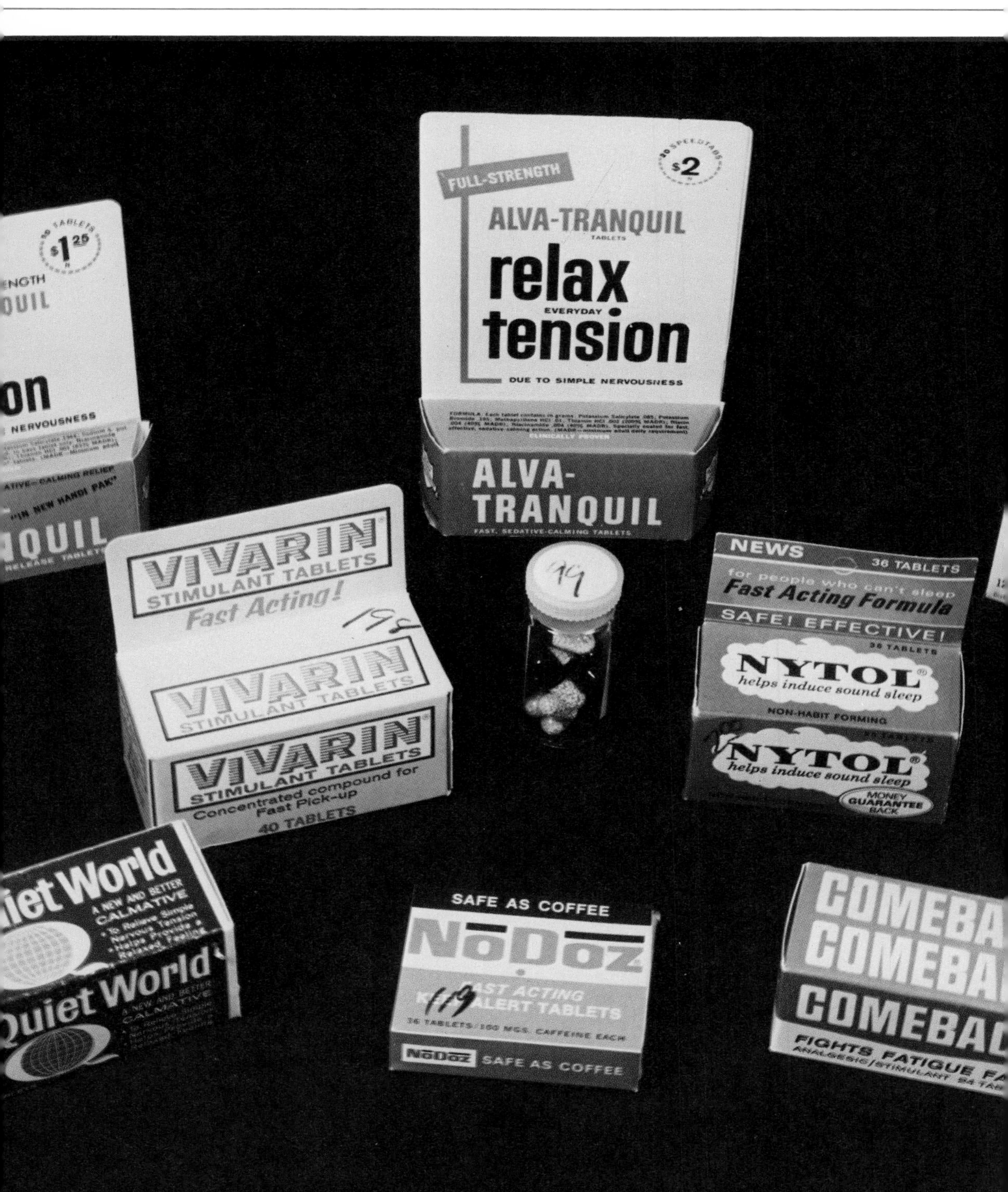

7

Drugs

A drug may be defined as any substance introduced into the body to change the way the body systems work. The word "drug" comes from the Dutch word *droog,* meaning "to dry," and probably came into use because most early drugs were made from dried plant tissues. There are thousands of chemicals that can be classified as drugs, and scientists are constantly developing new ones. Many of these compounds are obtained by medical prescription and over-the-counter sales. They are intended to prevent, improve, or cure various undesirable physical or mental conditions. Over-the-counter sales are huge for such compounds as aspirin, cough drops, mouthwashes, eye and nose drops, fungus remedies, acne cures, laxatives, vitamins, tonics, and reducing and sleeping pills. In addition to these there is a wide array of prescription drugs which can be purchased legally only if prescribed by a physician.

Drugs are produced from a great variety of plants, many of which have been known for thousands of years. Primitive man used dried foxglove leaves to treat heart ailments long before scientists discovered that they contain digitalis. A small amount of this chemical causes the heart to beat more strongly. Other plant drugs used for centuries include atropine, belladonna, cascara, castor oil, cocaine, curare, morphine, and opium.

Almost every part of animals has been used by early man for medicinal purposes. Doctors began removing glands from animals in the late 1800s and extracting *hormones* to combat various diseases such as asthma and *diabetes mellitus*. Physicians commonly treat goiter, a disease of the thyroid gland, with thyroxine obtained from the thyroid glands of cattle and hogs. Adrenocorticotrophic hormone (ACTH) from the pituitary glands of sheep, cattle, and hogs, has been used to treat arthritis, leukemia, and rheumatism. Cortisol, produced by the adrenal cortex, has been used to combat arthritis, asthma, and inflammatory conditions of the skin and eyes.

Antitoxins, made by injecting various domestic animals with the bacteria of certain diseases, are used as disease-fighting substances. Antibodies developed in the animal's blood are injected as a serum into victims of diphtheria, tetanus, scarlet fever, and other diseases.

One of the greatest advances in modern medicine has been the development of synthetic drugs. Many of these are identical to those obtained from plant and animal tissue, and some are even better.

DRUG ABUSE

drug abuse

Drug abuse occurs when chemicals are taken, usually by self-administration, for purposes other than those prescribed by medical and social practice. The drug abuser takes drugs to obtain a pleasurable emotional or physical sensation rather than to effect a medical cure for some ailment. Taking drugs in large amounts often results in physical and/or psychological damage, usually accompanied by antisocial behavior.

addiction

When drugs are administered over a long period of time, they may produce a dependence generally called *drug addiction*. The user usually experiences a strong need for the substance if he is deprived of his supply, and the nature and strength of this addiction depend on the kind and amount of drugs used.

Physical dependence results when the body develops a dependence on the drug and is unable, for some unknown reason, to function well without it. If the drug is suddenly withdrawn, distressing or even violent reactions may occur. The severity of the withdrawal symptoms indicates to physicians how bad the drug habit is and the therapy needed to effect a cure.

Psychological dependence develops when a drug is relied upon to provide a strong emotional effect. During withdrawal the abuser experiences violent emotional reactions rather than a physical craving. This dependence, however, can be as strong and compelling as a physical need and is equally effective in leading a person to rely on drugs.

Drug abuse among youth is actually only a part of the nationwide drug problem. While many adults accuse young people of showing a lack of responsibility in turning to drugs, the adult world itself has become incredibly drug oriented. Advertising programs across the country lose no opportunity to urge the buying public to "pop" pills for every conceivable malady from the slightest sign of a common cold to nervous tension and sleeplessness. About 20 percent of the drugs prescribed by physicians are stimulants, tranquilizers, or sleeping pills, and studies have shown that the average medicine chest of families in California contains 30 drugs, 80 percent of which were bought without a prescription.

extent

How widespread is drug abuse among our youth? Studies indicate that as many as 50 percent of the students in some colleges have tried LSD or marijuana. High schools report similar percentages. These studies have limitations, however, since they do not involve the entire country and because many students refuse to cooperate in research programs even when they are not required to identify themselves. The information collected showed that while the abuse of mind-influencing chemicals may not be chronic or excessively heavy, it has increased substantially among both teenagers and adults within the past few years. This increase has been accompanied by an alarming rise in drug abuse among children still in grade schools.

KINDS OF DRUG TAKERS

tasters

Drug abusers fall into three categories. The first is the experimenter or taster. These individuals have actually tried some drug, usually marijuana, one or a few times and have refrained from further contact.

casual

The second type is the irregular user. These people take drugs once in a while for the fun of it, or when they are under stress. They usually do not suffer physical ill effects from the drug, and if psychological dependence develops it is usually mild. The average young person in this category does not really have a drug problem, unless of course he is arrested or becomes ill or dies from an accidental overdose. It should be pointed out, however, that irregular users can become addicted.

drug heads

Drug heads, the third type of drug takers, are individuals for whom drugs have become the all important concern of life, those who are so addicted that they cannot function without drugs and whose capacity for effective living is drastically disturbed often for long periods of time. These people have developed a strong physical and psychological dependence on drugs and find it almost impossible to "kick the habit"; they rarely do so on their own.

There is little question that the overwhelming majority of youthful drug users fall into the first two groups. While there is a distinct possibility that even an experimenter can be seriously harmed by just one dose, the number of youths mentally crippled by chronic drug use is not large in spite of what alarmists tell us. The drug problem, however, is a serious one and even one drug fatality is a great tragedy. There is strong evidence suggesting that some drug users were emotionally disturbed even before they started taking drugs and probably would have turned to some other form of self-destruction, such as crime or accident-prone behavior, if drugs had not been available. Teenagers have always had to deal with the problem of "teenage conformity," the pressure to be like the others and to do the "in thing." This tyranny of conformity often makes a youth feel "out of it" if he does not experiment along with the others. This exposes an emotionally disturbed person to the beginning of a long nightmare.

INHALANTS: TAKE A DEEP DEADLY BREATH

inhalants

The mind-influencing capabilities of inhalants have been known since the middle of the nineteenth century when scientists discovered that breathing a gas called *nitrous oxide* or the fumes of *ether* puts people to sleep. These drugs become a godsend to doctors because operations could be performed that were previously impossible. The effects of these gases were far from new. They had been used for years by people to produce a "giddy" state of mind.

The anesthetic properties of nitrous oxide, originally known as laughing gas, were discovered by Horace Wells when he observed that an individual high on the substance cut himself deeply without knowing it. Early medical students used it at "ether frolics" to get high.

ether frolics

glue sniffing

Drug abuse through inhalation grew rare when ether and laughing gas fads died away, but glue sniffing has recently become widespread.

Glue contains mind-influencing drugs and other inhalants which are highly volatile at room temperature; in other words, they easily change to a gas. Upon inhalation the vapors quickly pass from the lungs into the bloodstream and within seconds are carried to the brain. These vapors of glue commonly contain such drugs as *toluene, naphtha, benzene,* and *carbon tetrachloride*. Different kinds of glue are used, but airplane glue is the most popular.

Glue is sniffed after a handkerchief is soaked in it, or after it is poured into a plastic or paper bag which is then placed over the nose and mouth; it can also be inhaled from a heated pan. A habitual sniffer may require many tubes of glue a day, but glue is cheap and a large habit can be supported on very little money.

The immediate effects of sniffing last only about 1 hour, but it can influence behavior for a day, or even longer. A user may feel happy, drunk, and often very powerful. A delusion that nothing can harm him often so clouds his judgment that he may seriously endanger his life. Pleasant daydreams may flood his thinking. Among the emotionally disturbed, aggressive behavior often results, and in a few cases violent crimes such as rape and murder have been committed by sniffers under the influence of glue.

The real glue head usually comes from an unhappy family. He does not feel emotional security and starts sniffing glue early in life, sometimes as early as five. He develops symptoms that persist even when he is not high. The inhalants of glue are very irritating to the mucous membranes of the body. The eyes are always bloodshot and the lining of the nose, throat, and lungs becomes inflamed. Coughing is frequent, and the breath has a peculiar odor. The glue head always has a bad taste in his mouth; he lacks energy and has little appetite.

Glue sniffing over long periods can cause severe damage to the liver and kidneys and permanent damage to the brain. Lack of coordination, tremors, loss of memory, and epileptic seizures are sometimes seen among glue users, indicating a permanently damaged nervous system. Sudden death from glue sniffing is rare, but it does occur. Once death has struck, however, it makes little difference whether the victim was an experimenter or an addict. Glue sniffers have passed out with glue-filled plastic bags over their heads and suffocated. Youngsters have been killed by gas explosions while sniffing, by freezing their throats while breathing Freon 12 from aerosol cans, by convulsions, and by the sudden filling of the lungs with fluid.

Death can also come slowly in the form of a lingering sickness. Benzene, a common inhalant, causes damage to bone marrow and shuts off the production of red blood cells. Benzene abusers have become severely anemic, and death can result if the marrow fails to recover. Carbon tetrachloride, found in cleaning fluids, fatally damages the liver, an organ of vital importance in a great many processes.

Increasing public awareness during the late 1960s led to an outcry for legislation to help curb the problem. Laws have been passed in some states making glue, shoe polish, lighter fluid, and other household products a crime to possess for purposes other than their intended use. Some glue now contains oil of mustard, a substance that gives the abuser an unpleasant jolt instead of the high he expects. With new laws and a better educated public, the practice of glue sniffing may eventually fade from the drug scene.

AMPHETAMINES: UP TO THE CLOUDS

amphetamines

The drugs that are the most readily available to abusers are the *amphetamines* (Fig. 7-1), which typically stimulate the brain and nervous system, raising the blood pressure and decreasing the appetite. Taken in excess they give an intense feeling of excitement, self confidence, and power. Because they make the user unusually alert, awake and active, amphetamines are often called "pep" pills. When the effects wear off, the user may return to normal or he may feel a letdown and be tired and depressed. The drug does not magically eliminate fatigue but simply permits the body to use reserve energy which may suddenly be exhausted and a blackout or near-collapse can result. The dejected feeling that follows often promotes the desire to "pop" another pill.

Amphetamines were misused by soldiers during and after World War II, by truck drivers trying to stay awake, by students cramming for examinations, and by housewives needing a lift. At first, amphetamines were

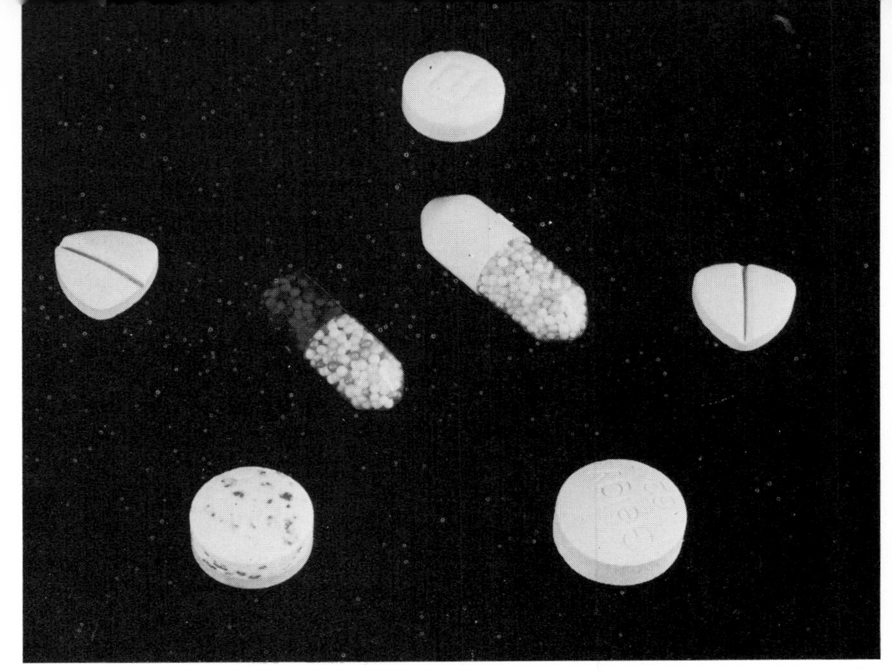

Fig. 7-1. Some of the amphetamine pills.

thought to be safe and nonhabit-forming, and this incorrect information spread until the use of these compounds became a worldwide problem.

Scientists have found that amphetamines stimulate the release of *norepinephrine* from nerve endings and concentrate it in the brain. This speeds up the metabolic rate and heart action.

Similar to many drugs, amphetamines are overprescribed and overproduced. Enough of these drugs are produced in the United States every year to supply every person with about 40 pills each. Half of these drugs find their way into the illegal market where enormous profit is made through the sale of pills to drug abusers.

narcolepsy

Stimulants are now usually prescribed for *narcolepsy* (an overwhelming urge to sleep), depression, and weight control. Many doctors, however, feel that amphetamines should not be given for mild symptoms because they can lead to abuse. For this reason many of the amphetamines that were once available through over-the-counter sales to relieve nasal congestion have been replaced by less stimulating drugs. They are used in the treatment of certain kinds of epileptic convulsions, childhood overactivity, and Parkinson's disease. The latter is a malady of the nervous system characterized by shaking of the hands and muscular stiffness.

Parkinson's disease

The most popular amphetamines include: amphetamine, Benzedrine, dextroamphetamine, Dexedrine, methamphetamine, Methedrine, and Fetamin. Preludin, a relative of the amphetamines used to control obesity, is alarmingly abused throughout the world today. Barbiturates are sometimes combined with amphetamines to reduce some of the unpleasant stimulating effects. The best known of these is Dexamyl.

Anyone going beyond the experimental stage soon builds up a tolerance to pep pills, and larger doses are required to obtain the desired effect.

The body does not develop a tolerance for the other properties of amphetamines, however, and increased dosages make the abuser progressively more nervous, irritable, unstable, and sleepless. Heavy doses may cause temporary toxic psychosis (mental derangement) accompanied by auditory and visual hallucinations. The individual grows restless, fearful, and irritable and may suddenly assault his friends or explode into homicidal or suicidal behavior. Physical signs of an overdose include chills, dry mouth, nausea, headache, vomiting, diarrhea, enlarged pupils, a change in blood pressure (either up or down), chest pains, and abnormal heart action. This can lead to a collapse of heart and breathing action, convulsions, bleeding in the brain, and death. Amphetamine users may seem to have a temporary increase in sexual drive but soon become so exhausted and deranged from the highs that sex becomes unimportant.

symptoms

The characteristic mental state of a heavy amphetamine abuser resembles that of a person suffering from the severe mental illness *schizophrenia*. His thinking becomes deranged and he has frightening visual and auditory hallucinations. Convinced that people are out to do him harm, he withdraws, becoming highly suspicious of everything (*paranoid*). Ordinary objects are misidentified as crouching animals ready to leap out at him, and he may assault imagined enemies. He is a hazard to anyone on a highway should he attempt to drive in this condition.

Worse yet, abusers often realize they are going crazy and attempt to medicate themselves to keep their worlds from crumbling. They take barbiturates, tranquilizers, or heroin to obtain relief from the overwhelming feeling of helplessness. In the last stage of psychosis, they are completely out of touch with reality.

While most experimenters try drugs for no particular reason other than curiosity, abusers had emotional difficulty even before starting on the drug scene. They tend to be immature, cannot bear frustration, and feel deeply ineffective and insecure. Amphetamines make them feel effective and capable of reaching goals that are far beyond reach.

There is evidence that the worst addicts were the most mentally unstable before they started on drugs and that drugs simply uncovered illnesses that were waiting for the right stress to bring them into the open. Psychiatrists believe that this is why so many abusers remain chronically disturbed even after long treatment. Some doctors believe that permanent damage to the brain results from prolonged amphetamine abuse.

Amphetamines do not produce physical dependence as do narcotics. As a result, withdrawal does not have to be gradual but a patient may become seriously depressed and attempt suicide when he is hospitalized and the intake of drugs is stopped. Tranquilizers, such as *Thorazine,* can be used to reduce mental stress, and a nutritious diet can restore the patient to good health.

Drug abusers usually return to the drug scene even after the most intensive psychotherapy. Once psychological dependence is established, the percentage of cures is extremely low.

SEDATIVES: DOWN AND OUT

sedatives

barbiturates

Sedatives have long been used as cures for insomnia, one of the most common complaints. Sedatives belong to a large group of drugs designed to relax the central nervous system. The best known of these are the barbiturates (Fig. 7-2). They are made from barbituric acid and were first manufactured in 1846. Since then a great many new sedatives have been produced, and each in turn has been proclaimed for its effectiveness and safety. Each has the same danger and abuse potential.

Barbiturates are usually taken by mouth, but they may be injected into a vein or muscle. They range from fast-acting *pentobarbital* (Nembutal) and *secobarbital* (Seconal) to slow-acting *phenobarbital* (Luminal) and *butabarbital* (Butisol). The fast-acting drugs are the ones most often abused.

depressants

Because barbiturates slow down the activity of the brain and other organs, such as the heart, and reduce the rate of breathing, they are called depressants. They are most commonly used as a cure for sleeping problems. Their depression of abnormal brain activity makes them useful in the treatment of epileptic seizures. Several medical conditions arising from emotional stress, such as ulcers and high blood pressure, also respond to barbiturate treatment.

Barbiturates are used to put patients to sleep before an anesthetic is administered. This prevents panic on the operating table. Injections of barbiturates such as *Amytal* can be used to help patients recover memories that have been lost following severe emotional shock.

Recent surveys show that sedatives are among the most popular drugs. Twenty-five percent of the prescriptions written by doctors are for barbiturates. Each year about 1000 companies produce enough of these pills to supply every man, woman, and child in the United States with more than 20 doses. An enormous quantity of these drugs reaches illegal channels and is sold to addicts at great profit.

Fig. 7-2. Barbiturate capsules carried by local drugstores.

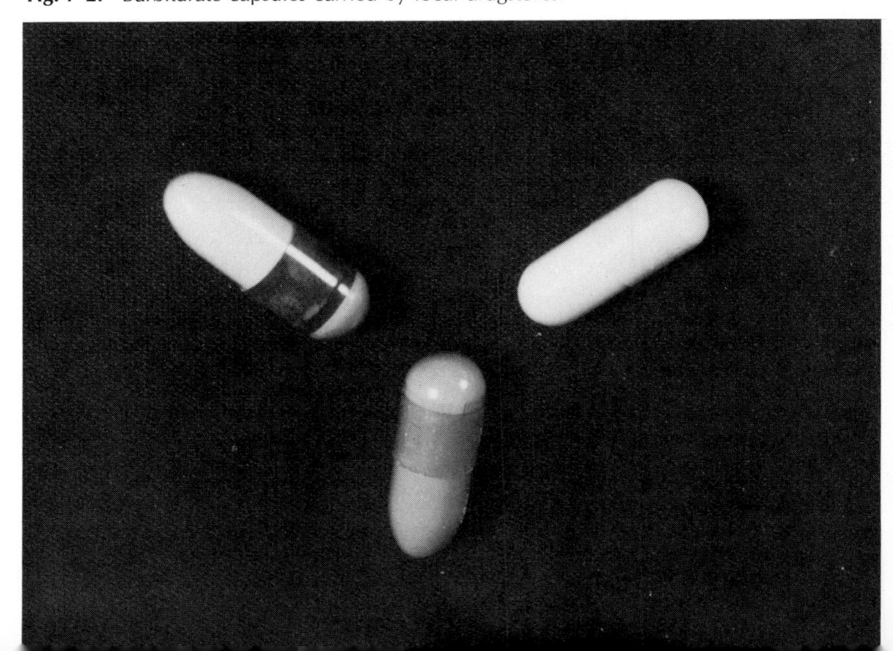

The barbiturate abuser is looking for more than a good rest. Barbiturates provide the "down head" with some of the same effects as alcohol, only stronger. He experiences a drowsy, dreamy feeling of happiness referred to as *euphoria.* The drug reduces his anxiety about himself and enables him to come out of his "shell." He feels less guilty about sex. Often traits that he normally controls, such as anger, aggression, irritability, and bad temper, come out when he is stoned on sedatives.

A barbiturate-dependent person tends to be of the same personality type as other drug addicts. He is psychologically weak and immature; he cannot face up to the reality of everyday living. The parents of drug heads are often confused, inconsistent, and weak. They spoil their child, at the same time they reject him.

A person high on barbiturates appears to be drunk. He staggers, his speech is slurred, and he often loses his balance and falls. His mood is infantile, goofy, and slow. He becomes drowsy and may pass out.

tolerance

The continued use of barbiturates rapidly leads to the development of a tolerance for the drug. Increased doses must be taken to obtain the desired effect. Worse yet, both psychological and physical dependence continue to build, and the user may graduate to heroin in an attempt to satisfy his craving. Heavy use of barbiturates may result in permanent damage to the brain.

Compared with heroin addiction, the barbiturate habit is cheap. A drug head can buy pills for a nickel to a dollar. His real world slips away as his personality, personal hygiene, and social habits decay. He lives in perpetual fantasy and finally becomes psychotic. His confusion grows as imaginary voices call out to him. Things that are not there seem real. This fearful and confused state, delirium, demands immediate and continued medical care.

withdrawal

The first day of withdrawal finds the abuser becoming increasingly nervous, weak, and nauseated. When night comes, he cannot sleep and he sweats profusely as his blood pressure drops. Epileptic-type seizures may begin on the second day. If treatment is not given at this time, he may die. During the third through the seventh days he becomes delirious and experiences panic, confusion, and terrifying hallucinations. The worst withdrawal symptoms pass in about a week, but the illness is still painful and potentially deadly. A doctor's care is essential. The addict who tries withdrawal alone is actually gambling with his life. Gradual withdrawal under the care of a doctor is accomplished by reducing the daily dose by about 10 percent each day. Sometimes the daily reduction is even less. After withdrawal is completed the patient may feel irritable and sleepless for several more days. Such individuals require long-term psychiatric care which they often resist.

An overdose of barbiturates often results in accidental or suicidal death. Statistics indicate that each year about 3000 Americans, either accidentally or deliberately take their own lives with these drugs. Abusers unintentionally kill themselves when they become so confused they cannot remember how many pills they have taken.

ALCOHOL: FLOAT WITH THE TIDE

alcohol

While part of the drug-taking population considers alcohol an "inferior trip," it continues to be the most popular, most easily obtained, and most abused drug in the English-speaking world. In the United States alone, more than 10 billion dollars a year are spent on alcoholic beverages, and more than 70 million Americans are regular drinkers. Some of course drink much more than others. Drinking is considered an enjoyable way to relax by people who are mentally and physically capable of handling it, but those who cannot often wind up living in a nightmare world.

Alcoholic beverages are fluids that contain ethyl alcohol in intoxicating quantities. The alcoholic content may range from as little as 2 percent in some beers to as much as 69 percent in absinthe. Ethyl alcohol, a product of the fermentation of fruits, grains, or vegetables, is a clear liquid with little odor and a stinging taste.

Although there are other alcohols besides ethyl alcohol, they have never been popular and some are extremely dangerous. Methyl alcohol (wood alcohol), for example, causes blindness and often leads to death.

Most alcohol enters the bloodstream through the stomach and small intestine. Absorption is faster when the stomach is empty. About 10 percent of the alcohol entering the bloodstream is excreted unchanged by the kidneys. Most of the other 90 percent is oxidized by the body and converted into energy. This in no way qualifies alcohol as an adequate food. Alcoholics are notoriously deficient in vitamins and proteins.

effect

Alcohol affects the body much the same as sedatives and anesthetics; it acts as a depressant on the brain and nervous system. The first areas depressed by alcohol are the centers that control emotions and thinking. Inhibitions and conscience emotions are relaxed, and the drunk feels less restricted by accepted codes of moral and sexual behavior. Sexual performance, however, is not improved by alcohol.

Drinking is usually accompanied by a warm sensation caused by increased blood circulation through the stomach and vessels near the surface of the skin. This seems to have led to the misconception that alcohol helps keep a person warm on cold days. Alcohol actually causes increased heat loss through sweating and by increasing circulation near the surfaces of the skin.

A person under the influence of alcohol is often deluded by the thought that he is mentally sharp and alert, but tests show that even small amounts of alcohol decrease body functions. While familiar tasks may be carried out with reasonable efficiency, complicated problems that arise swiftly leave the drinker bewildered. This is why anyone who has been drinking is such a menace on a highway.

The drinking of alcoholic beverages causes more damage than all the other drugs combined. It creates the greatest drug problem among teenagers because it is so easily available and accepted by the vast majority of society. More than half of all fatal automobile accidents in the United

States involve drivers who have been drinking, and about 30 percent of the pedestrians killed on streets and highways are people who have been drinking. Alcohol is a contributing factor to about 40 percent of all other violent deaths, including homicides and suicides.

alcoholics

It is estimated that there are between five and six million alcoholic adults in the United States today, making alcoholism our fourth most critical health problem. Mental illness, heart disease, and cancer cause more deaths. Nearly 250,000 more Americans become alcoholics every year. People cannot live in isolation, and every alcoholic seriously hurts five or six others in his family, as well as his business associates. Added to these enormous costs are desertion, mental and physical cruelty, and a high divorce rate among alcoholics. Children with alcoholic parents are often delinquent, often suffer emotional illness, and often turn to crime. The cost of these social maladies must be paid by the general public. Ineffectiveness and absenteeism among alcoholics cost industry several billion dollars a year, and this enormous sum is passed on to the public in the form of higher prices. There is no question that the cost of alcoholism to the American society is staggering.

Although people drink for a great many different reasons, experts point out that the alcoholic suffers from a behavior disorder which causes him to drink to the point where his physical and emotional health are damaged. Some people may react to unusual stress by "getting stoned" but may stay sober for months or even years afterward, perhaps drinking on a lower level. People who drink constantly, however, similar to other kinds of drug addicts, usually show an abiding interest in alcohol, and their "binges" do not seem to be related to any particular stress. They continue drinking until sickness or lack of money temporarily puts them "on the wagon."

There are psychologists who believe that every alcoholic is a bitter, depressed person who was overindulged, spoiled, and coddled as a child. When such a person meets frustration, he reacts with unusual resentment and feels overwhelming helplessness. He demands love and devotion but cannot obtain enough to satisfy his craving. Filled with despair and frustration, he strikes out at his world and himself with a whisky bottle.

Obviously, individuals build up a strong psychological dependence on alcohol. Although tolerance also develops, it takes longer than it does for other drugs such as barbiturates and narcotics. A chronic alcoholic may require three to four times as much alcohol as an occasional drinker to feel the same effects. Some physical dependence also develops, and many drunks feel physical discomfort if they suddenly go on the wagon.

Alcohol irritates the mucous membranes of the throat and stomach, and there is a high incidence of bleeding ulcers among drinkers. Because alcohol provides calories to the body and diminishes the appetite, alcoholics are likely to be poorly nourished. They are subject to infectious diseases such as tuberculosis and pneumonia, resulting in part from poor personal care. Deficiencies in vitamins, proteins, and minerals among drinkers result in a marked shrinking and scarring of the liver, a disease called cirrhosis. Other organs also suffer.

delirium tremens

Chronic alcoholics, usually over 30, often experience *delirium tremens*. It may occur after heavy drinking, or following the sudden stopping of drinking when their money runs out. It is characterized by fright, restlessness, and confusion. Terrible visions occur; spiders and bugs seem to crawl over the body, and monstrous animals confront the agonized victim as he shakes and sweats uncontrollably. Delirium tremens is a medical emergency, and about 10 percent of people afflicted with it die of convulsions, heart failure, or other complications.

MARIJUANA

marijuana

Marijuana is a drug made from the flowering tops and leaves of the Indian hemp plant *Cannabis sativa* (Fig. 7-3). It grows in the milder climates around the world, thriving in such places as Mexico, Africa, India, and the Middle East. Marijuana grows throughout much of the United States; however, plants grown in tropical areas produce the highest quality drugs. The psychocentric properties are found in the sticky resin exuded from the flowering tops and leaves.

The word marijuana is said to be a corruption of *mariguango*, a Portuguese word meaning intoxicant. In the United States the drug is also known as pot, tea, grass, weed, Mary Jane, and by other names.

Fig. 7-3. Marijuana and its derivatives. (1) Branch; (2) hashish; (3) pressed pollen; (4) manicured marijuana; (5) seeds; (6) rough marijuana. *(Courtesy Carolina Biological Supply Company.)*

THC

The active ingredient in marijuana is a chemical called tetrahydrocannabinol (THC). Its concentration determines the potency. Seeds and stalks contain very little THC, but they are often used in the marijuana trade.

hashish

The drug commonly referred to as marijuana is made from the whole plant and is considered to be of low quality. *Bhang,* slightly stronger than marijuana, is made from the leaves and flowering shoots of the plant. A drug called *ganja* is made from small leaves and resins and is about three times as strong as marijuana. *Hashish* or *charas* is made from dried resin and is five times as potent as marijuana. The drugs used in the United States are usually made from dried leaves and flowers. They are crushed and rolled into cigarettes, called reefers or joints, which produce a harsh sweet smoke.

The History of Marijuana Use

history

No one knows how long men have been using marijuana, but the first record of its use dates back to 2737 B.C. when the Chinese emperor Shen-Nung prescribed it for beriberi and as a psychic liberator, among other things. It was later used by Chinese physicians as a surgical anesthetic, but the Chinese appear to have found the drug undependable and continued to use opium for such purposes.

The hemp plant spread to India where its psychic properties were more appreciated than in China. Early Christian missionaries looked upon marijuana as a curse, but writings about 1000 B.C. told of its mind-expanding qualities and its value in treating disease. Theology students took it before reading the scriptures, and holy men enhanced their meditation through its use.

By A.D. 500 the weed had spread throughout much of Europe and to Arabia and Africa where the smoke from marijuana was often inhaled as the plant burned in the campfire. By this time North African users had developed a pipe which cooled and removed some of the irritants from the smoke. Napoleon's army, returning from Egypt, is most likely responsible for the spread of the hemp weed into western Europe during the early part of the nineteenth century. The first account of marijuana in western literature was written by the British physician W. B. O'Shaughnessy in 1840.

in the United States

Mexican laborers brought marijuana into the southern regions of the United States shortly after the turn of the century, and gradually the demand for cannabis spread up the Mississippi River valley and throughout the United States.

The general use of marijuana was outlawed by the Federal Marijuana Tax Act of 1937, and this was followed by strict laws and enforcement in every state. A sharp increase in use was reported in the 1960s, and arrests for misuse have more than doubled since that time according to the President's Commission on Crime. While the extent of marijuana use in the United States is not definitely known, health authorities estimate that

5 million Americans have used the drug at least once, and other estimates run as high as 20 million.

Marijuana Chemistry

There are over 30 cannabinoids in marijuana, many of which are inactive. Most of the active substances can be extracted from the crude resin with ether and other solvents. The most active ingredient is *l*-delta-THC, a substance that produces the same effect as the crude drug. There are several forms of this chemical, some of which are inactive. When certain of these inactive forms are smoked, however, they are converted to active *l*-delta-THC. This explains the higher activity of smoked marijuana.

The Effects of Marijuana

effects

Contrary to the claims of many users, marijuana has been shown to affect recent memory adversely. Under the influence of the drug, people are unable to reproduce material that they have recently read and have trouble recalling words as well as ideas.

A user high on marijuana finds it harder to make decisions that demand clear thinking, and he is more likely to follow the suggestions of others. Performing any task that demands thinking and good reflexes is difficult. For this reason it is dangerous to drive while under the influence of the drug. Recent investigations showed that high doses of the drug caused serious reactions in every person tested. According to the National Institute of Mental Health, for some unknown reason psychotic reactions sometimes occur in some persons from smaller dosages.

Scientific research has shown that a dose of marijuana equal to one American cigarette can make a person feel excited, silly, and gay. After a four-cigarette dose the user notices changes in perception and reports brighter colors and a keener sense of hearing. After a 10-cigarette dose he experiences visual hallucinations, sees objects in odd shapes, and has delusions or believes things not based on reality. He may initially feel great joy, but his mood may swing to one of extreme anxiety; he may become deeply depressed and feel extremely uneasy, panicky, or fearful.

Marijuana and Addiction

dependence

Authorities now consider marijuana in terms of dependence rather than addiction. It is not a narcotic and therefore does not produce physical dependence as do such drugs as heroin and morphine. This means that the body never becomes dependent on the continued use of the drug, and it does not appear to develop a tolerance for it either. Therefore larger and larger doses are not necessary to produce the same effects. Withdrawal from marijuana use does not produce physical sickness. Many investigators believe, however, that a user can develop psychological dependence if he takes the drug on a regular basis.

Recent studies have shown that more than 80 percent of narcotic addicts from city areas had previously used marijuana. Scientists agree, however, that few of the thousands of people who use marijuana have gone on to harder drugs. While no direct cause-and-effect link between marijuana use and narcotics has been found, informed observers in the drug field point out that persons predisposed to abusive use of one drug are likely to graduate to stronger kinds. Another factor that helps lead marijuana users to harder drugs, and one that is often cited in arguments for the legalization of marijuana, is the fact that its use may expose a person to a variety of drugs through contacts with drug pushers and other users.

What Are the Risks?

legal aspects

Breaking marijuana laws can have effects far more serious than users realize. Because continued use of the drug acts to change the personality, young users often find their education interrupted. Increasing numbers of American youth, estimated in the thousands, are arrested each year for use or possession of the drug. While this in no way acts to curtail the problem, it has serious implications among young Americans. Breaking laws dealing with marijuana and subsequent conviction can lead to fines up to $20,000 and 40 years in prison with no suspension of sentence, probation, or parole. Conviction for a felony further complicates their lives since anyone with a police record must meet special conditions to renew or obtain a driver's license. It can also prevent a person from entering certain professions such as law, medicine, or teaching, and can make it difficult for him to obtain a responsible position in business or industry. Special hearings must usually be held before he can qualify for a government job. Students should be aware of the social and legal facts before becoming involved with drug use.

risks to health

Risks to health are pointed out by experts on human growth and development. Although the more subtle effects of the drug are difficult to assess, its use is known to change the personality and to reduce growth and development. It is at best a seriously questionable practice for anyone to experiment with drugs of any kind during a period in his life when he is in transition to adulthood. It has long been known that youth have enough of a task adjusting to life and establishing values without engaging in activities that upset the delicate developmental equilibrium.

The statement is often made that marijuana is "medically safe," but this has not been supported by scientific evidence. The serious consequences of tobacco smoking were not known until hundreds of years after its introduction, after thousands of smokers unaware of the dangers had died from ailments now known to be directly related to the use of tobacco. Medical science does not know enough about the effects of marijuana at the present time to predict its dangers accurately, but the American Medical Association's Council on Mental Health feels there is sufficient evidence to show that it is harmful when used over extended periods.

NARCOTICS

narcotics

The term narcotic refers to opium and pain-killing drugs made from opium such as *heroin, morphine, paregoric,* and *codeine*. These and other opiates are made from the juice of opium poppy fruits (Fig. 7-4).

Opium was used by the Egyptians as far back as 1500 B.C. and since that time abuse in various parts of the world has resulted in a great many serious and costly social problems. The first laws restricting the use of opium in the United States were passed by San Francisco in 1875 in the form of a city ordinance, and since then the drug problem has grown with the population. While the use of opium itself is not a significant problem, the use of morphine, heroin, and other narcotics obtained from the poppy has been extremely difficult to control.

When the narcotics user becomes addicted, his body requires repeated and larger doses of the drug. Once the habit is established, larger and larger doses are required to produce the same effect because the body develops a tolerance for the drug.

Withdrawal symptoms always occur when an addict stops taking narcotics. He may sweat, shake, have chills, develop diarrhea and nausea, and suffer leg cramps and abdominal pains. Modern treatment is available, however, to help the addict through these painful and dangerous stages of withdrawal. This does not make withdrawal less serious. Scientists have uncovered evidence that physical addiction may last for longer periods of time than previously suspected.

Fig. 7-4. This type of seed pod is produced by members of the poppy family from which opium is derived. Seeds are shed through the pores near the top.

As with various other drugs, heroin use also results in psychological dependence. In other words, taking drugs becomes an urgent emotional need, and the addict uses it to escape facing the realities of life. Unfortunately, the narcotic can be more of an escape than expected since injections of large or unintentional pure doses often end in death.

The Effects of Heroin

The first reaction that most people feel after taking heroin is a reduction in tension and a relief from worry and fear. This high is usually followed by a period of inactivity which borders on stupor. Because heroin depresses certain areas of the brain, hunger is reduced, as are thirst and the sex drive. Feelings of pain are reduced. Because addicts seldom feel hunger, hospital rehabilitation often must include treatment for malnutrition.

heroin

Users take heroin by mixing it in a liquid solution and injecting it into a vein (Fig. 7-5). Addicts report that, once in the system, "It makes troubles roll off my mind" and "It makes me sure of myself." It apparently takes the edges off reality. The effects of the drug, however, are influenced by the user's personality, the size and frequency of the dose, how the drug is taken, and many other factors.

Fig. 7-5. Heroin injection technique. The crystals in the spoon are melted by heat from a match or cigarette lighter and the liquid injected with a 1-milliliter syringe. Persons wishing to avoid tell-tale arm scars may give the injection between the toes or under the tongue.

Who Uses Heroin?

The U.S. Public Health Service has made studies showing that heroin is used chiefly by young men of minority groups in ghetto areas, although use is increasing among middle-class groups. More than half of the 60,000 known heroin users live in New York State, mostly in the city of New York, according to the Bureau of Narcotics and Dangerous Drugs. Recent findings show that the majority of these addicts are under the age of 30. It was estimated as of May 1971 that 1.5 million Americans have taken heroin at least once.

Some older people who take narcotics regularly to relieve pain become addicted, as do people who have easy access to drugs. Doctors, nurses, and druggists can become hooked on heroin. As one might suspect, however, this type of addict is known to have emotional difficulties similar to those of other regular drug users. People who take drugs to relieve pain, however, are usually emotionally stable.

Most users admit that once on drugs the main objective of life involves obtaining a continuing supply. This powerful drive frequently prevents addicts from holding jobs or continuing their education. Furthermore, the heroin habit is too expensive to be supported by ordinary labor. Addicts are often sick one day from withdrawal and sick the next day from an overdose. Statistics show that their lives are shortened on the average by 15 to 20 years. Because of their single-minded desire to obtain more drugs, addicts often turn to crime and are usually in trouble with their families and constantly in trouble with the law.

Most addicts were in some kind of trouble with the law even before they started taking drugs. Once addicted to heroin, however, they usually become involved in crime, in part because it costs so much to support their habit and also because they must buy the drug from criminals. An addict may have to spend up to $100 a day to satisfy his craving for heroin. While involvement in crime is not a direct effect of the narcotic itself, it is the only way a user can obtain that much money. His crimes almost always involve thefts of anything he can lay his hands on, even the property of his own family, and not crimes of passion or violence.

Penalties for Illegal Use of Narcotics

laws

The Harrison Act of 1914 established federal penalties for the illegal use of narcotics. This act makes illegal possession of narcotics punishable by fines and imprisonment with sentences ranging from 2 to 10 years for the first offense, 5 to 20 years for the second, and 10 to 20 years for further violations. The illegal sale of narcotics can lead to fines of $20,000 and sentences of 5 to 20 years for the first offense, and 10 to 40 years for further offenses. Anyone who sells narcotics to persons under 18 is refused parole or probation for all offenses including the first. If the drug is heroin, the pusher can be sentenced to life in prison.

The Harrison Act has served as a model for the development of state narcotics laws. Both federal and state judges traditionally impose severe sentences for all narcotics violations.

The drug problem is one of the greatest threats to our cities because it is the half-crazed addict driven by an uncontrollable drug need that so often keeps 98 percent of the city population in terror. In New York City addicts make up only about 2 percent of the population, but this handful of sick people makes life nearly unlivable for the others. People are afraid to walk the streets at night; parks have become places to be dreaded rather than enjoyed; and rich and poor alike cower behind barred doors at night. The addict becomes violent for only one reason—to obtain large amounts of money each day to pay for his fixes. This amounts to an enormous profit for drug traffickers. The injustice of this illegal business is incredible —in light of the fact that a kilogram of heroin is worth approximately the same as a kilogram of potatoes. This amount of heroin is sold for as much as $50,000 on the streets of our cities. There is little question that as long as these enormous profits are to be made a whole galaxy of smugglers, processors, and pushers will remain in operation. Each child, youth, and adult these social parasites can "hook" means $5000 more a year in illegal profit. Until narcotics laws are made that take the profit out of narcotics, they will be self-defeating.

Paradoxically, every gain in the war on narcotics raises the price, creates new pushers and more addicts, and crime increases. Many experts feel that narcotics should be made legally available to all addicts. This would take the profit out of illegal traffic and create fewer addicts. When there is no profit in making an addict out of a teenager, there will be no pushers around schools talking students into trying drugs. The addict's need for money would greatly diminish and with it muggings, burglaries, and robberies. Needless to say, by supplying addicts with drugs we would need to open up avenues of rehabilitation as never before.

Medical View on Addiction

As with other types of drug addiction, the narcotics user is a sick person, according to medical authorities. He has little chance of recovery without treatment for physical addiction and withdrawal sickness. He must have continued help to keep him away from drugs after he has successfully withdrawn.

The most difficult part of his rehabilitation comes when he is released from the hospital. It is not extremely difficult for doctors to take him off the drug and restore his health, but a user is often subjected to a great deal of pressure to take up the habit again. Taking drugs has become part of his life; it involves his friends and the kind of job he can find. Worse yet, he has the type of personality that makes him prone to taking drugs and often he does not want to begin a fresh start in life or enjoy normal pleasures.

Various rehabilitation methods are being tested, but the search for cures has been an illusive business. Rehabilitation means the rebuilding of emotional, mental, physical, social, and vocational life. For many addicts it takes remodeling of all these aspects of their personalities to keep their lives from being wasted. This is an enormous task because it

rehabilitation

requires drastic changes and addicts are seldom able to put the required effort into it.

The maintenance of community clinics is one of the current experimental techniques used to help addicts. A user can go to these clinics regularly in his own neighborhood to take a drug that blocks the high produced by a shot of heroin. Addicts who have stayed away from the drug for many years report that continued treatment and close supervision after they had been cured were mainly responsible for their rehabilitation.

New York City maintains a self-help program, called Daytop Village, run by ex-addicts. Their "no-nonsense" treatment of new patients involves frank and open discussions several times a week. A gain in status and privileges is earned by hard work, honesty, and by staying away from drugs. This treatment, lasting a year, is proving to be quite successful.

Because the rebuilding of a life requires so much effort, special services, and programs, opportunity for help has been limited. The Narcotics Addict Rehabilitation Act of 1966, however, is designed to give addicts a better chance to kick the habit. It gives some addicts, those not charged with a crime, a chance for treatment rather than imprisonment. This law provides that a complete range of services be made available to addicts in their own communities.

LYSERGIC ACID DIETHYLAMIDE (LSD)

LSD

ergot

A powerful man-made drug, LSD, was developed in 1938 from one of the ergot alkaloids. Ergot is a rust fungus which grows on rye, a common grain plant. It is extremely powerful. A single ounce is enough to make 300,000 average doses.

Classified as a mind-affecting drug or hallucinogen, LSD is noted particularly for producing bizarre mental reactions and remarkable distortions in touch, smell, hearing, and vision. To a person under the influence of the drug, walls may appear to move; colors appear stronger and unusually brilliant. Sensory impressions may be translated or blended into each other. For example, music may be interpreted as a color, and colors may appear to have taste. One of the most common but confusing reactions to LSD is the feeling of strong opposite emotions at the same time. The user may feel depressed and elated, relaxed and tense, and happy and sad at the same time. Normal feelings of boundaries between the body and space become distorted, giving users the notion that they can fly or float with ease. Accidental death can and has occurred as a result of users leaping out of high windows or from other high places to their death. Also, because of the feeling that he cannot be harmed, a user may drive or walk in front of moving vehicles.

Taking LSD often results in panic. The user may fear he is losing his mind and grow frightened because he cannot stop the effect of the drug. He may suffer from paranoia, a mental disorder characterized by continuing elaborate delusions of persecution or grandeur. He may become increasingly suspicious of everything, feeling someone is trying to harm him or regulate his thinking. This feeling usually wears off after about 72 hours.

Unfortunately, recurrences occurring days, weeks, or even months after the drug has been used are not uncommon. When this happens, the user is helpless to stop the strong effects and may fear he is going insane. The effects vary with the individual, however.

Certain users think that LSD can heighten their senses and make them more creative. Studies have shown, however, that works of art, such as paintings and writings, produced by drug users fail to lend credulity to this point of view. Works produced by individuals after they have used LSD is noticeably poorer than before.

Reactions to LSD range from great worry and deep depression to borderline and severe mental derangement. Overwhelming worries and fears, according to medical experts, can be disturbing enough to cause acute and long-lasting mental illness.

There is evidence that LSD affects the chromosomes, the tiny threads carrying hereditary information in every cell in the body, which are the genetic links between generations. Scientists have reported that chromosomal damage or changes result when LSD is added to tissue cultures. Also, cells taken from users have shown unusual chromosomal breaks. This can cause abnormalities in offspring, cause cancer, and shorten life expectancy.

Fetal damage in pregnant rats and mice given the drug has also been reported, and human birth defects have been observed in the babies of mothers who have taken LSD. Such evidence is arousing concern of scientists and laymen alike.

Because LSD is a dangerous drug it is regulated by the Bureau of Narcotics and Dangerous Drugs of the U.S. Justice Department. Strict penalties are provided for anyone who illegally produces, sells, or disposes of dangerous drugs. Convictions can result in fines of $1000 to $10,000 and imprisonment for up to 5 years. Adults selling or giving drugs to anyone under 21 can receive 10 to 15 years in jail and be fined up to $20,000. Mere possession of this dangerous drug can result in a fine of $1000 to $10,000 and a sentence of 1 to 3 years in jail; some state laws are even more severe.

Suggested Readings

Abel, E. L. "Marihuana and Memory. Acquisition or Retrieval?" *Science*, 173(4001):1038–1040, 1971.

Greenberg, H. *What You Should Know about Drugs and Drug Abuse*. New York: Four Winds Press, 1970.

Houser, N. W. *Drugs*. New York: Lothrop, Lee and Shepard, 1969.

Skakkebaek, N. E., and E. Philip. "LSD in Mice: Abnormalities in Meiotic Chromosomes." *Science*, 160(3833):1246–1248, 1968.

Snyder, S. H. *Uses of Marijuana*. New York: Oxford University Press, 1971.

Trout fingerling girdled by a soft-drink can pull tab. The introduction of fish, although often successful, is fraught with a variety of hazards. *(Courtesy Michigan Department of Natural Resources.)*

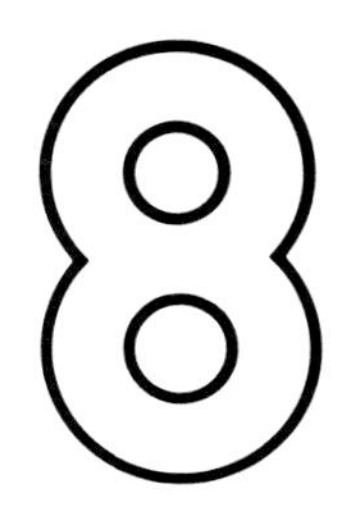

Introduced Species

An understanding of the potential for disaster and also the possible benefits resulting from man moving plants and animals to new areas first requires a consideration of the natural distribution of organisms. It may be logical to ask why a species that succeeds after being brought to a new area was not already there.

ANIMAL MIGRATION

migrations

The kinds and numbers of organisms are not static in time or in space. Although the process of evolution is slow, new species are constantly evolving as others become extinct. A species may occur only in a limited area because it is new and has evolved tolerance only to a restricted set of environmental conditions. A species on the verge of extinction, however, may be experiencing a steady reduction in its range as the habitats in which it formerly flourished are gradually eliminated by climatic or other changes. Between the birth and death of a species, it may have occupied a wider area and tolerated a greater variety of environmental conditions.

Barriers to Migration

barriers

In addition to the restrictions imposed by its genetic makeup, a species may be confined to a particular area because of geographic barriers. These range from high mountains, keeping a particular population within an enclosed valley, to the expanse of ocean surrounding a continent.

continental drift

In the geological past the major landmasses of the world were much closer than they are today. A single supercontinent, *Pangaea,* is thought to have split about 250 million years ago into a northern section, *Laurasia,* and southern section, *Gondwana* (Fig. 8-1). Animals at that time ranged widely since there were few barriers to their migrations. However, Laurasia fissured into what we now know as Eurasia, Greenland, and North America. Rifts in Gondwana carved Australia, southern Asia, Africa, South America, and the Antarctic. Slowly but measurably, the continents drifted apart like gigantic arks, isolating plant and animal populations.

Links between the continents have formed periodically and allowed animals to move from one landmass to another. Even today, ice bridges sometimes serve the same function. The formation of a land bridge between North and South America (Fig. 8-2) at various times in the past

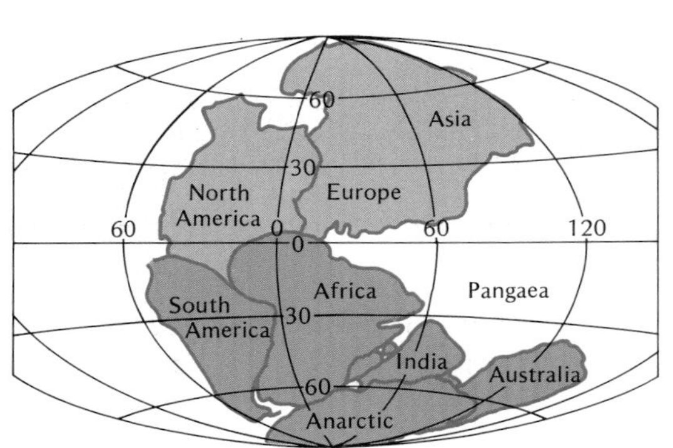

Fig. 8-1. The supercontinent of Pangaea as it is thought to have been about 225 million years ago. The first separation by continental drift formed the landmass of Laurasia from what would be North America, Europe, and Asia. Gondwana comprised the rest of the exposed land surface. (Redrawn from *The Sciences,* 10(11), 10, 1970.)

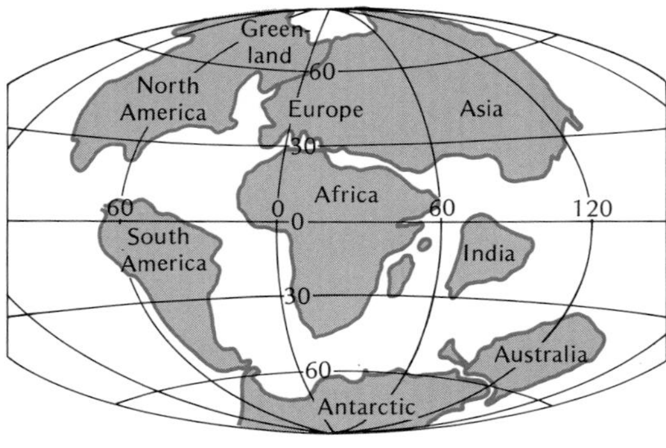

Fig. 8-2. Configuration of the continents at a time, about 65 million years ago, when North and South America were separated. (Redrawn from *The Sciences,* 10(11), 10, 1970.)

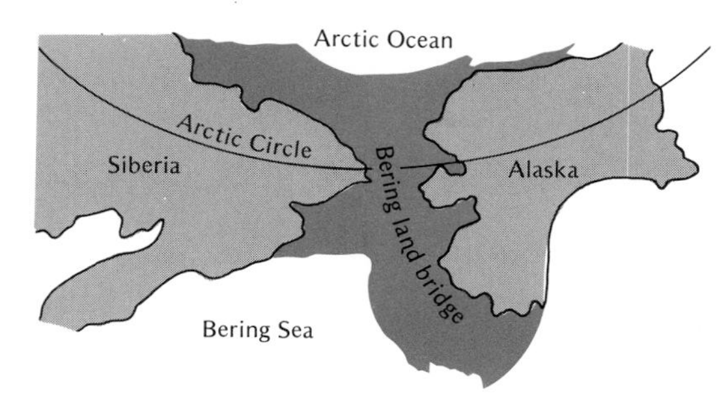

Fig. 8-3. The Bering land bridge. An exposed land connection occurred between Siberia and Alaska during three periods of the early Pleistocene epoch: 70,000 to 50,000 years ago, 32,000 to 28,000 years ago, and 22,000 to 15,000 years ago. The last of these three periods is the most likely one for the migration of man into North America. (Adapted with permission of The Macmillan Company from *Man and the Environment* by A.S. Boughey. Copyright 1971 by Arthur S. Boughey).

Bering land bridge

and between Siberia and Alaska during Pleistocene times permitted movements of animals from areas of former isolation. The opossum, porcupine, and armadillo moved into North America from the south; South America was invaded by foxes, dogs, cats, otters, and deer from the north. The Bering land bridge, as the Siberia–Alaska land connection is called, enabled deer, bison, bears, rhinoceroses, mammoths, cats, and even men to migrate from Asia to North America (Fig. 8-3). Dogs, camels, tapirs, and horses moved across into Asia. Another bridge, present during the late Mesozoic era, connected southeastern Asia and Australia. Marsupials migrated into Australia and have been isolated there for the last 100 million years. This bridge was closed by a rising sea level by the time placental mammals appeared on earth. The intermixing of species has occurred slowly, but by the time modern man arrived on the scene the post-Tertiary plant and animal redistribution was almost stabilized.

Zoogeographic Realms

disjunct distributions

Some plants and animals occupy very limited areas; others are widespread. Alligators live along the southeastern coasts of the United States from South Carolina to Louisiana. They also occur in southeastern Asia. Tapirs have a similar disjunct distribution: Malaya, Sumatra, and Borneo; and Central America and northern South America.

In the Northern Hemisphere north of 45° North latitude, many plant species are distributed around the world. They are limited, however, to a narrow north–south belt. This applies in particular to those whose seeds or spores travel readily by air currents. Whatever the distribution of a species, one or more factors limit its range. These may be physical (mountains, deserts, bodies of water), climatic (temperature, rainfall, wind, light), or biological. The last-mentioned category includes parasite–host relations as they restrict the parasite, and the availability of suitable food.

Wallace

The peculiarities of animal distribution led Alfred Wallace (1832–1913) to divide the world landmasses into what he believed represented natural zoological areas. Each major area, which he termed a realm, contains distinctive groupings of animals that are different from those of other areas. These are outlined in Table 8-1.

TABLE 8-1
Zoogeographic Realms

Realm	*Area*	*Typical animals*
Nearctic	North America from the Mexican highlands northward, Greenland	Bison, caribou, musk ox, pronghorned antelope, mountain goat, muskrat
Palearctic	Europe, Asia south to the Himalayas, Persia, Afghanistan, Africa north of the Sahara	Fallow and roe deer, wild boar, hedgehog, wild ass, reindeer
Ethiopian	Africa (Sahara and southward), Madagascar and adjacent islands	African elephant and lion, hippopotamus, giraffe, horned antelope (many species), ostrich, gorilla, chimpanzee, guinea fowl, lemur in Madagascar, zebra
Oriental	Asia south of the Himalayas (Ceylon, India, Borneo, Malay Peninsula, Java, Philippines, Celebes)	Indian elephant, peacock, jungle fowl, gibbon, macaque, orangutan, tarsier, tiger
Neotropical	West Indies, Central and South America, low coastal area of southern Mexico	Alpaca, llama, anteater, most hummingbirds, sloth, curassow, rhea, tapir, toucan, vampire bat
Australian	Australia, New Zealand, New Guinea, Pacific Ocean islands	Monotremes (duck-billed platypus), most marsupials, bird of paradise, Australian lungfish, emu, lyre bird, cockatoo

Man-Made Migrations

invasions

In the absence of intrusions, the organisms within an ecosystem achieve a kind of dynamic equilibrium which may be easily upset. Now, as never before, plants and animals are invading new territories, not because of land bridges or climatic changes, but because of the wanderings of man. Wherever he goes, man takes plants and animals with him—some intentionally, some otherwise. This has occurred since man first domesticated animals. The development of rapid transportation and the increase in shipping within the last 100 years, however, has accelerated the introduction of foreign species into every region.

Many of man's attempts to improve on nature by moving plants and animals to new homes have been beneficial, but others have led to disastrous consequences. The majority of his importations have failed to become established as self-sustaining populations because the niches required were not there or because the invaders were unable to compete with native species.

INTRODUCTION OF NEW SPECIES

When a species is released into a suitable new habitat that contains abundant food and no natural enemies, it multiplies rapidly until some aspect of the environment causes a leveling of the population. The population level reached may be so high that considerable damage results before the interloper is brought under control. The following examples illustrate this situation.

The European Rabbit

rabbits

The *European rabbit* has a long history of detrimental introductions. Originating in the Mediterranean area, it was introduced into the British Isles at the time of the Norman invasion, about A.D. 1000. It multiplied rapidly in the islands and, although the overabundance caused serious problems, the consequences of rabbit introductions were either not understood or ignored.

In 1859 a dozen pairs of European rabbits were released on an estate in Australia in the province of Victoria (Fig. 8-4). By 1865 the rabbit population had skyrocketed to an estimated 10,000 in spite of the fact that about 20,000 had been killed. Expansion continued and in just 15 years this prolific forager, introduced as a game animal, had become a grave problem. Five rabbits ate as much forage as one ewe, so as the rabbit population increased the sheep population decreased. Many efforts were made to slow the expanding rabbit population and stop them from spreading into new areas. Fifteen hundred miles of rabbit-proof fence were constructed. Rabbits were trapped, bountied, dug out, ferreted, and poisoned, with little success. The rabbit population density reached 1000 rabbits per square mile over an area of 1 million square miles.

Fig. 8-4. The European rabbit in Australia. *(Courtesy Commonwealth Scientific and Industrial Research Organization, Division of Wildlife Research, Canberra City, Australia.)*

virus

The population was finally brought under control by a second introduction. A species-specific virus disease, discovered in South America, has little effect on South American rabbits but was found to be lethal to the European rabbit. The disease, myxomatosis, was inoculated into some captive rabbits which were then turned loose. After World War II and through the early 1950s, the disease spread through the population, killing millions of rabbits and ridding the Australian farmer of his worst scourge. This disease, similar to others, may lose its virulence and be ineffective in a few years, and the rabbit hosts may eventually develop immunity to the disease. For the time being, however, the rabbit problem has been solved.

The Gypsy Moth

gypsy moth

The French astronomer Leopold Trouvelot brought some eggs of the *European gypsy moth* to the United States in 1869. He intended to use the caterpillars in breeding experiments to improve silk production. Some escaped, however, and to date government agencies have spent over 100 million dollars trying to bring these potential money makers under control. Concentrations of up to 10 million insects per acre have been reported in a 2-million-acre tract extending from the New England states to Pennsylvania. Specimens have even been found in California and as far south as Florida.

Eggs are deposited in July, usually near the cocoon since the adult female seldom flies, and hatching takes place the following spring. Most eggs do not hatch if they are exposed to temperatures lower than −20°F. Because the larvae are very light and covered with long hairs, they can be carried many miles by the wind, and very great distances if they become lodged on vehicles.

Gypsy moth larvae feed on the leaves of many kinds of broad-leaved plants, particularly oaks. Trees frequently die when they are defoliated during the dry season. From early May to mid-July, vast quantities of feces fall from infested trees, lodging on houses, cars, people, or anything left near trees.

Widespread applications of pesticides have been ineffective in controlling the spread of the moth. Unfortunately, the indiscriminate use of pesticides has often led to even greater infestation since natural enemies are killed along with the pest.

sex attractant

The gypsy moth is vulnerable, however, because of the nature of its mating devices. The female, a much weaker flier than the male, seldom ventures far from the cocoon from which she emerged. Males are attracted by an odor produced by the female which apparently can be detected in minute quantities at great distances. This odor, first obtained by grinding female moths, but now artificially synthesized, can be used to attract males to traps where they can be destroyed. One advantage this method has over chemical poisons lies in the inability of the moth to develop a resistance to its own sex attractant. Such a resistance would have the effect of disrupting the life cycle and be equally useful from man's point of view. Also, the sex attractants are specific. Only the intended victim is affected.

Fig. 8-5. The Norway rat in America. This species invades homes in the United States, particularly those in the cities. *(Courtesy U.S. Department of Agriculture, Office of Information.)*

The Norway Rat

rats

Where man goes the Norway rat, a prolific, destructive camp follower, is soon found. Originating in central Asia, this disease-carrying pest arrived in Europe with the returning crusaders in the 1600s and ran down ships' mooring lines onto American docks about 1775 (Fig. 8-5). Since then the rat, a stowaway in countless cargos, has spread throughout the world. Unable to compete successfully in the wild, it has adapted well to the environments of man. A versatile appetite enables it to satisfy its nutritional requirements with anything from garbage to the most sophisticated human foods. Houses, farm buildings, dumps, warehouses, and any poorly constructed and poorly cared for structure provide suitable living quarters, as long as man is in the vicinity.

Few animals are as prolific as the Norway rat. A female begins its breeding life at the age of 3 months and produces five litters per year which average eight young per litter. In less than 1 year one female could produce 250 descendants. Fortunately, this seldom occurs.

cost

The cost of supporting a rat population is enormous. About 2.5 billion dollars are lost annually as a result of damage to property and food supplies by this noxious invader. Historically, the rat has had a devastating impact on humans through its remarkable capacity to harbor diseases and parasites that carry the diseases to man. Trichinosis, tularemia, bubonic plague, typhus, and food-poisoning organisms are characteristic of the disease reservoir of these animals. People sleeping in rat-infested dwellings are often infected with disease after being bitten, usually on the nose or toes, and unattended babies have been killed by rat attacks.

Although they seem indestructible, rats are not without natural enemies. Snakes and weasels often enter their burrows and kill them. Foxes, cats, hawks, and owls feed on rats, although often with little effect on a population. Since rats are dependent on niches created by human activity, however, man alone has the power to control them. Poisons and traps have been tried with varying degrees of success. The most effective way to control the rat, however, is through the proper storage of food, disposal of garbage, and the construction of rat-proof buildings.

The Carp—A New Fish for America

carp

About 1870 reports were received in the United States about a superfish which was very tame, could be kept in ponds, grew rapidly, ate practically anything, and was delicious to eat. Introduction of this fish, the carp, seemed too good to ignore. Funds were appropriated, and a shipment of fish arrived in the United States in the winter of 1876–1877, all of which were dead. More fish were ordered and a second shipment, alive and healthy, was received from Germany in the spring of 1877. It contained 345 fish, of which 227 were of the partly scaled or mirror variety; the remaining 118 were common scaled carp which were destined to write fish history in North America.

The carp reproduced rapidly and were soon available for distribution. By 1879, 12,000 fish had been introduced in 25 states and territories. Congressmen were anxious to obtain shares for their states so over 250,000 were divided among 298 congressional districts. Only two districts were "neglected." The planted carp soon spread to many waterways, helped along by floods, washouts, and the accidental opening of dams.

By 1900 complaints about the fish began to be made. The carp was found to be inferior as a table food. Lake Erie fishermen were worried that their catches would soon contain only worthless fish. Grubbing and mixing of bottom sediments by the carp was killing aquatic vegetation needed by waterfowl and destroying the spawning grounds of native fish. Furthermore, the carp adapted readily to a changing environment, tolerating high temperatures, agricultural and industrial pollution, and a low oxygen level. While the activity of man was making lakes and streams too foul for native fish, it was creating a more suitable habitat for the carp.

There is little hope of eradicating the carp, and efforts to control populations are proving to be both expensive and difficult. The state of Wisconsin uses nets up to ¾ mile long, capable of hauling 100,000 pounds of fish to shore, in an attempt to alleviate the problem. Rotenone poisoning is widely used in lakes and some rivers, but this chemical kills all species and replanting is usually necessary. There is some hope that a species-specific toxicant can be developed which will kill only the carp. Such chemical treatment, however, would be less than 100 percent effective, and the few carp that survive could quickly repopulate waterways. A female carp can lay two million eggs in one season.

It has been nearly 100 years since the scaled carp was introduced into the United States, and now a cousin, the grass carp, also called the white amur, is being planted in some southern states. A few have escaped into the Mississippi River. By proceeding up this drainage system, the fish could invade the waters of numerous states.

Fig. 8-6. A netting operation to remove carp from a lake. *(Courtesy Michigan Conservation Department.)*

grass carp

The grass carp is a large fish which feeds mostly on aquatic vegetation. Because of their voluminous appetites they may be a solution to the weed problem that plagues canals and rivers in the southeastern United States. The weed causing most of the trouble is water hyacinth, accidentally introduced from Brazil. Millions of dollars are spent each year in attempts to control this beautiful but troublesome flowering plant which forms the solid mats of vegetation that clog southern waterways. Some northern states are also considering introduction of the grass carp to control rapidly growing vegetation in highly productive lakes.

Although the grass carp appears to be a solution to the weed problem, past experience tells us that the introduction of a new species without extensive investigation often leads to disaster. We could conceivably end up with two serious problems instead of one.

New Zealand Introductions

The havoc man has created by introducing foreign species is extreme in such places as New Zealand where a virgin paradise has been changed beyond any chance of recall. While much of New Zealand is still beautiful, to the ecologist looking for remnants of vanished primeval splendor, it resembles a once lovely fairyland ravished by bulldozer and landscape developer. Until settlers arrived, New Zealand had always undergone natural changes through glaciation, climatic cycles, and the gradual shifting of populations of plants and animals. The Polynesian people, who occupied the islands during the 400 years before Captain Cook's visit in 1769, had scarcely altered the land. They had introduced a Polynesian cat and a breed of dog, neither of which was aggressive.

Cook

Serious damage began when Europeans appeared on the scene with their own domestic animals and camp followers. Captain Cook introduced several aliens to New Zealand, including pigs, sheep, goats, cabbage, potatoes, and turnips, during his second visit in 1773, and still others in 1777. Whalers, explorers, and merchantmen made a practice of keeping animals on board ship as a fresh meat supply on long voyages. They liberated many of these animals on islands that appeared capable of supporting them, with the hope of producing a future supply of fresh meat, and sometimes with good intentions of helping the natives.

The blackbird and the song thrush were later introduced, as were the brier rose and the blackberry. Birds, feeding on the fruits of these almost ineradicable plants, were responsible for their widespread distribution throughout the islands.

rabbits

Rabbits were introduced as a food source, but by 1870 had become a serious problem. Carnivorous mammals, such as the weasel, were then introduced to control them. The result was overpopulation of both rabbits and carnivores.

As more settlers arrived during the nineteenth century, bringing with them the most cherished reminders of their homeland, the introduction of alien forms accelerated. Attempting to make these beautiful South Sea Islands replicas of England, they introduced 130 species of birds, 30 of which became established. The red deer and fallow deer were introduced from Europe. North America furnished moose, elk, mule and white-tailed deer. Sambar, rusa, and sika deer and wild sheep were brought from Asia. Emperor Franz Josef sent chamois as a gift, resulting in a New Zealand population of these mammals as great as in their native home in the European Alps. The Australian opossum, introduced in 1858 as a potential fur bearer, is the worst pest in New Zealand today. They are so abundant that electric power poles must be sheathed in metal to keep opossums from climbing them and short-circuiting the power lines. In all, 48 species of mammals have been introduced, 25 of which have become established.

The populations of such domestic mammals as goats, pigs, sheep, horses, and imported horned grazing animals grew enormously as they destroyed forests which had developed through countless ages. More than half of the original forests of New Zealand have been converted into almost worthless desert grasslands. Destruction of forests was followed by soil erosion, and once-navigable channels are blocked with mud.

plants

More than 400 species of plants have been introduced into New Zealand. Although not all have proved to be detrimental, many have spread in rank profusion. Rather small and unimportant in its native land, the European watercress grows 14 feet long and has stems 2 inches in diameter in New Zealand, where it chokes river channels with heavy vegetation.

New Zealand has been ravaged by burst after burst of ecological change, and the effects of introductions will continue for a long time to come. It can never be returned to its primeval state. Although the evidence is not as obvious, the same thing is happening all over the world, on islands and continents, on mountains and deserts, and in rivers and lakes. Exotics that have previously been unable to penetrate foreign environments have been willingly and even anxiously helped over geological barriers by man.

SOME BENEFITS

The Ringneck Pheasant

pheasant

A few introduced species have been successful in that they have provided a desirable addition to the array of native species without being excessively disruptive. The ringneck pheasant, a handsome bird with excellent game qualities was introduced into North America (Fig. 8-7). Ten centuries before Christ this bird was taken from Asia to Europe and to Britain. About 1790, pheasants were shipped from England to the East coast of America, and others were later brought over from China and Japan.

The state of Oregon imported 26 ringneck pheasants in 1881. The introduction was so successful that 11 years later a hunting season was established and 50,000 birds were shot on the opening day. Success stories such as this led to further introductions throughout the northern states and southern Canada. Although they failed to become established in many places, they were highly successful in several areas. South Dakota hunters took approximately 20 million pheasants between 1919 and 1940.

Relatively new agricultural land on which grains are grown, particularly corn, provide the best habitat. Over the years the increased use of mechanical equipment, the clearing of fence rows, the draining of low places, and the earlier mowing of hayfields have caused a decline in the number of ringneck pheasants.

Fig. 8-7. The ringneck pheasant has been successfully introduced into many areas. (*Courtesy Michigan Department of Natural Resources.*)

The German Brown Trout

Similar to the ringneck pheasant, the German brown trout has a long history of successful introductions. Originating in Europe, the fish was already widespread by 1883 when it was introduced into the United States.

brown trout

Very sly and difficult to catch, the brown trout stands up well under fishing pressure. Habitat changes due to logging, stream erosion, and increased development along river banks have produced aquatic conditions more favorable for brown trout and less tolerable for brook and rainbow trout. The brown trout can tolerate higher water temperatures, more turbidity, and some of the pollution common near high-population centers.

Reintroduction of Animals—A Second Try

The alteration of habitats through overcropping, overgrazing, burning, logging, draining swamps and coastal tidelands, elimination of fence row shrubs, and the use of more land by people has resulted in the disappearance of many kinds of wildlife from their original ranges. When a change in human pressure or stages in natural succession reestablish a favorable habitat, displaced species may reoccupy an area by migrating from surrounding regions, or man may take them back.

beaver

Much of North America was explored during the development of the fur trade. Among fur-bearing animals none has been so diligently sought over such a wide area as the beaver. By 1900 most beavers had disappeared from their former range. However, protection, replanting in suitable areas, and selective harvest when population density permitted, have reestablished the beaver in many of the localities it had previously occupied (Figs. 8-8 and 8-9). Michigan now issues a limited number of permits to take a specified number of beavers.

Fig. 8-8. Moving trapped beavers from an undesirable habitat. (*Courtesy Michigan Department of Natural Resources.*)

Fig. 8-9. Release of beavers in a favorable area. *(Courtesy Michigan Department of Natural Resources.)*

A similar program for the wild turkey has been successful in the United States (Fig. 8-10). This large game animal was so plentiful in colonial times that it sold for 1¢ per pound. By 1920 overhunting, deforestation, and the loss of the chestnut (a prime food source) had eliminated the turkey from three-fourths of its natural range. The state game department of Pennsylvania employed the capture-and-release method to reintroduce breeding stock into appropriate habitats. The number of wild turkeys has increased to over 50 thousand in the state. From this stock turkeys have been reintroduced into several other states. As the number increases in these states, it may be possible to offer hunters an open season as is done in Pennsylvania.

wild turkey

Fig. 8-10. A wild turkey in Michigan. *(Courtesy R. McNeill, Ferris State College, Big Rapids, Michigan.)*

Much the same story could be told for the elk, white-tailed deer, and pronghorn antelope. All these animals were adapted to certain niches, but as a result of overharvesting or changes in the habitat they were eliminated. With protection, partial or complete, they easily extended their range by dispersion, or readily accepted transplantation to suitable habitats within their former range.

Crop Introductions

Careful selection of examples can create the impression that most of man's gambles in moving plants and animals to new areas have been ecological disasters. Indeed, many cases can be cited in which the results have been considerably less than desirable. In the area of crop introductions, however, man has been right more often than wrong. Elimination of all introduced species from the American dinner table would impose an era of plain living. Of the 80 leading crops grown in the United States, only 10 are native.

U.S. Department of Agriculture

The U.S. Department of Agriculture has a long history of activity in importing and testing new crops and new varieties of crops already grown in the United States. The Department was established by Congress on May 15, 1862, but did not achieve cabinet status until 1889. Between the time of its establishment through 1889, the Department distributed 48,591,026 free seed packets in what was intended to be a testing program. Klose (1950) suggests that political abuse of the system often resulted in valuable material being sent to persons not qualified to test them.

wheat

As indicated in Table 8-2, many varieties of wheat were tried in the United States. In 1865 the U.S. Department of Agriculture imported 65 varieties from Great Britain, Russia, Prussia, France, Chile, and China. Many were not suitable, but some made outstanding contributions to the quality and quantity of the wheat grown in the United States. Because of the quantity of wheat planted, an increase in the average yield of wheat of 1 bushel per acre would, at the present time, add about 67 million dollars to farm income.

pineapple

The travels of the pineapple illustrate the circuitous route taken by some crops in arriving at their present locations. The pineapple originated in Brazil and Paraguay. The Guarani Indians carried the pineapple northward in their conquest of northern South America. Columbus found pineapples on the island of Guadeloupe when he landed there on his second voyage (1493) and took them back to Spain. From Spain the pineapple was taken to all parts of the world. For many years English and continental gardeners grew pineapples in greenhouses. The development of large tropical plantations and improved shipping methods, however, made the raising of pineapples under glass an unprofitable business. Today four-fifths of all pineapples are grown in Hawaii, and most of them are marketed in the continental United States.

TABLE 8-2
The Sources of Some Common Crop and Ornamental Plants

Plant	*Imported from*	*Introduced into*	*Introduced by*
Corn	West Indies	Spain	Columbus and others
Barley	Spain	West Indies	Columbus
		United States (central states)	Charles Mason (1854)
Wheat	Spain	West Indies	Columbus
Siberian	England	New Hampshire	Pre-1792
Pacific blue-stem	Australia	California	Pre-1850
White winter	England	Oregon	Pre-1855
Red fife	Poland via Scotland	Wisconsin	David Fife (1860)
Turkish flint	Turkey	Kentucky, Tennessee	U.S. Patent Office
Algerian flint	Oran, Algeria	Virginia	U.S. Patent Office
Cape	Cape of Good Hope	United States	Matthew Perry
Durum (hard)	Russia	United States	U.S. Dept. of Agriculture (1864)
Rye	Germany	United States (central states)	Charles Mason
Oats	Sweden, Lithuania, Russia, Prussia, Scotland, Denmark, England	United States	U.S. Dept. of Agriculture (1860–1870)
Sugarcane	Spain	West Indies	Columbus
Sorghums (16 varieties)	Africa via England	Southern United States	Leonard Wray
Dwarf corn	South America via Spain	United States high valleys	Charles Mason
Kiushu rice	Japan	Louisiana, Texas	U.S. Dept. of Agriculture
Sweet potatoes, beans, peppers	West Indies	Spain	Columbus
Muskmelon, lemon, orange	Spain	West Indies	Columbus
Navel orange	Brazil	California	U.S. Dept. of Agriculture
Soybeans	Japan	United States	U.S. Dept. of Agriculture (1900)

Table Continues

TABLE 8-2
Continued

Plant	*Imported from*	*Introduced into*	*Introduced by*
Grimm alfalfa	Germany	Minnesota	W. Grimm
Alsike clover	Sweden via Scotland	Ohio	M. B. Batcham
Buckwheat	France	United States (central states)	Charles Mason (1854)
Turnips (26 varieties)	England	All states	U.S. Patent Office
Opium poppy	Asia	United States	U.S. Patent Office
Chinese yam	France	United States	U.S. Patent Office (1856)
Fig	Mediterranean area	North and South America	Spanish explorers (1520)
Olive	Syria, Greece	West Indies, California	Spanish explorers
Camphor tree	China	Southern United States	Robert Fortune
Cork oak	France and Spain	Southern United States	U.S. Patent Office
Avocado	Mexico	Florida	Henry Perrine (1833)
Mango	Southeast Asia	Brazil, Africa	Portuguese
Banana	Southeast Asia	Throughout the tropics	Many travelers
Travelers' tree Crown-of-thorns Royal Poinciana	Madagascar	Florida, California	Many travelers
Pelargonium Gladiolus Fringed hibiscus Castor oil plant	Africa	Widely cultivated	Many travelers
Tulip	Turkey via Vienna	Holland	Clusius

opium

The opium industry in the United States never became firmly established, but serious efforts were made. Opium imported for medical use was often adulterated. Experimental plantings in Jefferson County, New York, showed that the cash return per acre from opium poppies exceeded that of any other crop then grown. Opium poppies were also planted in Vermont, Connecticut, and Kansas, but the extensive hand labor required for maximum yield resulted in production costs that were too high.

Early American statesmen were sharply aware of the agricultural problems facing a growing nation and pressed for the introduction of new food

Franklin

crops. Benjamin Franklin, probably the busiest man of his time, promoted agriculture both at home and abroad, sending many plants to his friends in America during his visits abroad. Among the plants he sent are two that are of economic value today, rhubarb and Scotch kale. George Washington, Thomas Jefferson, and many others in their time, imported agricultural plants and seeds.

plant hunters

Plant hunters, such as Robert Fortune, Joseph Rock, David Fairchild, and countless others, have changed the green face of the earth, adding enormously to the variety on many tables and greatly beautifying the landscape. Three hundred years ago there were no pineapples in Hawaii, no coffee in Brazil, and Amazon barons had a monopoly on rubber. The cinchona tree, moved from the jungles of South America to distant plantations, changed history. The quinine extracted from its bark enabled man to explore and occupy the malaria-infested tropics. Wheats from Russia and the Mediterranean area, oats from Scotland, rye from Germany, flax from New Zealand, and persimmons from Japan are only a few of the hundreds of plants brought to America, many of which have been drastically changed through plant breeding. By crossing varieties gathered from many places, modern plant breeders produce many new high-yielding plant strains which help to raise food production in countries around the world.

SAGA OF THE GREAT LAKES

The waters of the Great Lakes and its various tributaries contain an interesting and extensive fish fauna which includes representatives of most of the North American fish families. Because this entire drainage system was once covered by glaciers, fish now present invaded the area from other regions and followed the receding ice northward. The retreating ice cap left ponds of meltwater along its southern edge, producing a favorable environment for fish. Dammed to the north by the ice front, the ponds grew, eventually spreading over an area larger than that covered by the present Great Lakes. With the disappearance of the ice, the northern end of the lakes region rose, and this gigantic water reservoir tilted southward, spilling outward in several directions. This overflowing allowed fish to reenter and extend their range throughout the Great Lakes region (Fig. 8-11). The postglacial sites of origin of Great Lakes fish were principally the Middle Atlantic coastal plain, the upper Mississippi River valley, and the United States Northwest.

With the exception of the St. Lawrence River, all natural drainages from glacial times have ceased to exist. Niagara Falls, an awesome 190-foot cataract included in this water course, has presented an insurmountable obstacle to any would-be invaders for thousands of years. Geographically isolated, the Great Lakes fish followed their own path of evolution, developing unique ecological communities based for the most part on organisms within a closed system.

Within the past 150 years, numerous canals have been dug which have

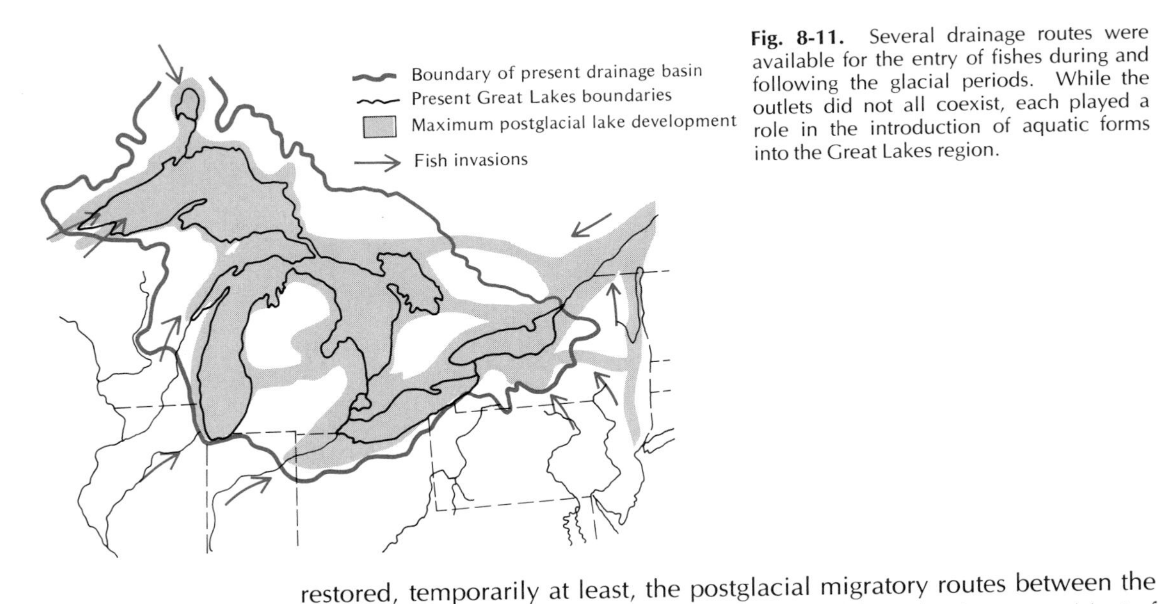

Fig. 8-11. Several drainage routes were available for the entry of fishes during and following the glacial periods. While the outlets did not all coexist, each played a role in the introduction of aquatic forms into the Great Lakes region.

restored, temporarily at least, the postglacial migratory routes between the Great Lakes and the upper Mississippi River and the Atlantic coast. Most of these canals, some of which are no longer in use, have been largely ineffective in producing faunal changes in the upper lakes. Niagara Falls would have remained the ultimate natural barrier between Lake Ontario and the other Great Lakes were it not for such navigable connections as the Welland Canal.

Welland Canal

The Welland Canal began in 1827 as a ditch connecting small rivers, but by 1932 it was a 27-mile-long waterway with eight locks, allowing ships up to 700 feet long to navigate between Lake Ontario and Lake Erie. Cargoes bound from distant ports could now be lifted easily and quickly the 326 feet from the lower to the upper lakes. Along with the ships, unknown to the builders of the canal, went two aquatic invaders which were to upset drastically the delicate ecological balance of the Great Lakes fish.

The Land-Locked Lamprey

lamprey

The *sea lamprey,* which enters rivers and streams to spawn, has always been known for its ability to adapt entirely to fresh water. Its passage into Lake Erie had been possible since 1828, but it did not take place until the Welland Canal was deepened between 1913 and 1918. When first introduced into Lake Erie, the lamprey did little damage, possibly because the water was too warm to encourage its development. Inhibited by adverse environmental conditions, it took several years for the lamprey to penetrate the other lakes. Between 1940 and 1955, however, it spread to Lake Huron, Lake Michigan, and finally to Lake Superior. In these cold waters, without natural enemies, lampreys multiplied explosively and, until the last few years, made devastating attacks on the fish population. In the wake of this invasion, the whitefish and lake trout fisheries collapsed.

Fig. 8-12. Sea lamprey on a river spawning run attempts to negotiate a boulder at the base of a waterfall. (*Courtesy Carolina Biological Supply Company.*)

The lamprey, little changed from prehistoric times, has a long slender body similar to that of an eel. Unlike the eel, however, the lamprey is not a true fish. Its spinal column is composed of a cartilaginous *notochord* rather than a backbone, and it has no paired fins or jaws as fish do. Its mouth is a round sucking organ, a feature that enables lampreys to climb even vertical surfaces such as wet lock walls and dam faces (Fig. 8-12). These and other features place the lamprey among the most primitive of vertebrates.

Lampreys spawn in clear streams in the spring. The spawning pair digs a shallow nest in the gravelly stream bottom, using their mouths to carry and move stones. After depositing eggs and milt, the adults die. The young hatch several days later and drift downstream until they reach some quiet backwater or the stream mouth where they burrow into the bottom and spend their larval stage. The larvae, called *ammocoetes,* feed on microscopic organisms which they strain from the oozy bottom. During the summer and fall of the fifth year they are transformed into adultlike lampreys ready to begin a destructive life. They lose the fleshy hood overhanging their mouths and develop a true sucking disc lined with a set of formidable teeth; thus they are transformed into vampires of the watery world.

As adults, young lampreys move into open lakes and prey on whitefish, lake trout, and other large lake fish. They feed by attaching themselves to their prey, rasping an opening through the skin, and sucking out blood and other body fluids. Some fish may survive such an attack, but an unsightly scar remains (Fig. 8-13). In just two decades these parasites killed most of the lake trout in the Great Lakes region.

In 1955 the United States and Canada formed the Great Lakes Fishery Commission to study, among other things, the lamprey problem. These and other pest fish were of international concern not only because both countries bordered the Great Lakes but also because lampreys migrate up

Fig. 8-13. Lake trout showing scars where lampreys had been attached. *(Courtesy Michigan Department of Natural Resources.)*

streams to spawn, crossing state and national boundaries. The commission was actively involved in efforts to control lamprey populations.

larvacides

weirs

Scientists have learned, after years of unremitting effort, that the only truly vulnerable part of the lamprey life cycle is the larval stage. Traps and electric *weirs* designed to serve as barriers to spawning adults proved ineffective (Fig. 8-14), but by 1958 specific poisons had been discovered that killed the larvae during their 5-year growth period. Scientists tested over 4000 chemical compounds, several of which were found to kill larval lampreys without harming fish. With these discoveries the slow process of restoring the Great Lakes fisheries finally began.

The Prolific Alewife

alewife

The Welland Canal served as a watery corridor for still another Great Lakes invader. All the upper Great Lakes showed an alarming increase in the plankton-feeding *alewife* between 1931 and 1954. The alewife originally lived in the Atlantic Ocean where it migrated up freshwater rivers to spawn. Over a long period some of them adapted completely to the fresh water of the St. Lawrence waterway and Lake Ontario where they spawned in shallow areas rather than in clear rivers and streams.

In the 1950s and early 1960s, alewives multiplied so rapidly that they comprised 9/10 of the weight of all the fish in the upper Great Lakes. Enormous dieoffs occurred, and dead fish piled up in shallow water and on beaches (Fig. 8-15). Shoreline property owners attempted to collect

Fig. 8-14. This weir across a small stream was intended to kill lamprey going upstream to spawn. (*Courtesy Michigan Conservation Department.*)

and bury the dead fish one day, only to be faced with greater numbers of dead fish the next day. The awful stench of decaying fish was accompanied by swarms of houseflies, flesh flies, and maggots. The resort industry along Lake Huron and Lake Michigan faced a serious economic crisis as tourists found other vacation spots or went home early to avoid the mess.

Alewives were essentially without predatory control; they had crowded out the original plankton-feeding fish and their numbers were limited only by their food supply.

Fig. 8-15. Alewife dieoff along the shores of Lake Michigan during July 1972. This indicates that the coho salmon have not completely controlled the alewife population.

The Coho in the Great Lakes

coho

With the lamprey under control, fish biologists were not confident that lake trout could reproduce in sufficient numbers to control the alewife. After careful study they decided to introduce coho salmon into the Great Lakes. The coho is extremely adaptive, has a strong homing instinct, and in its native waters along the West coast feeds on fish similar to the alewife. Eggs were supplied by the state of Washington and hatched in Michigan; the fry were released into streams and allowed to migrate into the lakes. The results were astonishing. Many of the released coho gained 7 pounds in a mere 4 months. After spending 2½ years in the Great Lakes, cohos of 20 to 30 pounds were common (Fig. 8-16). In a few years the alewives were under control, the dieoffs fewer, and lake trout began to appear in greater numbers.

During the spring and summer, maturing cohos are scattered miles offshore and fishermen, using their own or chartered boats, enjoy excellent salmon fishing. In the fall the salmon crowd along the shoreline and near the mouths of the parent streams before making their spawning runs. Enormous numbers of fishermen congregate in these areas (Fig. 8-17) and riotous behavior has been reported as being commonplace. Greed manifests itself among many of these "sportsmen." Fishing lines have been entangled and cut by the propellers of passing boats, resulting in fights and abusive language. Seven fishermen drowned and hundreds of boats were swamped one afternoon in 1967 because fishermen refused to heed the storm warnings issued by the U.S. Coast Guard.

Fig. 8-16. Coho salmon from Lake Michigan. Many of these fish have mercury concentrations that render them unsuitable for human consumption. (*Courtesy* Manistee News Advocate.)

Fig. 8-17. A case in which the predator population may exceed that of the prey. (*Courtesy Michigan Department of Natural Resources.*)

When the salmon enter the streams, the behavior of the fishermen worsens. Private property is not respected and trespassing is common. Fish refuse and other litter are frequently disposed of without regard for the environment. Fish are snagged with improvised equipment instead of being enticed to strike a lure or bait. Fishing laws have been changed because it was impossible to control the fishermen and to stop them from intentionally snagging fish.

Suggested Readings

Allen, D. L. *Our Wildlife Legacy.* New York: Funk and Wagnalls, 1962.

Camp, W. H., V. R. Boswell, and J. R. Magness. *The World in Your Garden,* Washington, D.C.: National Geographic Society, 1957.

Elton, C. S. *The Ecology of Invasions by Animals and Plants.* New York: Wiley, 1963.

Hubbs, C. L. and K. F. Lagler. *Fishes of the Great Lakes Region.* Cranbrook Institute of Science, Bloomfield Hills, Michigan, 1949.

Klose, N. *America's Crop Heritage.* Ames, Iowa: Iowa State College Press, 1950.

Laycock, G., *The Alien Animals.* Garden City, N.Y.: Natural History Press, 1966.

Nussey, W. and H. Reuter, "African Round-up." *Science Digest,* 69(6):42–48, 1971.

Scott, J. A. "Great Lakers Are Back." *Michigan Natural Resources,* 40(3):24–27, May–June 1971.

"Water Hyacinth Pernicious." *American Biology Teacher,* 32(8):479, 1970.

Weber, W. J. "A New World for the Cattle Egret." *Natural History,* 81(2):56–63, 1972.

The cotton-boll weevil has forced diversified farming in many areas of the South. Because of this pest, cotton cannot be grown year after year in the same fields. *(Courtesy U.S. Department of Agriculture, Office of Information.)*

Pests, Plagues, and Poisons

arthropods

Three-fourths of all animals living on the earth today belong to a group known as the arthropods. Members of this group have an external skeleton, jointed legs, and a segmented body. Lobsters, lice, spiders, mites, butterflies, and beetles are common examples. Those with three body regions (head, thorax, and abdomen) and three pairs of legs are classified as insects. Most insects have two pairs of wings attached to the thorax, the middle body region; some (flies) have one pair, and others (fleas, bristletails, silverfish, and sucking lice) are without wings during all stages of their life cycle. Insects compose 90 percent of the arthropods and are represented on the earth by more than 700,000 different kinds.

insects

A billion billion insects thrive on our planet and yet only a few, perhaps 3000 kinds, are harmful to man. The other 99.9 percent are either useful to man or of no concern. Many are essential.

pollination

Insects aid in the pollination of many flowering plants, both wild and cultivated. Pollination is the transfer of pollen grains from the stamen of a flower to the stigma of the pistil. This process is essential in most flowering plants to initiate the development of the embryo in the seed and the surrounding fruit. Tubes grow from the pollen grains germinating on the stigma to the ovules in the base of the pistil (ovary). The pollen tubes carry sperm nuclei which fertilize the eggs in the embryo sacs of the ovules. A fertilized egg becomes an embryonic plant in the developing seed. The pistil becomes the fruit.

Special Cases of Pollination

Certain clovers, such as alfalfa, red, and sweet, have small, tightly closed flowers. The heavy weight of the bumblebee is needed to open these flowers and expose the stamens and pistil.

When alfalfa and other legume planting increased in western Canada, Washington, Idaho, and Utah during the 1930s and 1940s, the abundance of wild bees assured high seed yields. Cultivation destroyed ground-nesting bees, however. This, coupled with an increase in *Lygus* bugs, resulted in a steady decline in seed yield. Effective control of *Lygus* and restoration of bee nesting sites has greatly improved yields. In some western irrigated valleys where bumblebee populations are high, alfalfa now produces more than ½ ton of seed per acre.

The fig (*Ficus*) has been associated with man more than 5000 years. Egyptian documents dated about 2700 B.C. (fourth dynasty) refer to the fig, and it is mentioned many times in the New and Old Testaments. Fig plants which have escaped from cultivation grow in rocky crevices around the Dead Sea and along the Jordan River.

The peculiar flower and fruit structure of the fig, known as a synconium, makes pollination difficult. A hollow, spherical receptacle bears flowers on the inside. Figs may be self-pollinating, parthenogenetic, or require pollination from another fig plant. The Smyrna fig, one of the most highly valued varieties, bears only pistillate flowers and thus produces no pollen. Wild caprifigs, sometimes more-or-less cultivated, produce pollen that fig wasps carry to Smyrna fig tree flowers. The wasp enters the synconium through a small opening and distributes pollen unintentionally as it selects a place to lay its eggs. The wasp lays a cluster of eggs and then dies, usually before escaping from the developing fig fruit. Both the wasp and the eggs are digested and absorbed as the fruit grows. No wasp larvae develop in Smyrna figs.

The fig wasp is successful, however, in completing its life cycle in the caprifig fruits. The caprifig produces two types of flower clusters. Some have only staminate (pollen-producing) flowers; other flower clusters are pistillate. A fig wasp enters one of the flower clusters at random. If it enters a pistillate flower cluster first, it lays eggs but the unpollinated flowers do not develop and the larvae have no food. If the wasp enters a staminate flower cluster, no eggs are laid because the receptacle wall is not suitable. After becoming covered with pollen, the wasp leaves the staminate flower cluster and tries another. When a pistillate flower cluster is found, the wasp lays eggs and, as it walks around over the flowers, rubs pollen onto them. These flowers, in the caprifigs, develop into fig fruits and the wasp larvae feed on the seeds and fruit wall.

This relationship between wild figs and commercial figs was known to the Greeks several centuries before Christ. Although not fully understanding the details, they hung caprifig branches on the trees, producing the good figs. In California imported Smyrna fig trees failed to produce until the caprifigs and the fig wasps were introduced in 1899.

This relationship between the caprifig wasp and the Smyrna fig illustrates an interesting evolutionary quirk. The Smyrna fig depends on the wasp for the pollination necessary for setting fruit and repays the wasp by death and destruction of its eggs. This is not the usual symbiotic relationship between plant and pollinator. The main factor here is the inability of the wasp to distinguish among the three kinds of flower clusters, caprifig staminate, caprifig pistillate, and Smyrna pistillate. If the wasp did "learn" to differentiate among these, however, it might more readily recognize the difference between staminate and pistillate flowers and thus avoid the staminate ones because it never lays eggs in them. This would be self-defeating for the wasp and harmful to the caprifig. Natural selection would tend to eliminate those lines of wasps in which such an ability developed. They would have few or no offspring.

The flowers of plants pollinated by insects may be fragrant, showy, or both. The pollen grains are waxy and tend to clump. In contrast, wind-pollinated species produce inconspicuous flowers and dry, light pollen. Corn, pine, oak, and hickory are examples of wind-pollinated plants. Willow seems to display the characteristics of wind-pollinated plants. Staminate flowers are on one plant and pistillate flowers on another. The relatively inconspicuous flowers develop before the leaves. The myriad of bees, flies, beetles, and wasps that can be seen working on both staminate and pistillate flowers of the willow, however, indicates a role for insects in the pollination of these plants.

The destruction of dead animal bodies is in part brought about by many kinds of insects but especially by beetles. Burying beetles dig soil from under the dead bodies of small animals and thus lower them into the ground. The beetles then lay their eggs in the carcass. Carrion beetles, many of which are brilliantly colored, complete their life cycles in decaying animal bodies.

Zamia

Zamia, a cycad of Central America and the Caribbean Islands also grows in Florida as far northwest as the Suwannee River. Its large taproots served as a source of starch for the Seminoles. (These roots should *not* be eaten raw.) Two beetles are closely involved in the reproduction of Florida *Zamia*. Just before pollen is shed in the spring, the staminate cone enlarges rapidly by the elongation of the internodes between the microsporophylls. As the microsporangia burst and the pollen is shed, small gray beetles, resembling larger editions of the confused flour beetle, are found working in the tissues of the staminate cone. In all probability these particular beetles do not play a role in the life cycle of *Zamia* (Fig. 9-1). They do bring about the rapid deterioration of the cone after pollen is shed. *Zamia* appears to be wind pollinated.

Fig. 9-1. *Zamia*. The life cycle of this primitive cycad is closely associated with the activities of certain beetles.

TABLE 9-1
The Vector Problem in Disease[a]

Disease	*Host*	*Vector*	*Cause*
Curly top	Sugar beet	Leafhopper	A virus
Dwarfing	Rice	Leafhopper	A virus
Streak	Corn	Leafhopper	A virus
Yellow spot	Pineapple	Thrip	A virus
Mosaic	Beet, cucumber, sugarcane, bean	Aphid	Viruses
Yellows	Peach	Leafhopper	A virus
Stewart's disease	Corn	Corn flea beetle	A bacterium
Cucurbit wilt	Cucumber	A beetle	A bacterium
Fire blight	Pear, apple	Bees and leaf-hoppers	A bacterium
Soft rot	Potato, cabbage	Flies	A bacterium
Blight	Chestnut	Beetles	A fungus
Dutch elm	American elm	Bark beetles	A fungus
Blue stain	Red pine	Bark beetles	A fungus
Fusarium wilt	Cotton	Grasshopper	A fungus
Downy mildew	Lima bean	Bees	A fungus
Tapeworm	Man and dog	Dog flea	*Dipylidium*
Tapeworm	Man and rat	Rat flea	*Hymenolepis*
Tapeworm	Man, other ani-mals	Water flea	*Diphyllobothrium*
Lung fluke	Man	Crayfish, crab	A flatworm
Elephantiasis	Man	Mosquito	A roundworm
Guinea worm	Man	Water flea	A roundworm
Malaria	Man	Mosquito	A protozoan
African sleeping sickness	Man, other ani-mals	Tsetse fly	A protozoan
Equine sleeping sickness	Horses (mild in man)	Mosquito	A protozoan
Chagas sleeping sickness	Man and rodents	Assassin bug	A protozoan
Texas cattle fever	Cattle	Tick	A protozoan
Bubonic plague	Man	Rat flea	A bacterium
Typhus (three kinds)	Man and rodents	Lice, flea, mite	*Rickettsia*
Yellow fever	Man, monkeys, rodents	Mosquito	A virus
Amebic dysentery	Man, other ani-mals	Flies	A protozoan
Equine encepha-lomyelitis	Horses, mules, burros, and many others	Mosquito	A virus

[a] Compiled from data in D. J. Borror and D. M. DeLong, *An Introduction to the Study of Insects*, pp. 662–664, 669–673. New York: Holt, 1964.

The seed-bearing cones of *Zamia* grow at ground level on separate plants. A large seed cone may have 90 to 115 seeds. Each seed is covered by a fleshy aril which attracts a large, black scarabid beetle. This beetle is related to the sacred scarab of ancient Egypt. In Florida the beetle buries *Zamia* seeds in vertical tunnels. Sometimes the seed is jammed in the tunnel, and sometimes it is placed in a short lateral pocket. The buried seeds are the ones that germinate, and they may be buried at depths of up to 18 inches. Seldom do seeds germinate that simply lie on the surface.

LOSSES ATTRIBUTABLE TO INSECTS AND OTHER PESTS

The battle lines between man and insects become hazy because of the great number of beneficial insects. Nevertheless, a problem does exist. Insects are well adapted to survival and can be expected to outlast man on earth. Their tremendous reproductive capacity enables them to make a rapid population recovery after a calamity.

losses

In spite of the increased use of pesticides, crop losses have remained about the same in the United States over the last 20 years. Grasshoppers cost the United States about 20 million dollars a year. The European corn borer toll is much higher—75 to 100 million dollars. The small grain green bug (*Pyrausta nubilalis*) siphons off another 75 million dollars. The National Cotton Council calculated that the 1950 cotton losses due to insects amounted to about 900 million dollars. Tables 9-1 and 9-2 list some of the more common plant and animal pests.

TABLE 9-2
Some Insect Pests[a]

Disease	*Victim*	*Site of Action*
Ox warble fly larva	Cattle	Under the skin
Bot fly larva	Sheep	In the nose
Bot fly larva	Horses	Intestinal tract
Screwworm	Cattle, goats, and so on	Under the skin
Louse (chewing)	Birds and mammals	Skin
Louse (sucking)	Mammals	Skin
Corn borer	Corn	Stalks
Corn earworm	Corn	Ear
Chinch bug	Cereals, other grasses	Stems and leaves
Boll weevil	Cotton	Seeds, boll fibers
Alfalfa weevil	Clover	Leaves, growing tips
Engraver beetle	Various trees	Cambium layer
Timber beetle	Various trees	Wood

[a] Compiled from data in D. J. Borror and D. M. DeLong, *An Introduction to the Study of Insects*, pp. 662–664, 669–673. New York: Holt, 1964.

CHARACTERISTICS OF A GOOD PESTICIDE

To feed the present world population, man must hold crop losses due to insects to the absolute minimum consistent with long-range environmental protection. A whole array of chemical pesticides has been used lavishly in the past with little concern for their ultimate effects or their impact on other than target organisms, even on man himself.

Because the use of poisons for insect control will be needed for some time, even though in reduced quantity, we should consider the characteristics of an ideal pesticide. Although a control measure meeting all the objectives may be difficult to find in specific cases, some such controls do exist and the search for more can be expected to pay dividends in reducing the poisoning of our planet. The following criteria are suggested as goals for the development of pest controls. A good control is:

1 Effective only against the target organism.

2 Stays where it is applied; is relatively immobile.

3 Has a short effective life; is readily biodegradable to nontoxic substances.

4 Requires a minimum of instruction and machinery for its safe use.

THE EVOLUTION OF PEST CONTROL

Nature has its own biocides, and man has learned to use several. The balsam tree produces an antibiotic which gives it some protection against the linden bug. Paper toweling made from balsam tree wood has some insecticidal properties which trace to these natural terpenes (unsaturated hydrocarbons found in the essential oils of plants). The material the balsam tree produces is selective against the family of insects that includes the most destructive pests of cotton. The great variety of such chemicals that exist in nature makes this a fertile field for the discovery of biocides without the devastating side effects of DDT and related poisons.

nicotine

Nicotine, produced in the roots of tobacco plants and transported to the leaves, is the basis for Black Leaf 40, the standard louse killer used in the henhouses of America. If lice have never developed an immunity to nicotine, some other insects have. The tobacco worm is addicted to a diet of tobacco leaves. In spite of what seems to be a built-in biocide, tobacco is subject to a variety of diseases. Tobacco breeders search constantly for genetic strains resistant to cyst nematodes, root knot nematodes, Granville wilt, fusarium wilt, powdery mildew, anthracnose, black shank, frogeye leaf spot, rattle virus, tobacco etch, wildfire, mosaic virus, streak virus, and many other diseases.

pyrethrum

Several species of *Chrysanthemum* produce the insecticide pyrethrum. This forms the basis for extracts known as pyrethrins which are the active

Fig. 9-2. Wood eaten by termites is digested by symbiotic protozoa living in the termite digestive tract. Without these protozoa, the termite would starve. Termites eat the interiors of joists, floors, sills, and studding of houses. They have been known to tunnel from floors into the legs of tables that were seldom moved. (*Courtesy Carolina Biological Supply Company.*)

Fig. 9-3. Cockroach. This insect has remained relatively unchanged through the last several million years. Sprays used in some restaurants and groceries to kill cockroaches may drift onto exposed foods. (*Courtesy Carolina Biological Supply Company.*)

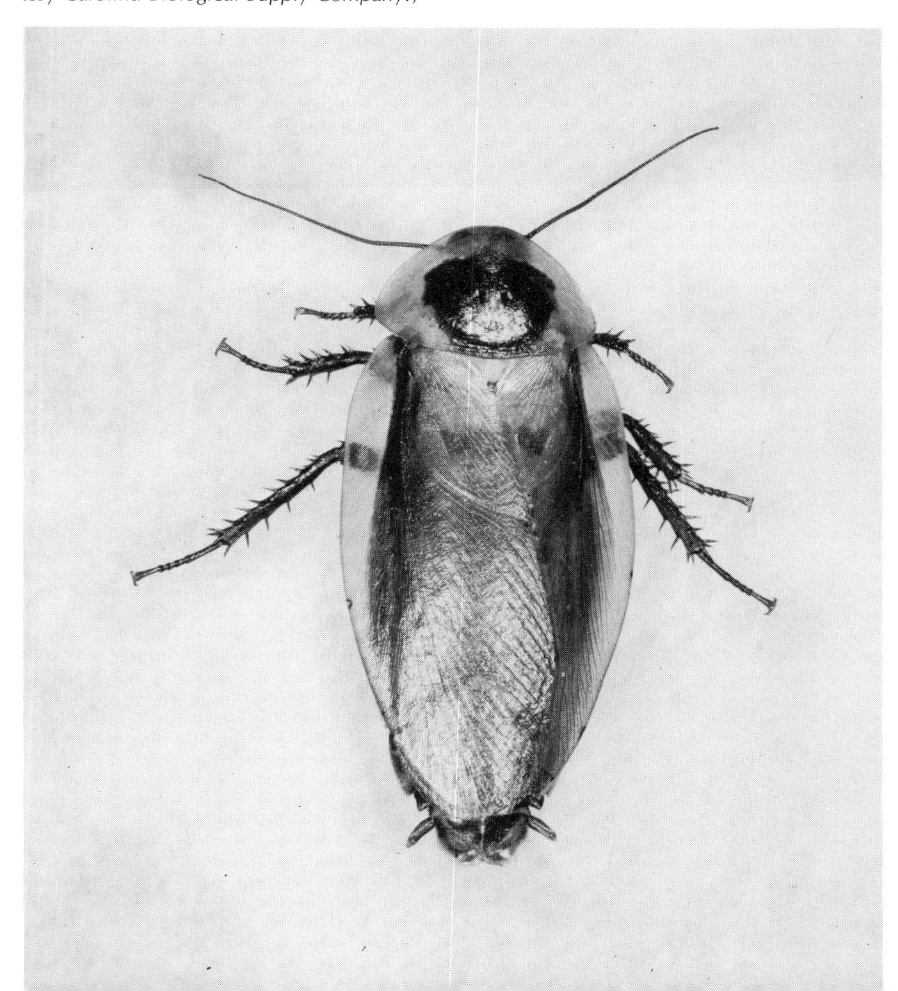

ingredients in several insecticides marketed for the home. The most widely used species, *Chrysanthemum cinerariaefolium,* is native to southeastern Europe but is now grown extensively in Africa and Japan. Dried flowers yield 3 percent crude pyrethrum, most of which is concentrated in the flower ovaries. Under cultivation, 500 pounds of flower heads per acre may be harvested. Grain stored in elevators is protected from weevils by pyrethrum. Pyrethrum is also used in aerosol sprays around the home and in dusts for garden produce. Pyrethrum is not toxic to warm-blooded animals.

locusts

Locusts have ravaged cultivated and wild plants since the time man first recorded the events of his everyday life and, we may assume, long before (Fig. 9-4). Except for locusts (grasshoppers that appear in catastrophic numbers), however, insects have emerged as a threat to the world food supply only within the last 200 years. Prior to this most agricultural areas consisted of small farms growing a variety of crops. Under such conditions insect outbreaks might decimate local plantings but usually not whole regions. This contrasts sharply with the corn belt in Iowa, the Kansas wheatfields, or the cotton plantings in southern California. When confronted with such extensive plantings of susceptible plants, insects and disease organisms multiply explosively and extensive damage results.

arsenate of lead

rotenone

The early use of chemicals in efforts to control insects included pouring kerosene on ponds to destroy mosquito larvae. Chewing insects were treated to a diet of stomach poisons, often arsenate of lead. Sucking insects are not affected by poisons that must be swallowed and required a contact spray. Nicotine and rotenone were used widely. The latter, the product of a South American shrub (*Lonchocarpus nicou*) and a similar Far Eastern plant, has been used extensively for killing fish. This was used to dispose of trash fish prior to stocking ponds with bluegill and bass, and by South American Indians to obtain food. South American Indians cultivate the plant by sticking live stems into the ground. About half take root and grow into mature plants. The roots are dug up after 2 or 3 years, cut

Fig. 9-4. Lubber grasshopper on corn leaf. During a heavy infestation, corn stalks may be completely stripped of leaves. *(Courtesy Carolina Biological Supply Company.)*

into suitable lengths, and dried. Similar to pyrethrum, rotenone does not affect warm-blooded animals.

DDT

In 1939 the second phase of insecticide development was ushered in when the Swiss entomologist Paul Muller discovered the insect-killing powers of a rather simple chemical which had been sitting on the shelves for many years. This substance, DDT, had been synthesized by a German graduate student, Othmar Zeidlar, in 1874, but its potential for destroying life was not recognized at that time.

The United States entered World War II unprepared for the host of parasites and other diseases that would be encountered by a military campaign in the tropics. A crash program was initiated to locate all the cinchona trees in the high Andes of South America. At that time, quinine, made from the bark of three species of *Cinchona,* was the only reliable drug that cured or at least alleviated malaria. The concern of the military could be understood. Malaria had contributed to the failure of the French to build the Panama Canal, to defeats suffered by the Confederate Army, to Hannibal's failure against Rome, and it played a role in the fall of Troy. Then, in 1942 DDT was brought to the United States and World War II became the first war in which more soldiers died from combat wounds than from disease.

PARADE OF POISONS

classification

The spectrum of chemical controls used to assure man a better life fall, for the most part, into six general categories: chlorinated hydrocarbons, organophosphates, carbamates, arsenicals, plant derivatives, and chemicals for rodent control. Of these, the first two constitute the greatest threat to the environment (Fig. 9-5).

Fig. 9-5. Man's efforts to control organisms that compete for his food and timber crops have resulted in his poisoning vast areas of the earth. (*Courtesy Michigan Department of Natural Resources.*)

Fig. 9-6. Spraying onions in the Rio Grande Valley near Las Cruces, New Mexico. *(Courtesy U.S. Department of the Interior, Bureau of Reclamation.)*

chlorinated hydrocarbons

Chlorinated hydrocarbons are synthesized by substituting chlorine atoms for hydrogens on carbon chains. They are intended primarily for use against insects but affect a broad range of cold- and warm-blooded animals. Insects are more susceptible to external contact with these sprays than mammals because insect cuticle absorbs the toxin more readily than mammalian skin. Among the chlorinated hydrocarbons, extensive use has been made of DDT, endrin, and dieldrin.

DDT has been used longer, and its far-reaching effects are better understood than those of most pesticides. Although greatly restricted in use, at least in the United States, DDT, similar to mercury and lead, will be with us for a long time. A few facts about DDT indicate the hazards of introducing this and other synthetic poisons into the biosphere. DDT in particular, and chlorinated hydrocarbons in general, have four characteristics that compound the problems arising from their use.

characteristics of DDT

1. We cannot restrict the effect to the pest we want to eliminate. DDT is a broad-spectrum killer and destroys life from the unicellular algae to the highest level of animals.
2. DDT is much more soluble in fat than in water and thus tends to concentrate in living organisms. The cell membrane contains fat, thus any living cell can absorb DDT. Water dissolves very little DDT; 1.2 parts per billion is saturation. As rapidly as DDT dissolves in water, it moves into the aquatic plants and animals living in the water.

3 DDT does not stay put. It has been found in insects in the High Sierras, in seals off the coast of Scotland, and in penguins of the Antarctic, all remote from areas of spraying. DDT has been carried around the world on dust particles and it codistills with water. When water evaporates, DDT goes with it, thus the hydrological cycle carries DDT. A reduction in DDT in the soil of a sprayed field does not necessarily mean that the poison has been rendered harmless. It may simply have moved into another area—the drinking water, the streams, the lakes, or the air.

4 The organic molecules that occur in nature are sources of energy and simpler materials necessary to sustain life. They have been with us for a long time, and natural selection, the driving force of evolution, has resulted in adaptation of some organisms to utilize each of these compounds. Few natural compounds remain long without being degraded to simpler materials by the metabolism of living organisms. This represents the filling of ecological niches. Synthetic pesticides, in contrast, are new materials in the environment. The time they have been with us is far too short for the evolution of organisms that might depend primarily on synthetic pesticides for their energy and raw materials. The fact that some organisms do have enzyme systems that can break down some of the new synthetic molecules is a fortunate coincidence. The enzymes of few decomposers attack chlorinated hydrocarbons. The half-life of many pesticides in this group has been variously estimated at 10 to 15 years. The DDT molecule is broken down slowly—it lasts a long time.

These characteristics, lack of selectivity, affinity for living systems, mobility, and stability, render chlorinated hydrocarbons a serious threat to whole ecosystems.

Fig. 9-7. Spraying lettuce in the Rio Grande Valley near Valencia, New Mexico. *(Courtesy U.S. Department of the Interior, Bureau of Reclamation.)*

PESTICIDES AND THE LIVING SOIL

soil

As discussed in Chapter 4, soil is more than crushed rock. From the standpoint of fertility, soil animals play an important role. An analysis of the impact of pesticides on the soil ecosystem was presented in detail by Edwards (1969).

soil animals

The top 3 inches of soil contain the majority of soil animals. These animals belong mainly to four groups. The smallest in size but the largest in number is the protozoa. Then come the roundworms (nematodes), the earthworms (annelids), and the arthropods. The 3-inch layer of soil under 1 square meter of surface may contain one million arthropods and many times that number of nematodes and protozoa. The total biomass of such a layer may amount to as much as 1 pound. Most of the biomass consists of earthworms and protozoa. The earthworms in an acre of good pasture can easily equal the weight of a steer (about 1200 pounds). Earthworms, nematodes, wood lice (crustaceans), mites, millipedes, termites, springtails, fly maggots, and beetle larvae all contribute to soil conditioning.

Pesticides are applied to soil because the action of some members of the soil community is contrary to man's interests. Nematodes attack a variety of crops. Eleven or more species parasitize the roots of citrus trees. Nematodes destroy Easter lily bulbs in the fields of western Oregon. Root knot caused by nematodes occurs in many species (Fig. 9-8).

Soil animals differ widely in their reactions to particular pesticides. Among the most susceptible are predatory mites and fly larvae. Earthworms and certain nematodes have a high degree of immunity.

Fig. 9-8. Fumigation of soil for control of golden nematodes. *(Courtesy U.S. Department of Agriculture, Animal and Plant Health Service.)*

The Pesticide Syndrome

Cotton ranks high among the crops most dependent on chemical protection. A critical look at what has happened to this crop in some places may help us to avoid similar mistakes elsewhere. Doutt and Smith (1971) outline the impact of chemical control on the cotton crop. According to these authors, what happened in the Canete Valley in Peru illustrates the successive stages of cotton farming.

The Canete Valley, similar to about 40 other coastal river valleys in Peru, relies on river flow for agricultural water. The farm valleys are separated from each other by a desert. This isolation makes these valleys ideal areas for studying the effect of any agricultural practice.

Prior to 1920 cotton farming in the Canete Valley was at the subsistence level. The main crop was sugarcane; corn, beans, and potatoes were also grown. At the subsistence stage of cotton farming, as also practiced on the poorer lands of Afghanistan, Uganda, parts of Equador, and other places, only a minimal effort is made to protect crops. Most of the harvest, usually less than 200 pounds per acre, is used by local weavers rather than being exported.

During the 1920s farming in the Canete Valley shifted from sugarcane to cotton. This began the exploitation phase. About two-thirds of the valley was planted with cotton. During the late 1940s, organic chlorides were added to the usual arsenicals and nicotine sulfate sprays. This, coupled with the practice of growing ratoon cotton (allowing a second and third crop to come from sucker growth at the base of the old stem, thereby reducing the labor of replanting) gradually built up the pest population to a severe outbreak in 1949. The cotton yield dropped from the 1943–1948 averages of 415 to 526 pounds per acre to 326 pounds per acre. This precipitated the next phase of cotton farming—the crisis phase.

The crisis phase in cotton growing and in the cultivation of many other crops is a well-marked series of events. As insect or other pests become more difficult to control, the time interval between sprayings is shortened and heavier applications are used. In the Canete Valley this had the initial effect of doubling the 1949 yield. In spite of warnings from ecologically oriented entomologists, organic insecticides, particularly DDT, toxaphene, and BHC (lindane), were applied in increasing amounts. The greater yields fostered the belief that, because a little insecticide did a little good, more would do more good. Trees were cut to reduce the hazards of aerial spraying. This greatly reduced the number of tree-nesting birds, many of which lived primarily on insects. Insect predators that attacked the pests were eliminated along with other useful insects. Spraying had to start earlier each year because the pests appeared earlier in the season.

The crisis phase culminates in the development of insect resistance to chemical control. In 1952 in the Canete Valley, BHC failed to control aphids. In 1954 toxaphene was no longer effective against leafworm on tobacco. By 1955 and early 1956, cotton-boll weevils reached very high numbers. Six other insects, which had always been present but in limited numbers, became serious pests. This did not happen in nearby valleys where organic insecticides were not used or were used in small quantities.

continued

The 1955–1956 season in the Canete Valley illustrates the disaster phase of complete reliance on chemical control of crop pests. Rising costs for chemicals and their application, coupled with lower yields, first eliminate farmers operating on marginal lands. Finally, even the best land no longer yields a profit. In the Canete Valley the 1955–1956 crop was almost a total loss. The cotton yield dropped to less than 300 pounds per acre in spite of spray schedules which had been reduced from 2-week to 3-day intervals.

Canete is not an isolated case. Similar experiences occurred in the Chinch and Pisco Valleys south of Canete.

The final phase of crop protection—integrated control—was instituted at Canete and in several other valleys in Peru. Integrated control is a system of crop protection designed to modify any of the environmental conditions that allow insects to reach pest status. Full use is made of all natural enemies and any other factors that contribute to pest mortality. At Canete cotton production was prohibited on marginal land. Natural enemies, primarily predator insects, were restocked from other areas. Cotton was not planted 2 years in succession in the same field. The use of synthetic organic insecticides was greatly reduced and put under government control. Organic chlorines and phosphates were largely replaced by arsenicals and nicotine phosphates—the chemical controls that were in use before the crisis developed.

The success of integrated control at Canete indicates a solution for other areas of pest-ridden agriculture. As a result of a many-pronged attack replacing total reliance on organic pesticides, the severity of the crop pest problem diminished rapidly. The six new pests reverted to their former innocuous status. The population of the chief pests declined, and pest control costs were lower. In 1 year the cotton yield was back to 468 pounds per acre. In later years yields ranged from 644 to 922 pounds per acre. These equal or far exceed the best produced during the time of heavy application of insecticides.

Reestablishment of a hard-hit soil animal population may be slow. The toxicity of the pesticide and its persistence in the soil affect the recovery cycle. In tests DD mixture (1,2-dichloropropene and 1,3-dichloropropene) eliminated the springtail population in a test plot and then disappeared from the soil in 30 days. Recolonization of the plot by springtails did not begin until 4 months after the DD application, however. In contrast, the herbicide simazin is less toxic to soil animals but lasts longer in the soil. Simazin eliminated 70 percent of the springtails in a 30-day period but lasted 5 months. By the time the herbicide had disappeared, the springtails had been restored to 90 percent of the original population density. In relation to effect on the total soil animal population, however, more drastic upsets come from the more persistent chemicals.

fumigants

Fumigants, gases injected directly into the soil, kill almost all soil animals in the upper 10 inches of treated soil. The gas penetrates into the smallest air spaces between soil particles. Fumigants last a few weeks at most. Animals, such as springtails, carried into the area by wind, may produce explosive populations. The predator mites that could hold them in check have been eliminated.

earthworms

From man's point of view, the earthworm is the most valuable soil animal. Earthworms assist agriculture by aerating the soil and by decomposing the plant debris that reaches the soil. Few pesticides reduce their numbers; phorate, carbaryl, and two chlorinated hydrocarbons, heptachlor and chlordane, are exceptions. Earthworms present a pesticide hazard to their own predators because they concentrate these chemicals in their tissues. Aldrin and dieldrin have been found in earthworms at 10 times the concentration in the soil in which the worms were living. Organophosphates, fortunately, do not seem to concentrate in earthworms and, more and more, organophosphates are replacing chlorinated hydrocarbons.

Certain organisms, although drastically reduced in number initially, may increase in number under the influence of attempted chemical control. This is a general principle in the ecology of pesticides which applies more to herbivores than to their predators. In one test over a 5-year period following a single application of DDT, a springtail population increased 400 percent. At the same time the population of mites dropped to 5 percent of that at the start of the test. The predatory mite exerts a natural control on the springtail population. In this respect organophosphates have essentially the same effect as DDT.

The loss of soil animals on cropland may be offset by cultivation and application of fertilizers. The most serious effect of soil animal destruction occurs in forest soils as a result of widespread spraying to combat tree parasites. Loss of the animals of the forest floor slows or stops soil formation.

RESISTANCE TO PESTICIDES

New types of plants and animals evolve because some members of a population, a little different from the others, are better able to survive. Thus they have more offspring, some or all of which have the same survival advantage. In most cases natural selection operates on small differences which confer only slight advantages. The population of a species changes gradually over long periods of time. A disaster that very few can cope with produces a rapid change in the frequency of some genes in the population, however. These may be genes that were of no survival value in the predisaster environment. Suddenly, in the new situation, they make the difference, and the individuals that survive pass these genes on to their offspring.

When man enters the picture with chemical killers, he replaces natural selection in the evolution of that which he seeks to control. If the makeup of the target population varies in individual resistance to the pesticide, then some of the intended victims escape. These may make up a minute fraction of the original population, but they are left to produce the next generation. With such a heavy selection pressure, the genotype promoting resistance quickly becomes the most prevalent one. The species becomes immune to the poison. In reality, some individuals were immune when the poison was first used and these produced the next generation. Rather than developing immunity, a shift takes place in the percent of certain genes in the population.

resistance

The more intense the application of pesticides and the more extensive the treated area, the more rapidly resistance develops. Only the highly resistant are left for breeding stock. The susceptible strains are no longer present.

By 1961, 137 species of insects and nematodes had developed known resistance to biocides. Six years later resistance was present in 224 species. Of these, 127 are pests of field or forest crops, or of stored foods. The remaining 97 are important to public health or to the health of animals. Insects may be resistant to one specific insecticide or to several kinds. The 224 species included 135 no longer controllable by dieldrin, 91 immune to DDT, and 54 resistant to organophosphates. (These figures total more than 224 because some insects are immune to more than one of these three categories.) A few—the housefly, the cotton leafworm of Egypt, and the cattle tick of Australia—are immune to all three. Each year new names are added to the list of resistant species. Aldrin formerly controlled western corn rootworm effectively but a pocket of resistant worms developed in one locality in Nebraska. In 5 years these resistant worms spread into seven states.

Resistance has been shown in animals other than insects. Strains of mice nearly twice as resistant to DDT as a control population were produced in a laboratory in just nine generations. Resistance in natural populations of fish and mammals is also a matter of record.

Pesticide resistance has several ecological effects. These depend in part on the mechanism by which the resistance is achieved. In some species the ability to store more pesticide in the body confers immunity. In such cases these individuals are more effective concentrators of the poisons and therefore a greater hazard to the animals that prey upon them. As a consequence, any tendency on the part of the target insect to show this kind of immunity increases the harm done to the biological controls present before the spraying. Insect predators missed during the spray application now find their food more and more contaminated (Fig. 9-9).

Vertebrates in a sprayed area are unlikely to develop resistance as rapidly as the pest or other nontarget invertebrates. Generation time is part of the problem. Some insects might produce 6 to 10 broods in a season and a great many more offspring per brood than their bird predators.

The fact that the kill rate from an initial application of a pesticide is often very high suggests that the gene or genes responsible for resistance made up a small portion of the gene pool. If so, prior to the spraying these genes may have had other effects which were detrimental, or they may have been linked with genes that were disadvantageous. If they were beneficial other than in reference to the pesticide, they would have been present in a greater portion of the population, and the pesticide would have killed far

Fig. 9-9. Spraying American elm trees in a vain effort to eradicate the Dutch elm disease. *(Courtesy Michigan Department of Natural Resources.)*

fewer. This holds some hope for recovery of control where insects seem to be developing immunity. Generally, the reduction of insecticide pressure results in a decrease in the proportion of resistant individuals. Susceptible ones replace them for the same reasons that accounted for the predominance of the susceptibility prior to the use of the pesticide. Thus if we change our tactics from all-out war aimed at complete extermination of pests to obtaining control at a level we decide we can live with, even though some damage occurs we may buy the time needed to develop effective biological controls.

BIOLOGICAL CONTROL

The previous discussion of pesticides placed emphasis on the problems man has created in his efforts to produce more food by eliminating insect competitors for his crops. We now look at some of the possible alternatives to the suicidal practice of poisoning the earth (Table 9-3).

juvenile hormones

The use of juvenile hormones is one answer to the problem of insect control. Normal development of sexually mature adult insects from the final larval stage cannot occur if certain hormones are present. These hormones, produced by the juvenile stage, are necessary for larval growth but their production normally stops prior to the final adult stage.

The potential of these hormones as an insecticide was recognized about 17 years ago, and recent improvements in methods of synthesis have produced extremely potent materials. Less than 3 ounces of a synthetic material resembling a juvenile hormone is sufficient to kill all insects within a

2½-acre plot. Although effective only against insects, these hormones still have one of DDT's disadvantages. Within the insect group they are broad-spectrum killers and not selective for any one species or even one group of species.

TABLE 9-3
Examples of Biological Control

Pest	*Location*	*Control*
Insects		
Citrus black fly	Central America	*Eretmocerus serius*, a parasite of the black fly, from Malaya
Cottony cushion scale of citrus	California (accidental introduction in 1868)	Vedalia lady beetle from Australia
Red scale insects on citrus	California	A parasitic wasp from China
Oriental fruit moth	United States (peaches)	*Agathis diversus*, imported from Japan
Squash bug	United States	Tachnid fly larvae
Gypsy moth	United States	Sex attractant lures males to a trap
Screwworm	Southern United States and the Caribbean	Sterilized males produce no offspring
Spotted alfalfa aphid	California and elsewhere	Mowing field in strips maintains high predator insect population
Grape leafhopper	California	Planting blackberries (predator wasp winters in the eggs of another leafhopper which occurs only on blackberries)
Plants		
Prickly pear cactus	Australia	A moth imported from Argentina
Klamath weed	Western United States	A beetle
Pamakani (a weed)	Hawaii	Gallflies reduce the growth of the stem
Alligator weed	Southeastern United States (97,000 acres in eight states)	Flea beetle from South America
Puncture vine	Introduced into the United States from the Mediterranean	Weevil stage of two beetles from the Mediterranean
Tansy ragwort	California, Oregon, Washington	Cinnabar moth

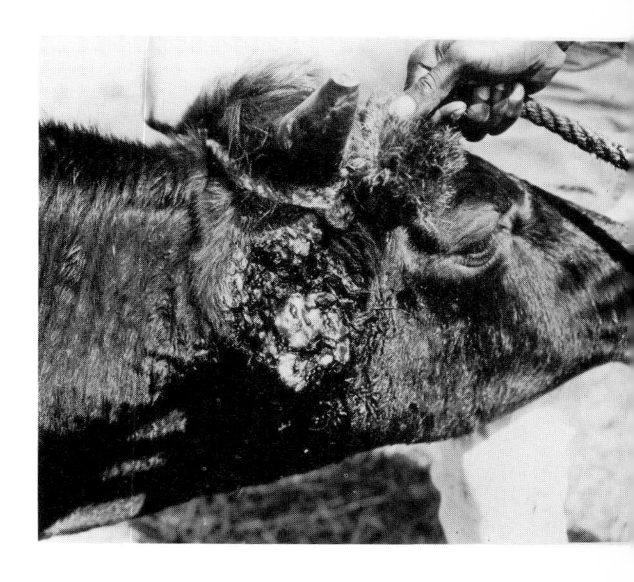

Fig. 9-10. Screwworm infection in the ear of a steer. An untreated fully grown animal may be killed in 10 days by thousands of maggots feeding in a single wound. *(Courtesy U.S. Department of Agriculture, Office of Information.)*

Some examples of biological control in addition to the use of juvenile hormones illustrate their advantage over chemical poisons. Chief among these advantages is the specific nature of the control—only the intended pest is affected.

screwworm

The screwworm (the maggot stage of a fly) infects cattle, horses, sheep, and goats in southern Florida and the southern parts of the states bordering Mexico. An adult female lays up to 300 eggs in a wound (or even in a tick bite) of the host. Within 24 hours the eggs hatch and the larvae begin feeding. The activities of the larvae cause bloody sores which attract more flies to lay more eggs. Soon the animal is host to thousands of flesh-eating maggots. A short life cycle, with 10 to 12 generations each year, and the large number of eggs produced by each female, result in a rapid spread of the disease. Prior to effective control the annual loss from screwworm maggots (due to lower milk production, reduced value of hides because of holes, and dead animals) amounted to about 25 million dollars annually in the United States (Fig. 9-10).

The application of a biological control method termed the sterile male technique eradicated the screwworm in Florida, Texas, and parts of the Caribbean. In the southeastern United States, starting in January 1958, two billion sterile flies were released over an 18-month period. The pupae had been sterilized by exposure to 2500 roentgens of gamma radiation from a cobalt-60 source. (Both male and female flies were sterilized because it was not feasible to try to separate them under conditions of mass culture.)

After irradiation the pupae were placed in paper bags in which they emerged as adult flies. The flies were distributed over Florida and parts of adjacent states from an airplane. The equipment used released a paper bag of flies every 2 or 3 seconds through a tube which automatically opened the bag (Fig. 9-11).

An adult female screwworm fly mates only once; males mate many times during their lifetime. When the ratio of sterile to fertile males is 9:1, 83 percent of the matings produce unfertilized eggs.

Fig. 9-11. Loading sterilized screwworm flies for release over an infected area. This technique lacks the harmful side effects that may occur when chemical poisons are used to control pests. *(Courtesy U.S. Department of Agriculture, Office of Information.)*

The original test of the sterilization technique was made on Curaçao, an island 40 miles off the coast of Venezuela. Curaçao was chosen because of its isolation and because constant easterly trade winds prevent new flies from reaching it from the mainland. Screwworm infection was very high among the 25,000 goats that roamed the island. A weekly release rate of 435 sterile males per square mile reduced the number of fertile egg masses collected at 11 sampling areas to zero in 7 weeks. Infertile egg masses were collected up to 14 weeks after the release program started. The weekly releases were continued for 22 weeks (Fig. 9-12). Conclusions seem obvi-

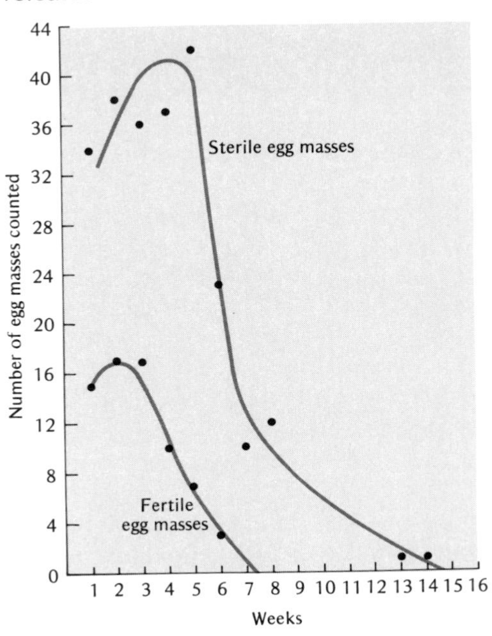

Fig. 9-12. Number of screwworm egg masses found each week after start of sterile male release program. Count taken from 11 groups of penned goats. (Based on tabular data from Baumhover, A. H., et al. *Journal of Economic Entomology*, 48: 462–466, 1955.)

Fig. 9-13. Moth sterilized with gamma radiation for release in California cottonfield. *(Courtesy U.S. Department of Agriculture, Animal and Plant Health Service.)*

ous: screwworm was eradicated from Curaçao. It took only a little more than 1 year to eliminate the screwworm fly in Florida with this method.

Mediterranean fruit fly

The sterile male technique is being applied successfully to the control of the Mediterranean fruit fly. This fly lays its eggs in the rind of citrus and other fruits. Even if the egg does not hatch, a rotten spot develops where the egg was laid and the fruit is unsalable. The Mediterranean fruit fly was introduced into Costa Rica in 1955 and spread rapidly northward to Guatemala and southward to Panama. The potential annual crop loss from this insect is about 80 million dollars.

The sterile male technique has not worked equally well with all species tested. Ideally, it should produce sterility without affecting the insect in any other way (Fig. 9-13). In some cases, however, irradiation reduces the mating drive and the sterile males have relatively little effect.

According to Huffaker (1971), efforts have been made to apply biological control to 223 species of insect pests. Half of these have been successful to a degree that made the effort economically worthwhile. Among families of insect pests, scale insects and mealy bugs (Coccoidea) were the most susceptible to biological control. Some significant degree of success was achieved with 50 of 64 species (78 percent). Among all other insects groups, success has been achieved in 44 percent of the biological control attempts.

bacterial control

Pest control through the spreading of a bacterial disease of the pest promises minimum harm to other organisms. The bacterium is effective against only one or a few target organisms. In the cases so far, this method has had the additional advantage of mass production and easy application of the control organism. The bacteria can be grown in large quantities in laboratories and spread by airplane. The Japanese beetle is susceptible to control by a bacterium.

Japanese beetle

The North American invasion by Japanese beetles was first noticed in New Jersey in 1916. At that time 12 beetles were found. Three years later 20,000 could be collected in a single day. In 40 years they had occupied most of the states east of the Mississippi River. Recently, several Japanese beetles were seen near San Francisco. The adults skeletonize the leaves of many kinds of plants and eat some fruits, notably peaches.

Intensive efforts were made to eradicate the Japanese beetle by massive use of chemical poisions. Spray schedules called for 400 pounds of lead arsenate per acre, 25 pounds of DDT, or 2 pounds of aldrin. At these concentrations the effects of the three are about the same. Dieldrin and heptachlor were also used, in amounts of 2 to 3 pounds per acre. These methods, in addition to quarantine, had little effect on the Japanese beetle but were hazardous to other organisms. According to Scott et al. (1959), treatment of areas near Sheldon, Illinois, temporarily eliminated ground-feeding birds. Rudd (1964) indicates that losses also occurred among muskrats, ground squirrels, and rabbits.

Bacillus popilliae, a spore-forming bacterium, provided the solution to the Japanese beetle problem. This bacterium attacks the larval stage and produces what is known as "milky spore" disease. The larvae turn opalescent white and die. Nontarget organisms are not affected. The bacterium produces resistant spores which last for years in the soil—until ingested by the larvae of the Japanese beetle. This type of control does not exterminate the beetle, but it has reduced the population to a level that man can live with.

A related bacterium, *Bacillus thuringiensis,* is effective against a broader spectrum of crop pests. Its victims include: alfalfa caterpillar, artichoke plume moth, bollworm (cotton), cabbage looper (many garden plants), diamondback moth (cabbage), European corn borer, budworm (tobacco), hornworm (tobacco), leaf roller (oranges and grapes), orange dog (oranges), and imported cabbage worm. In preventing diseases of trees and ornamental plants, *Bacillus thuringiensis* controls or helps control California oakworm, webworm, cankerworm, Great Basin tent caterpillar, gypsy moth, linden looper, salt marsh caterpillar, and winter moth.

The solution to many pest control problems lies in further research to find biological control methods that work. Employed in a system of integrated controls these procedures offer hope for improving the present precarious state of many agricultural crops. Figures 9-14 and 9-15 illustrate ways in which biological control works.

Fig. 9-14. Mormoniella wasp laying eggs on the pupa of the fly that it parasitizes. (*Courtesy Carolina Biological Supply Company.*)

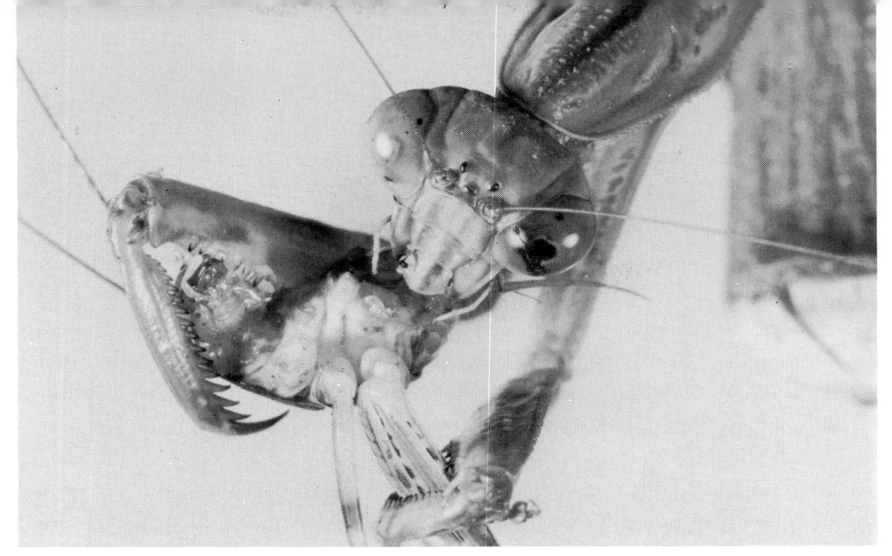

Fig. 9-15. Praying mantis eating a cave cricket. The praying mantis is an effective predator on many pest insects. *(Courtesy Carolina Biological Supply Company.)*

natural insect repellents

Plants have lived with insects much longer than has man, and natural repellents have evolved. Thyme planted near cabbage reduces cabbageworm damage. The Mexican bean beetle (the only destructive lady beetle) slows its activity considerably when rosemary is growing near by. Marigold is repulsive to nematodes in the soil and to several other pests.

Garlic has even greater potential in routing mice, moles, aphids, and Japanese beetles. A dilute oil made from garlic actually kills mosquito larvae and houseflies. The effective ingredients have now been synthesized.

Mustard seeds in water, viable or heat-killed to prevent germination, produce a sticky mucilage which attracts and traps mosquito larvae. The larvae then drown.

Biological control, sometimes simple, sometimes more complex, provides a solution to many pesticide problems. The application of man-made poisons can and must be greatly reduced. Hopefully, it will finally be eliminated.

MERCURY

mercury

Mercury occurs throughout the earth's crust but is by no means uniformly distributed. The average concentration of mercury in the soils and rocks of the world is 30 parts per billion. (At this concentration, 500,000 tons of earth would have to be processed to obtain 30 pounds of mercury.) In areas of volcanic activity and near hot springs, the mercury content may be much higher. Cinnabar, composed of mercury sulfide, contains 86.2 percent mercury. Almaden, Spain, has the largest mercury lode, and Idria, Italy, is a close second. High-production areas in California were named after these two cities (New Almaden and New Idria). California mercury production, along with that of Nevada and Idaho, accounts for more than 90 percent of the annual total output in the United States. Important mercury mines are worked in Alaska, Arizona, Oregon, and Texas.

production

In 1969 United States mines produced more than 2 million pounds of mercury valued at nearly 15 million dollars. This was far below the consumption of mercury by industry, however. United States imports of mercury from eight foreign countries that year amounted to 1,155,732 pounds. The recycling of waste mercury supplied 18 percent of the 1969 requirements, and the Atomic Energy Commission has been releasing mercury from stockpiled material in recent years.

Sources of Mercury Pollution

uses

Mercury has about 3000 industrial uses. These include the manufacture of pharmaceuticals, explosives, electrical fixtures, pesticides, paints, catalysts, caustic soda, chlorine, thermometers, pastes, glues, and sizing compounds. Montague and Montague (1971) state that a conservative estimate of the amount of mercury dumped into American waters during the last 40 years places the yearly average at 1 million pounds. In addition, considerable mercury is released into the air from several kinds of smelters. Refining operations for gold, copper, zinc, and tin are among the worst offenders. The 550 million tons of coal burned each year in the United States release at least 275 tons of mercury into our atmosphere. Some observers suggest that the amount may reach 1800 tons.

Mercury Poisoning

poisoning

Along the coastlines of the world, bays and estuaries play an important role in the lives and livelihoods of the people living near them. These waters produce a variety of shellfish (oysters, clams, mussels), along with crabs, lobsters, and fish. Seafood animals not taken in bays and estuaries may have spent a part of their life cycle there. Bays have been the dumping ground for the refuse of cities along their shores, and rivers flush the waste of the mainlands into the sea. Toxic materials from industry are particularly important to the health of people utilizing the produce of coastal waters. Mercury dumping especially has had tragic consequences.

Minamata

Manufacturing chemicals was definitely secondary to fishing in the town of Minamata on the most southern of the Japanese main islands. The chief product of the chemical plant was fertilizer, and its waste products included carbide, ammonium compounds, calcium phosphate, and sulfuric acid. The plant also produced a polyvinyl chloride plastic, and mercury was used in this process. Mercuric chloride was in the effluent the company poured into Minamata Bay. Before barbiturates, mercuric chloride was a favorite alternative for those who could not face reality.

The alarm that signaled something wrong at Minamata sounded in April 1953. Cats and birds that had regularly feasted on the debris from fish cleaning showed the first symptoms of mercury poisoning although the cause was not immediately recognized. Wild, screeching cats plunged into the sea to drown, and crows fell out of the air. Then the strange disease hit sporadically among the people. Children complained of sore areas and numb fingers and toes. Their angles of vision became constricted, and the world seemed to be at the end of a tunnel. Afflicted persons lost control of their muscles and suffered blindness and loss of hearing. Over a period of 9 years, somewhat more than 100 persons were affected; of these, 36 died. Of the 40 families afflicted between 1953 and 1956 by what came to be known as Minamata disease, fish or shellfish were a part of the daily diet of 25 families and eaten two or three times a week by the remaining 15.

The next outbreak of this strange disease took place in 1965 in Niigata, Japan. Symptoms of the Minamata disease were recorded for 120 persons; 5 died. The fish eaten contained mercury levels from 5 to 20 parts per million.

methyl mercury

A major part of the problem developed from the conversion of mercury from a much less toxic form, leaving as the source of pollution the deadly methyl mercury. In the United States failure to recognize that a serious threat existed was due partly to lack of knowledge concerning the transformation of inorganic mercury to methyl mercury. Many industries were known to be dumping mercury but it was considered a "safe" form. We know now that under the anaerobic conditions of bottom lake and bay sediments bacteria convert inorganic mercury to methyl mercury.

What Mercury Does

effects

The ability of methyl mercury to attack the central nervous system and dissolve brain cells has been well documented. In the Minamata and Niigata disasters, 22 women who had eaten poisoned fish while pregnant gave birth to defective children. The women themselves had not developed symptoms of Minamata disease. This indicates that the fetus is more susceptible to mercury than the adult. Mercury levels in the bloodstreams of new born infants have been found to be 28 percent higher than in the blood of the mothers. Autopsies on children at Minamata and Niigata showed typical brain damage produced by methyl mercury. Patches of brain cells had been dissolved and inert liquid filled the pockets.

A complicating factor in the mercury problem lies in the absence of clinical symptoms in spite of damage to the brain. When a few brain cells are killed, other cells nearby take over their function. This type of damage may go on for many years and not be detected. However, normal losses due to aging combined with that caused by mercury can be critical for elderly persons. From 15 to 20 percent of the body's load of methyl mercury lodges in the brain.

half-life

The principle of half-life can be applied to the duration of a given amount of mercury in an animal body. This refers to the time required to eliminate half the mercury if no more is taken in. A fish loses half its mercury load in 200 days and half the remainder in another 200. The half-life of mercury in the human body is 70 days.

Delayed Action

The evidence of mercury poisoning at Minamata Bay dated back to April 1953, yet no serious alarm was sounded in the United States until February 17, 1970, when the tragic story of a family near Alamogordo, New Mexico, was reported. The grain used to fatten hogs that were butchered and eaten came from sweepings from a grain elevator. The sweepings included seed intended for planting—seed that had been treated with a mercury fungicide. The mercury in the pork left four children blind and essentially unable to walk.

U.S. Department of Agriculture

The United States Department of Agriculture responded to the announcement by deciding suddenly that mercury was an imminent hazard and canceled registration of 17 mercury fungicides. They also asked the main supplier of mercury fungicides to recall its product from dealers' shelves; the supplier did not wish to do this.

Lake St. Clair

Mercury was already in the food chains, however. Lake St. Clair, between Lake Huron and Lake Erie, harbored fish with mercury levels reaching 7090 parts per billion. This was an area of commercial fishing. On March 24, 1970, the Ontario government closed its portion of Lake St. Clair to commercial fishing. Michigan soon closed the American side. Manitoba closed most of its lakes, and Ohio closed its portion of Lake Erie.

PCBs—A WARNING FROM THE TERNS

terns

Terns, compact editions of gulls, nest on rocky islands and deserted barrier beaches around the world. They have survived the storms that batter the coasts and have had their ranks decimated by the millinery trade. An early piece of conservation legislation protected terns shortly after 1900, but they must still yield territory to the encroachment of man.

Great Gull Island, 17 acres of rocks and sand at the entrance to Long Island Sound hosted a colony of 14,000 terns in the late 1800s. This tern colony terminated when construction of Fort Michie began in 1897. Continued use of the island for military purposes until 1949 kept the terns out. Then in 1949 the island was turned over to the American Museum of Natural History. The Linnean Society of New York drew the chore of making the island suitable for a badly needed tern nesting site. Commercial development along the coast was pushing the terns into oblivion. On Great Gull Island buildings were leveled and sand beaches reestablished. Then in 1955, 25 pairs of terns set up housekeeping. The colony increased to a population of 7000 birds, one of the largest in Long Island Sound. Two species of terns, common and roseate, along with red-winged blackbirds, barn swallows, spotted sandpipers, song sparrows, and starlings, rapidly filled the vacant ecological niches.

Weather delt the tern colony a blow during the summers of 1967 and 1968. Heavy rains and lack of fish combined to increase chick mortality. The 1969 hatch was studied intensively for signs of recovery. More than 2000 chicks were marked with plastic bands.

abnormal chicks

Among the chicks one common tern was found with a deformed beak. The upper mandible crossed over the lower in such a way that the "jaws" did not mesh. The survey also recorded two roseate tern chicks with missing flight feathers. Three deformed chicks among more than 2000 did not seem to be cause for alarm. A year later, 1970, the census showed 37 young terns with abnormalities. Some were normal at hatching and later lost flight feathers. Loss of flight feathers had not been seen before in the banding of thousands of terns along the Atlantic coast, in the Caribbean, and on Pacific islands. This was a new plague. Other abnormalities appeared. One chick had very small eyes, another no down, and a third was found with four legs. A variety of beak deformities was seen. The following year observations were essentially the same, and a new threat was seen: thin eggshells (Fig. 9-16).

PCBs

Analysis of the defective birds and of the fish they were feeding upon disclosed high levels of polychlorinated biphenyls (PCBs). PCBs, similar to DDT, concentrate in fat tissue and are even more stable than DDT. In fact, this stability accounts for the usefulness of PCBs in our present technological

Fig. 9-16. A pair of nesting herring gulls on Round Island, Lake Superior. These, along with terns and other fish-eating birds are endangered by pesticides. (*Courtesy Michigan Department of Natural Resources.*)

society. They function as insulating and heat exchange fluids in high-voltage equipment, and they reduce breakdown when added to paint, rubber, and plastics. Laboratory birds fed PCBs lost feathers.

A case of this nature tempts people to pass judgment on the basis of somewhat circumstantial evidence. As Hays and Risebrough (1971) point out in their Great Gull Island study, however, slight impurities in the PCBs due to the manufacturing process might be responsible. One of the impurities often found in PCBs, dioxin, produced the types of beak deformities found among the terns of Great Gull Island when injected in very small amounts into fertile chicken eggs. Dioxin, perhaps the most powerful teratogenic substance known, may now be in the oceanic food chains.

dioxin

What happened on Great Gull Island is not an isolated case. Disquieting reports have now been received of similarly deformed chicks in tern colonies of Massachusetts, Jones Beach in New York, and the Dry Tortugas Islands west of Key West, Florida. This disaster can be expected to spread to the shores of all industrialized nations.

Suggested Readings

Chisolm, J. J. "Lead Poisoning." *Scientific American,* 224(2):15–23, 1971.

Compton, G. "The Mad Hatter Disease: Mercury Poisoning." *Science Digest,* 69(3):61–66, 1971.

Cox, J. L. "DDT Residues in Marine Phytoplankton: Increase from 1955 to 1969." *Science,* 170(3953):71–73, 1970.

Doutt, R. L. and R. F. Smith. "The Pesticide Syndrome—Diagnosis and Suggested Prophylaxis." *In* C. B. Huffaker, Ed., *Biological Control,* New York: Plenum Press, 1971.

Edwards, C. A. "Soil Pollutants and Soil Animals." *Scientific American,* 220(4): 88–99, 1969.

Goldwater, L. J. "Mercury in the Environment." *Scientific American,* 224(5): 15–21, 1971.

Hammond, A. L. "Mercury in the Environment: Natural and Human Factors." *Science,* 171(3973):788–789, 1971.

Hammond, A. L. "Chemical Pollution: Polychlorinated Biphenyls." *Science,* 175(4018):155–156, 1972.

Hayes, H. and R. W. Risebrough. "The Early Warning of the Terns." *Natural History,* 80(9):39–47, 1971.

Huffaker, C. B., Ed. *Biological Control.* New York: Plenum Press, 1971.

Janicki, R. H. and W. B. Kinter. "DDT: Disrupted Osmoregulatory Events in the Intestine of the Eel *Anguilla rostrata* Adapted to Seawater." *Science,* 173(4002):1146–1148, 1971.

Lisk, D. J. "The Analysis of Pesticide Residues: New Problems and Methods." *Science,* 170(3958):589–593, 1970.

Montague, K. and P. Montague. *Mercury.* San Francisco: Sierra Club, 1971.

Pichirallo, J. "Lead Poisoning: Risks for Pencil Chewers?" *Science,* 173(3996): 509–510, 1971.

Rudd, R. L. *Pesticides and the Living Landscape.* Madison, Wis.: University of Wisconsin Press, 1964.

Scott, T. G., Y. L. Willis, and J. A. Ellis. "Some Effects of a Field Application of Dieldrin on Wildlife." *Journal of Wildlife Management,* 23(4):409–427, 1959.

Seba, D. B. "Pesticide Concentration in Seawater." *Science,* 171(3974):928, 1971.

Weiss, H. V., M. Koide, and E. D. Goldberg. "Mercury in a Greenland Ice Sheet: Evidence of Recent Input by Man." *Science,* 174(4010):692–694, 1971.

Wurster, C. F. and D. B. Wingate. "DDT Residues and Declining Reproduction in the Bermuda Petrel." *Science,* 159(3818):979–981, 1968.

The peregrine falcon, an endangered species. *(Courtesy U.S. Department of the Interior, Bureau of Sport Fisheries and Wildlife.)*

The Extinction of Species

INTRODUCTION

Many of the plant and animal species that once inhabited the earth have been lost. They are extinct. Many of our present-day species are in danger of extinction. These problems of survival have resulted primarily from changes in the habitats of the organisms that either have become extinct or are endangered. In many instances these changes have been the result of natural factors; in other instances the changes have been due to man's activities. The organisms are now, or have been, unable to adapt to these changes; therefore they have failed to survive, or at least to survive in numbers sufficient to insure the perpetuation of their species.

value of species

The preservation or conservation of species has relatively broad implications: scientific, esthetic, recreational, and economic. Scientifically, man needs to know how natural ecosystems function. He will never obtain this information if they are destroyed. Esthetically and recreationally, wilderness areas, particularly in mountains and along seashores, provide man's only escape from the pressures and cares of overpopulated, urban existence. These values are difficult to measure in terms of money, but they are nevertheless real, and, if we wish to continue to enjoy these areas, we must realize that they are irreplaceable. Many of the plant and animal species that man depended upon at one time for his survival are now extinct or endangered. In these instances man has been forced to adjust himself and his economy to these changes—changes that have often been difficult.

As a part of the larger picture, species preservation is the conservation of the genetic stocks, or gene pools, of rare species of plants and animals. The evolution of these species took thousands or millions of years. In one way or another, man has eliminated many species during the last several thousand years, and most of them within the last century. The mammoth and the mastodon were part of the North American fauna when man first arrived. They have been gone for only a few thousand years. Evidence indicates that man hastened their extinction. We know that man eliminated giant moas, birds much larger than ostriches, in New Zealand only a few centuries ago. We know, too, that man killed off untold millions of passenger pigeons in eastern North America within the last 125 years. He did almost the same to the American bison, reducing its number from millions to a few hundred in the space of a few decades and thereby upsetting the economy of the Plains Indians. Practically speaking, the gene pools of these extinct species are lost forever. These genes are no longer available for use in the breeding of new varieties. This is especially true where there are no close relatives of the extinct forms. Where close relatives do exist it is sometimes possible, through repeated crossing and selection, to recreate the extinct form. Some German scientists, after many years of work, created a facsimile of the extinct European golden cattle of prehistoric times by starting with genes still present in domestic cattle.

effect of man

While species extinction due to inability to adjust to changes in environments is an age-old phenomenon, it may not be ecologically wise for man to hasten the process by his own carelessness. In addition to the drabness of living in a monotonously uniform biological world having a few crop plants, a few kinds of domestic animals, and a few widespread weeds, there is also danger in uniformity. Disease organisms which are continually evolving new and more virulent forms could completely destroy a uniform variety or species which has no biological resistance to the attack. Moreover, apparently there is greater ecosystem stability, or homeostasis, in communities with a great variety of organisms. For example, it has been suggested that the very complexity of a tropical rainforest, with its great number of species occupying all kinds of ecological niches, tends to keep any one species from outbreaks in numbers which could disrupt the system. Insect outbreaks in such forests are almost unknown because there are always so many enemies and parasites ready to utilize the new food supply. In the relatively uniform and relatively new coniferous forests of the Northern Hemisphere, there are not always enough natural checks and balances to contain outbreaks of insects such as the spruce budworm or the pine bark beetle. Moreover, the uniform stands of the dominant trees in such forests often results in spectacular destruction. Consequently, we must preserve genetic, and thus ecological, diversity in a community if we want continued biological stability in the future.

diversity

Extinction, as a natural process, eventually results in the replacement of all species by better adapted, newly evolved forms. The horseshoe crab

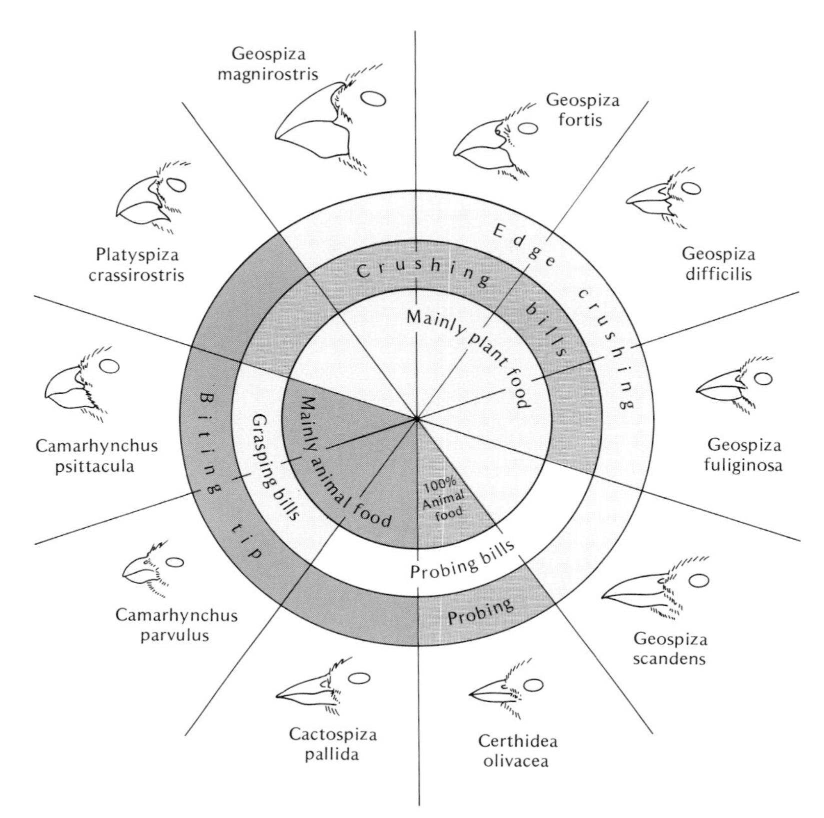

Fig. 10-1. Morphological and functional differentiation of the bills of the Galapagos finches. (Redrawn from T. H. Eaton, Jr., *Evolution*, 1970.)

which has remained unchanged for 200 million years, and the opossum which has changed very little during the last 75 million years, represent exceptions to the fact that most species remain unchanged only for relatively short periods of time. For example, the average bird species now lasts only for about 40,000 years. A species may become extinct and be replaced by other species, or it may gradually evolve into one or more new species and in this way become extinct. An example of the latter process is the evolution of the Galapagos or Darwin finches (Fig. 10-1).

birds

A single species of finch apparently found its way to the Galapagos Islands, 600 miles off the coast of Ecuador, perhaps a million years ago. Competition within the species, combined with an abundance of unfilled niches, allowed the variants that comprised the original species to evolve into several specialized forms—a seed eater, an insect eater, a fruit and berry eater, and a woodpecker type which, instead of developing a woodpecker-like tongue and beak, learned to use twigs to probe for insects. The original species no longer exists; it has been replaced by several new ones.

finches

The year 1600 has been selected as the reckoning date for modern extinction. This date was selected because it represents the first recording of adequate descriptions of species, particularly colors. It is a practical date because it coincides with the approximate beginning of man's special attack on nature. In 1600 there were approximately 4226 living species of mammals. Since then 36 have become extinct, and at least 120 are presently in danger of extinction. In 1600 there were approximately 8684 species of birds; 94 of them have become extinct, and at least 187 are presently on the endangered list. This means that about 1 percent of our mammals and birds have become extinct during the last 400 years and that over 2 percent are in danger of extinction. These figures are much larger for subspecies and races.

A.D. 1600

According to the U.S. Fish and Wildlife Service, nearly 40 different mammals and birds have disappeared from the American scene during the last 150 years. Half of them have become extinct since 1900.

There is a tragic finality in the term extinction. The extinction of a species is a direct and absolute contradiction of existence. It is not a part of life; it is a denial of it. An extinct species is one that has completely vanished, leaving no proof of its existence except, perhaps, a skin, skeleton, picture, name, or fossil.

Because extinction implies the absolute end of a species, there is a certain risk involved in using the term. In a few instances a species considered extinct by the entire scientific world has reappeared. Probably the most dramatic reappearance of an "extinct" species was the astonishing discovery in 1938 of the coelacanth, believed to have been extinct for 70 million years. Throughout recorded history, no one had been aware that this unusual "fossil" fish was still in existence. Since 1938 at least 15 specimens have been taken from the waters around South Africa, proving that the coelacanth is a living fossil, not an isolated freak of nature. Comparisons between the living specimens and fossil remains, which date back 325 million years, show that it has undergone few structural changes during this period of time.

coelacanth

The notornis of New Zealand is another example of a species once thought to be extinct but which is still alive. When the skeletal remains of this large bird were first discovered in a semifossil condition in 1847 by Walter Mantell, it was thought to be extinct. The bird was described as a member of the rail family, named *Notornis mantelli,* and forgotten. It refused to remain forgotten. In 1849 it made its reappearance on Resolution Island, near the southwestern end of South Island, New Zealand, at the precise moment a party of sealers, camped on the island, was looking for food. Even the sealers were impressed by the bird's indigo blue and deep green plumage and they saved the skin. It finally reached Mantell, who quickly established that it was indeed the notornis. Two years later, another specimen was found 40 miles from Resolution Island. A hundred years after the so-called fossil remains were first discovered, a living specimen was found on the shores of Lake Kohakatakahea in New Zealand. The New Zealand government declared the area (435,000 acres) around the lake a sanctuary in an effort to save the notornis. In spite of these efforts, fewer than 100 individuals are in existence today. They are seri-

New Zealand

Notornis

Extinct Wildlife

Many of our wildlife species (mammals, birds, and fishes) have become extinct. This has been the fate of numerous other species, both plant and animal. The extinction of species is a worldwide problem and no doubt has been occurring since the origin of life on earth. Some consideration of the kinds of species that do become extinct and the reasons for their extinction may be helpful in preventing this from happening to our existing species, especially those listed as endangered species (see page 231, "Endangered Species"). A list of some extinct wildlife species[1] (United States) follows. This list indicates the distribution, the approximate date of extinction, and the reason for extinction in each instance where known.

Plains wolf *(Canis lupus nubilus).* Great Plains; probably extinct in 1926; reasons: eliminated to protect livestock.

Steller's sea cow *(Hydrodamalis stelleri).* North Pacific, Bering Sea; extinct about 1768.

Eastern elk *(Cervus canadensis canadensis).* United States east of the Great Plains; extinct in 1880.

Merriam elk *(Cervus merriami).* Arizona; extinct in 1900.

Badlands bighorn *(Ovis canadensis auduboni).* North and South Dakota; extinct in 1910; reason: overhunting.

Labrador duck *(Camptorhynchus labradorium).* Northeastern North America; extinct about 1875; reason unknown.

Heath hen *(Tympanuchus cupido cupido).* Eastern United States; extinct in 1932; reasons: overhunting and loss of habitat.

Sandwich rail, moho *(Pennula sandwichensis).* Hawaii Island, Hawaii; extinct about 1893; reason: probably predation by introduced rats and mongooses.

Mauge's parakeet *(Aratinga chloroptera maugei).* Puerto Rico; extinct about 1892; reason: destruction of forest habitat.

Carolina parakeet *(Conuropsis carolinensis carolinensis).* Southeastern United States; extinct about 1920; reasons: overhunting and loss of forest habitat.

Oahu thrush *(Phaeornis obscurus oahensis).* Oahu Island, Hawaii; extinct after 1825; reasons: alteration of environment by modern man and probable avian disease from introduced birds spread by introduced mosquitoes; predation by introduced rats and cats hypothetical.

Laysan millerbird *(Acrocephalus familiaris familiaris).* Laysan Island, Hawaii; extinct between 1904 and 1923; reason: loss of habitat due to introduced rabbits which ate vegetation.

Laysan apapane *(Himatione sanguinea freethii).* Laysan Island, Hawaii; extinct soon after 1923; reason: destruction of the vegetation by introduced rabbits.

San Gorgonia trout *(Salmo evermanni).* Santa Ana River, California; extinct about 1935.

Harelip sucker *(Lagochila lacera).* Found in a few clear streams of the upper Mississippi Valley, Scioto River in Ohio, Tennessee River in Georgia, White River in Arkansas, Lake Erie drainage, and Blanchard and Auglaize Rivers in northwestern Ohio; not seen since 1900.

[1] From a listing compiled by the U.S. Department of the Interior, Bureau of Sport Fisheries and Wildlife.

Fig. 10-2. Puerto Rican parrot. A bright green parrot which is one of the rarest of the endangered species. It is confined to the Luquillo National Forest in Puerto Rico. *(Courtesy U.S. Department of the Interior, Bureau of Sport Fisheries and Wildlife.)*

ously threatened by animals brought to the island by man. The Puerto Rican parrot (Fig. 10-2) has suffered a similar decline as the Caribbean climax forests were destroyed.

Man must remember that, in spite of his ingenuity, he is able to live on earth because the whole physical and biological environment is still favorable to him. Surely he would like to keep it that way. Dinosaurs were adapted to their environment too, but they were powerless to do anything about the complex environmental changes, or to evolve genetically adapted forms, quickly enough to meet changing conditions; thus they became extinct. Many other animals and plants have become extinct too, some quite recently. In spite of our numbers, the same thing can happen to man because trigger factors in the environment sometimes act so subtly that we may not realize what is happening until too late. We can be sure that checks and balances in the earth's ecosystem will eventually result in some kind of ecological homeostasis, but man may not be a part of the system. It is up to us to understand our environment and to protect it as much as possible from the introduction of potentially dangerous factors. No other organism has ever had this chance.

CAUSES OF EXTINCTION

failure to adapt

There is evidence from paleontology indicating that extinction of species is commonplace in the geological time scale. Biologists estimate that there may now be as many as two million species of plants and animals in the world and that the total number of species that have existed since the dawn of life may be as high as 500 million. These estimates suggest that at least 99 percent of the species that have ever existed are now extinct. The vast majority of these species became extinct before man appeared on the scene. Although the precise causes of the extinction of such large numbers of species remain a mystery, their fate is contained in the phrase "failure of adaptation." The most spectacular example of wholesale extinction is that of the giant reptiles, or the dinosaurs, that lived between 225 and 75 million years ago.

dinosaurs

The name dinosaur refers to a group of extinct reptiles which included the largest land animals that ever lived. They thrived for 150 million years. Their fossils were first found in the rocks of the early Triassic period, about 225 million years ago, and were last seen in the deposits formed at the end of the Cretaceous period, about 75 million years ago. Dinosaurs were the dominant vertebrate land animals of their time, and their remains have been found on every continent except Antarctica. They were perhaps the most successful animal group that ever inhabited the earth.

No dinosaur survived beyond the Cretaceous period. Why they and other groups, such as flying reptiles and ichthyosaurs, became extinct while other groups, such as snakes, turtles, tortoises, crocodiles, and birds, continued to live and flourish is not completely understood.

It is known that the disappearance of dinosaurs was not caused by epidemic disease, catastrophe, or the destruction of dinosaur eggs by rapidly evolving mammals, but it was probably related to major physical changes which occurred at the beginning of the Upper Cretaceous period.

There were many widespread changes in geography at the end of the Cretaceous period. In North America and in Europe, the land was rising. In the western United States, the Rocky Mountains were slowly being raised and the great inland sea of North America was being drained. Rivers began to run more swiftly, and the homes of swamp dwellers disappeared as a new ice age developed.

Climatic changes also took place in the Upper Cretaceous period. The climate became cooler, and seasonal differences were intensified. These changes affected the food supply of the dinosaurs. At the end of the Cretaceous period, the majority of dinosaurs were herbivores. The changed climate stimulated the development of flowering plants, such as the oak, maple, elm, apple, and rose. Dinosaurs probably could have eaten these new kinds of vegetation, but the forests, bush country, and herbs created a new kind of environment, which, with fruits, was a great stimulus to mammals, birds, and insects. Also, much of the vegetation was leafless in winter, a factor that affected dinosaurs severely. The cumbersome, plant-eating reptiles probably found the winters increasingly intolerable, especially since they had no temperature advantage in shelter and hibernation. They probably declined slowly while the mammals increased in numbers and replaced them.

The decline of herbivores eventually led to the decline of carnivores. Since most of the late Cretaceous dinosaurs were large, they probably could not have withstood the long journeys necessary to find new food sources. It is significant that the last successful group of dinosaurs was made up of herbivores which led a browsing life very similar to that of our present-day rhinoceros.

Physiological factors of metabolism, egg laying, and egg hatching also probably added to the decline and extinction of dinosaurs. The demise of this large group of reptiles no doubt resulted from a combination of a whole host of factors rather than any one.

Extinction is a part of the process of evolution, and it continues as a natural process. In any period, including the present, there are doomed species—naturally doomed species, bound to disappear through overspecialization, incapacity to adapt to climatic change or the competition of others, or occasionally some natural disaster involving earthquake, volcanic eruption, flood, or the like. We can now add man to the list of causes for extinction. His influence upon species is discussed under the four major headings that follow.

Habitat Disturbance and Destruction

The principal cause of the loss of species is the alteration of the ecosystems in which they live. Such changes involve the modification, degradation, and sometimes total destruction of a habitat, usually by man, more particularly the felling of forests and the drainage of swamps for timber, farmlands, reservoirs, buildings, airfields, and many other purposes, even recreation. The black-footed ferret (Fig. 10-3), a predator of the prairie dog, cannot survive man's modification of the grasslands.

Fig. 10-3. Black-footed ferret. One of the rarest of the North American mammals. *(Courtesy U.S. Department of the Interior, National Park Service.)*

Fig. 10-4. California condor. One of the endangered species of North America. *(Courtesy U.S. Department of the Interior, Bureau of Sport Fisheries and Wildlife.)*

Madagascar

A classic example of habitat destruction is the island of Madagascar, which at one time was completely clothed in a rich forest. Its long isolation from the rest of Africa resulted in the preservation of an unusual group of primitive primates, most of which were tree dwellers. In recent years development of the island has destroyed over 80 percent of Madagascar's forests, leaving this unique group of primates clinging to the scattered remnants. Already 1 form is extinct, 4 more are rare, and 23 are threatened. Unless steps are taken to preserve some areas of undisturbed forest as refuges, this entire collection of primates will disappear.

Many of America's most magnificent bird species, notably the ivory-billed woodpecker, the whooping crane, the trumpeter swan, and the California condor (Fig. 10-4), began to disappear after their natural habitats had been invaded and destroyed by man. A few of them managed to find sanctuary in remote wilderness areas where they survived for awhile in small colonies. However, from the time they were forced to leave their original homes, they have held on to life by a slim margin.

ivory-billed woodpecker

The ivory-billed woodpecker has not been seen since 1952. The total whooping crane population is estimated at less than 75. The California condor has increased its population to almost 100. Miraculously, the trumpeter swan appears to have made a comeback and may yet regain a firm grip on life in spite of gloomy predictions to the contrary.

Unfortunately, a recovery such as that of the trumpeter swan is comparatively rare. Far more often it is too late to repair the damage, and the species are lost forever. The ivory-billed woodpecker was an alert and aggressive bird and probably could have resisted any predator if its food supply had not been destroyed. Its disappearance occurred almost simultaneously with the destruction of our swamp forests. If these birds had been able to retain even a small fraction of their former natural habitat, they might have survived.

Hunting

The demand for many animal products has decreased during the twentieth century as a result of the development of industrial technology and the subsequent manufacture of synthetic materials from wood, coal, and oil. In some cases, however, the favorable impact of this decrease on animal populations has been more than offset by the increased human population size and the increased market for luxury items in some countries. The blue whale and the green turtle, among others, are nearing extinction because of the animal products industries.

egret
ostrich

Even in the early part of this century, when animal populations were large and human populations small, short-lived fads for animal products brought several species to the verge of extinction. The craze for egret and ostrich feathers during the late 1800s and early 1900s almost succeeded in wiping out several species of these once abundant birds. In a single year, 1912, more than 160 tons of ostrich feathers were sold in France alone. Fortunately, man finally realized that it was possible to meet the demand through ostrich farms. This, coupled with a change in fashions, saved the ostrich. Similarly, the American egret was saved just in time by a change in fashion and by the action of groups such as the Audubon Society. The diamondback terrapin, an inhabitant of Atlantic coastal salt marshes, was saved in much the same way as the ostrich and egret. The diamondback's northern and southern subspecies were relished by gourmets after the turn of the century. In 1921, 6-inch turtles sold for as much as $90 per dozen. As for the ostrich, commercial breeding farms were established, but the real relief for the diamondbacks came when the fad expired during the Great Depression. Terrapin populations have now recovered in some localities, but these animals came very close to extinction.

terrapin

snow leopard

tiger

At present, there are few species that can resist the fur and hide industries the way the egret resisted the feather hunters and the diamondback terrapin resisted the restaurant demand. Snow leopard furs are still being advertised and sold, despite the fact that there are not enough snow leopards left in all the Himalayas to provide 100 coats. New tiger coats can be purchased in the United States despite the fact that there are now less than 4000 Bengal tigers in India and only about 500 tigers in all southeastern Asia, China, Siberia, and the Transcaspian region. The same holds true for virtually all large or showy varieties of cats. The leather industries have placed several reptiles, particularly crocodiles, alligators (Fig. 10-6), and turtles, in jeopardy in an attempt to satisfy the public's desire for reptile leathers and to make a quick profit.

Fig. 10-5. Bryant fox squirrel. A rare and endangered species. (*Courtesy U.S. Department of the Interior, Bureau of Sport Fisheries and Wildlife.*)

Fig. 10-6. American alligator from the Arkansas National Wildlife Refuge in Texas. (*Courtesy U.S. Department of the Interior, Bureau of Sport Fisheries and Wildlife.*)

Game hunting survives and flourishes as a sport in technologically advanced countries, particularly in the United States. If properly regulated, it can be an effective conservation practice. It serves to control the population sizes of animals such as deer, squirrels, and rabbits which tend to multiply excessively in the absence of their original predators. Hunting provides revenues to defray the cost of many governmental conservation activities. Most important, good hunters are familiar with natural environments; they recognize both their unique value and precarious balance and are often eager to work as a group to prevent the environment from being ruined.

polar bear

In spite of the good that results from game hunting, abuses do occur. A notorious example of this is the hunting of polar bears in the state of Alaska. The bears are spotted by plane and the "sportsman" is then landed at a convenient spot to await the arrival of the bear, which may be driven into his gunsight by another plane. The polar bear population in Alaska is rapidly dwindling to extinction. Fortunately, both Canada and Russia have taken steps to protect the polar bear in their territories. Similar abuses of good hunting practices still occur in Africa where big game is occasionally chased down and shot by hunters in planes. For those who insist on a trophy, several African national reserves and conservation departments issue restricted hunting licenses as a way of reducing local population surpluses of potentially destructive animals such as elephants.

zoos

orangutan

Public zoos are under constant pressure to exhibit rare and unusual animals. This pressure is transmitted to collectors and ultimately to animal populations throughout the world. A good case in point is the orangutan. Native to Borneo and Sumatra, this large ape is a favorite exhibit, but as the orangutan became rare the demand increased sharply and irresponsible animal collectors redoubled their efforts. The situation was aggravated by the placing of a premium on the young apes, usually collected only by killing their mothers. In an effort to save the orangutan in its native habitat, most zoos, to their credit, no longer purchase this or other animals in danger of extinction. In cases in which extinction seems imminent, however, a special effort is made to establish breeding stock in the safety of a zoo as was done for the Arabian oryx in Phoenix, Arizona.

research use

Medical research constitutes a potential danger to some wild animal populations, especially monkeys. The demand for animals for experimental purposes is increasing. It is perhaps ironic that as more and more products of potential danger to man are generated, more and more animals are required for testing programs. Many of these animals, rabbits, hamsters, guinea pigs, and rats, are bred especially for this purpose, but monkey metabolism is the most analogous to man's, and thus it is eagerly hunted in Central and South America by collectors. Although some effort is being made today to establish breeding colonies of the most desirable species to increase the supply and to assure greater uniformity and higher quality animals, it will be several years before the demand can be satisfied.

Introduced Predators

Predators have been introduced most commonly into areas colonized since 1600 where mammals, in particular, have been brought in to control the explosive populations of other introduced species. In some instances they have readily turned their predatory attention to the native fauna of the area. A calamitous example of the impact of predator introduction occurred in the West Indies sugarcane fields which were plagued with rats that destroyed much of the crops. The mongoose, a fierce, weasellike native of Asia was introduced to control the rat population. The mongoose attacked rats for a time but soon began to attack other ground dwellers. Amphibians, reptiles, and birds were preyed upon until several species became extremely rare.

mongoose

Disease-causing organisms and parasites, although not predators in the strict sense, may completely wipe out populations when introduced. Such introductions are of course unintentional. The American chestnut is an example of the destruction that can result. Originally, it was an important member of the forests of the Appalachian region of eastern North America. It had its share of parasites, diseases, and predators. Likewise, the oriental chestnut trees in China—a different but related species—had their share of parasites, and so on, including a fungus which attacks the bark of stems. In 1904 the fungus was accidentally introduced into the United States. The American chestnut proved to be susceptible to this new parasite. By 1952 all the large chestnuts had been killed. Their gaunt gray trunks are a characteristic feature of Appalachian forests. Chestnut trees continue to sprout from the roots, and such sprouts may produce fruits before they are killed back; but no one can say whether the ultimate outcome will be complete extinction or adaptation. For all practical purposes the chestnut can be removed from the list of successful species.

chestnut

Introduction of Other Species

When introduced into an ecosystem, nonpredator species, if successful; usually become competitors in the habitats of the indigenous species. They may become habitat destroyers (for example, the goats on Galapagos), or they may bring diseases to which native populations have little or no resistance. The damage may range from the loss of a few native species to a total collapse of the entire community.

Species introduced into favorable habitats reproduce in large numbers; their populations often literally explode. This is due primarily to a freedom from predators and parasites which gives them an unfair advantage in competition with native species. In 1884 a visitor returning to Florida from New Orleans brought back a water hyacinth, a floating plant native to Central America. The plant survived, reproduced, and spread

water hyacinth

rapidly. At last count it covered 90,000 acres of Florida's freshwater lakes and streams with an impenetrable mat of curly leaves and purple flowers. Despite their attractiveness, they represent a major obstacle to the passage of boats on inland waterways. Perhaps more important, they shade the naturally occurring water flora from the sun, and without this normally abundant and varied source of food, the community of herbivores and carnivores breaks down. In most hyacinth-covered ponds, the numbers of minnows, turtles, waterfowl, predatory fish, and alligators are greatly reduced. A possible solution to the problem involving the introduction of the grass carp was discussed in Chapter 8.

eucalyptus

In the marshy areas of the Canea Valley of Colombia, eucalyptus trees were planted to dry the land enough for sugarcane planting. Now the water table is so low that irrigation wells have become useless. Similar phenomena are occurring in the beleaguered Everglades, where an introduced ornamental tree, the melaleuca, is infiltrating and drying cypress swamps. The introductions of many species are unpredictable. Not until we understand and computerize all the interactions of an ecosystem will we be able to make introductions of new species with safety.

SOME EXAMPLES

the Nile Valley

The plundering of the land and its wild creatures is by no means a unique characteristic of modern man. Only the weapons are new. In A.D. 1300 the Egyptians made a futile attempt to preserve the forests of the Nile Valley by passing a law forbidding further destruction of this invaluable natural resource. The law, forcefully worded and duly recorded, was apparently ignored. Most of Egypt's forests were destroyed. The laws have continued to be ignored to this day; modern Egypt still has little concern for the land.

This disregard for the land, among those in authority as well as the general public, is also obvious in Egypt's treatment of its wildlife. Large numbers of quail and other birds, protected by international agreements, are illegally netted along the shores as they arrive from their exhausting trans-Mediterranean migration. That this practice is accepted by the authorities is evidenced by the fact that the birds are sold openly in the markets of Cairo, Alexandria, and other major cities.

ibex

Egypt's lack of concern for its natural resources took a decided turn for the better when Prince Kamal El Din Hussein established the Wadi Rashrash Ibex Reserve for the vanishing ibex at the beginning of this century. There was every reason to believe that the ibex and other wild animals would not only be saved from extinction but that they would increase sufficiently to be out of danger. Unfortunately, when King Farouk came to power, he saw no reason to deprive himself of this ready-made hunting area and promptly turned the entire reserve into his private shooting preserve. The ibex is now dangerously close to extinction, as are the slender-horned gazelle and the wild ass. Hunters and poachers have launched an all-out attack on the remaining herds by shooting them from automobiles.

gazelle

The gazelle of Saudi Arabia has also been pushed to the brink of extinction by mobilized hunting parties. At one time a kill of 300 animals in a single day was considered commonplace. The hunting has ceased because gazelles are so scarce that it is hardly worthwhile to pursue them.

ostrich

The Arabian ostrich has suffered much the same fate. Only a few remain in the southernmost part of their former range which once extended throughout Saudi Arabia. One of the last living specimens was killed in 1945. The cheetah, the leopard, the oryx, and the lynx are also rapidly dying off in Saudi Arabia.

giraffe

In Africa giraffes were almost wiped out by the British and German armies during World War II. Soldiers had instructions to shoot them on sight because they wrecked overhead telephone wires essential to army communication. Although capable of dodging tree branches with skill, giraffes were unable to avoid wires. Frightened herds often ran into the unseen wires, bringing all communications to a halt. As soon as the war ended they were given protection and, from all indications, are in no greater danger than other African wildlife.

India

Internal strife in India during the 1940s had grave repercussions on the wildlife of that country. By 1955 the status of the Kashmir stag was so precarious that its extinction seemed imminent; the last Indian cheetah was dead; and brow-antlered deer had been nearly destroyed in India. Admittedly, these species had been in danger for many years, but they had managed to survive.

The rash spirit of rebellion against the departing British apparently precipitated the sudden deterioration of India's enlightened attitude toward its wildlife. When the country was finally free of colonialism, a period of lawlessness swept the land and any reminder of British rule was a potential target. The game preserves and sanctuaries, among the finest in the world, were not spared. They too were attacked and plundered.

The damage in India was not irreparable. Shortly after order was restored, the preserves were put under strict government supervision and, in the years that followed, they were as effective as they had been before the British left. The brow-antlered deer was rediscovered in 1952 in India and, although it and the Kashmir stag are still critically low in numbers, their survival appears assured.

With the formation of the Indian Board of Wildlife in 1952, the Indian government joined with the very influential and widely acclaimed Bombay Natural Historical Society, the oldest wildlife preservation group in the world, to establish India as one of the leading nations in the field of wildlife preservation and conservation.

The British have every right to be proud of their conservation efforts in India. The picture in Australia was unfortunately quite different. The unique inhabitants of this island continent suffered great losses at the hands of early British settlers who killed off vast numbers of them and indirectly destroyed others by introducing foreign species. There can be no question that the most destructive mammal to arrive in Australia was *Homo sapiens*.

koala bear

The koala bear represents an example of an Australian species brought to the brink of extinction by man. These animals were slaughtered in large numbers for their thick, durable, and long-lasting fur which commanded a high price in the London market. Hundreds of thousands of skins were shipped annually. By 1920 the koala was becoming scarce, and by 1934 its survival was in danger. Shortly thereafter it was given protection, and by 1950 it had staged a remarkable comeback.

Many other Australian species were not able to stage such dramatic comebacks. Several exotic parrot species are now extinct as a result of the caged-bird market, the practice of burning grassland for cattle grazing, and the assaults of introduced species. The emu, the Tasmanian wolf, and Leadbeater's opossum are endangered.

Some North American species that have become extinct, or are in danger of extinction, are described in the sections that follow. The kit fox (Fig. 10-7), a native of southwestern United States, is an example of a species endangered because of poison bait campaigns.

Fig. 10-7. Kit fox. A rare and endangered species. (*Courtesy U.S. Department of the Interior, Bureau of Sport Fisheries and Wildlife.*)

Endangered Species

The U.S. Department of the Interior lists, as of October 1970, 101 species and subspecies of wildlife (14 mammals, 50 birds, 7 reptiles, and 30 fishes) as being threatened with extinction. The criteria used to determine when a species becomes endangered are contained in Public Law 91-135 (passed by the 91st Congress, December 5, 1969). The paragraph from this act indicating the criteria follows.

> Sec. 3. (a) A species or subspecies of fish or wildlife shall be deemed to be threatened with worldwide extinction whenever the Secretary determines, based on the best scientific and commercial data available to him and after consultation, in cooperation with the Secretary of State, with the foreign country or countries in which such fish or wildlife are normally found and, to the extent practicable, with interested persons and organizations and other interested Federal agencies, that the continued existence of such species or subspecies of wildlife is, in the judgment of the Secretary, endangered due to any of the following factors: (1) the destruction, drastic modification, or severe curtailment, or the threatened destruction, drastic modification, or severe curtailment, of its habitat, or (2) its overutilization for commercial or sporting purposes, or (3) the effect on it of disease or predation, or (4) other natural or man-made factors affecting its continued existence.

Selected examples from the official list of endangered species published by the U.S. Department of the Interior, Bureau of Sport Fisheries and Wildlife are listed below. This list was revised in October 1970.

Indiana bat *(Myotis sodalis)*
Delmarva Peninsula fox squirrel *(Sciurus niger cinereus)*
Eastern timber wolf *(Canis lupus lycaon)*
Florida panther *(Felis concolor coryi)*
Columbian white-tailed deer *(Odocoileus virginianus leucurus)*
Sonoran pronghorn *(Antilocapra americana sonoriensis)*
Aleutian Canada goose *(Branta canadensis leucopareia)*
Mexican duck *(Anas diazi)*
Brown pelican *(Pelecanus occidentalis)*
California condor *(Gymnogyps californianus)*
Hawaiian hawk (io) *(Buteo solitarius)*
Southern bald eagle *(Haliaeetus leucocephalus leucocephalus)*
American peregrine falcon *(Falco peregrinus anatum)*
Masked bobwhite *(Colinus virginianus ridgwayi)*
Whooping crane *(Grus americana)*
Hawaiian coot *(Fulica americana alai)*
Ivory-billed woodpecker *(Campephilus principalis)*
Red-cockaded woodpecker *(Dendrocopus borealis)*
Hawaiian crow (alala) *(Corvus tropicus)*
Bachman's warbler *(Vermivora bachmanii)*
Dusky seaside sparrow *(Ammospiza nigrescens)*
American alligator *(Alligator mississippiensis)*
Puerto Rican boa *(Epicrates inornatus)*
Texas blind salamander *(Typhlomolge rathbuni)*
Houston toad *(Bufo houstonensis)*
Shortnose sturgeon *(Acipenser brevirostrum)*
Piute cutthroat trout *(Salmo clarki seleniris)*
Colorado River squawfish *(Ptychocheilus lucius)*
Warm Springs pupfish *(Cyprinodon nevadensis pectoralis)*
Unarmored three-spine stickleback *(Gasterosterus aculeatus williamsoni)*
Watercress darter *(Etheostoma muchale)*
Blue pike *(Stizostedion vitreum glaucum)*

The Bison

bison

It is estimated that up to 60 million bison roamed the North American prairies in the late eighteenth century. One observer, standing at the summit of Pawnee Rocks in Arkansas, stated that he could see 6 to 10 miles in all directions and that the entire panorama was filled with buffalo.

For centuries the culture and economy of the Plains Indian were intertwined with the bison. He depended on buffalo meat as a dietary staple. From sinews he fashioned bowstrings. From bone he made tools and ornaments. Hides were used for bedding, clothing, and shelter. Rare albino hides were thought to be capable of healing a variety of ills. Even dried feces, known as buffalo chips, provided badly needed fuel.

For many years the prairie grasses, buffalo, and Plains Indians represented the major living components of a balanced ecosystem. The unmounted Indian made slight impact on the buffalo hordes with his bow and arrow. When the buffalo moved to a new range, the Indian quickly broke camp and followed, for the herd represented his food, clothing, and shelter.

When the Civil War ended, a ruthless campaign of bison butchery was launched which brought this animal to the brink of extinction. The U.S. Army apparently believed that subjugation of the fierce Plains Indians would be assured once the buffalo was exterminated. It has been estimated that during 1871 and 1872 about 8.5 million buffalo were slaughtered, about one-seventh of the peak population. Such butchery was facilitated by the westward extension of several railroads into the prairie country. The railroads employed professional buffalo hunters to provide their crews with food. Some hunters were able to kill as many as 200 animals in a single day. During the winter of 1872–1873, almost 1.5 million hides were shipped to eastern markets and sold for $3 per hide. Less than 1 percent of the meat was marketed, however. Over 100,000 ani-

Fig. 10-8. Bison from a herd at Stanwood, Michigan. Conservation of this species is only part of the plan. These animals are being used for genetic and medical studies.

mals were killed just for their tongues which were considered a delicacy. Once the tongue had been removed, the rest of the carcass was left to decay. Bones were collected and shipped by the ton to fertilizer plants in Kansas and Minnesota. Kansas plants paid 4.5 million dollars for bones over a 13-year period. The long facial hairs were used for stuffing mattresses, and a buffalo wool company was even established in the Red River valley. Inevitably the buffalo herds dwindled to scattered bands. By 1889 only 150 bison survived in the wild; in 1894 the last wild buffalo was shot by a rancher in Parke County, Colorado. Sound management practices have successfully built up the herd in several wildlife refuges (Fig. 10-8) from the 250 animals that still survived in captivity.

The Passenger Pigeon

passenger pigeon

The passenger pigeon was once the most abundant bird on earth. Early in the nineteenth century, the well-known ornithologist Alexander Wilson observed a migrating flock which streamed past him for several hours. Wilson estimated the single flock to be 1 mile wide and 240 miles long and composed of about two billion birds. Yet not one passenger pigeon is left today.

Several factors contributed to the passenger pigeon's extinction. Many potential nest and food trees (beech, maple, and oak) were chopped down or burned to make room for farms and settlements. The pigeons fed extensively on beechnuts and acorns; the single flock observed by Wilson could have consumed 17 million bushels per day. Disease may have taken a severe toll. Breeding birds were susceptible to infectious disease epidemics because they nested in dense colonies. It was reported in 1871 that a concentration of 136 million pigeons nested in an 850-square-mile area in central Wisconsin. Up to 100 nests were built in a single tree. Many pigeons may have been destroyed by severe storms during the long migration between the North American breeding grounds and the Central and South American wintering regions. One record states that a very large flock of young passenger pigeons descended to the surface of Crooked Lake, Michigan, after becoming confused by a dense fog. Thousands drowned and lay a foot deep along the shore for miles. The low reproduction rate was no doubt a factor in their extinction. Although many perching birds such as robins lay 4 to 6 eggs per clutch, and ducks, quail, and pheasants lay 8 to 12 eggs, the female passenger pigeon produced only a single egg per nesting. Reduction of the flocks to scattered remnants may have deprived them of the stimulus for reproduction. The bird's decline was hastened by persecution from market hunters. They slaughtered the birds on their nests. Every imaginable instrument of destruction was employed, including guns, dynamite, clubs, nets, fire, and traps. Over 1300 could be caught in one throw of a net. Pigeons were burned and smoked out of their nesting trees. Migrating flocks were riddled with shot. Over 16 tons of shot were sold to pigeon hunters in one small

Wisconsin village in a single year. Pigeon flesh was considered both a delectable and fashionable dish in Chicago, New York, and Boston. Almost 15 million were shipped from a single nesting area near Petoskey, Michigan, in 1861 and sold for 2 cents per bird. The last wild pigeon was shot in 1900. The last captive survivor died on September 1, 1914, at age 29, in the Cincinnati Zoo.

The Grizzly Bear

grizzly bear

The grizzly bear is easily distinguished from the much more abundant and widely distributed brown bear by its grizzled brown fur, shoulder hump, and concave face. Its huge front claws are effectively employed in digging rodents from their burrows and in ripping the flesh of a variety of animals, including, on occasion, colts, calves, and lambs. The grizzly originally ranged throughout western North America, from the Arctic to subtropical Mexico and eastward to the prairies. Although 11,000 of these animals wandered the wilds of Alaska in 1963, only a remnant survives in the contiguous United States, primarily in the mountainous regions of Wyoming, Montana, Colorado, and Idaho. In California, where it once was relatively abundant, the grizzly was eradicated in 1922. Livestock men have been an important contributing factor to the grizzly's present endangered status. Many a rancher who has found the torn carcass of a calf has waged a personal program of eradication employing poisons, traps, bullets, and dogs. Such environmental resistance has proved especially severe because of the bear's limited biotic potential (reproduction rate); the female gives birth to a single or twin cubs in alternate years. Restrictive hunting laws now provide some protection for the grizzly. In national wildlife refuges such as Glacier and Yellowstone National Parks, it was unlawful to kill the grizzly for many years until the ban was lifted in 1967. Future management for grizzly survival will probably include its removal from livestock areas to mountainous wilderness areas.

The Whooping Crane

whooping crane

The 4-foot whooping crane (Fig. 10-9) is the tallest bird in North America. Snow white except for its scarlet crown and black wing tips, it resembles a flying cross as it passes overhead, with its neck projecting forward and sticklike legs trailing. In the early nineteenth century, its breeding range extended from the prairie provinces of Canada south through the Dakotas to Iowa. It wintered along the Gulf coast from Florida to Mexico. In 1970 it was known to nest only in Wood Buffalo National Park in Alberta, Canada, and its wintering area was restricted to the Arkansas Wildlife Refuge on the Texas coast. Although their number is increasing slowly, only 55 wild and 8 captive whooping cranes survived in 1969.

The factors that have contributed to the endangered status of the

Fig. 10-9. Whooping cranes on the wintering grounds in South Texas at the Arkansas National Wildlife Refuge. *(Courtesy U.S. Department of the Interior, Bureau of Sport Fisheries and Wildlife.)*

whooping crane include: a low biotic potential—only two eggs per clutch; the appropriation of its prairie nesting habitat by ranchers and farmers; intensive hunting; and severe storms occurring during migration.

The Whale

whales

After centuries of excessive hunting on a highly organized scale, some species of whales have already been wiped out and others have become so scarce that their survival is in doubt. The most vulnerable target of all, because of its value, size, and increasing scarcity, is the blue whale. Often more than 100 feet long and weighing up to 130 tons, it has been reduced to a population of about 600, a level so low that even with complete protection it may slowly slip into oblivion.

Fig. 10-10. Tip of upper jaw of the finback whale illustrates the baleen filter plates which strain plankton from the ocean water. The pits at the end of the pointer are thought to function in the detection of odors in the water. *(Courtesy U.S. Fish and Wildlife Service.)*

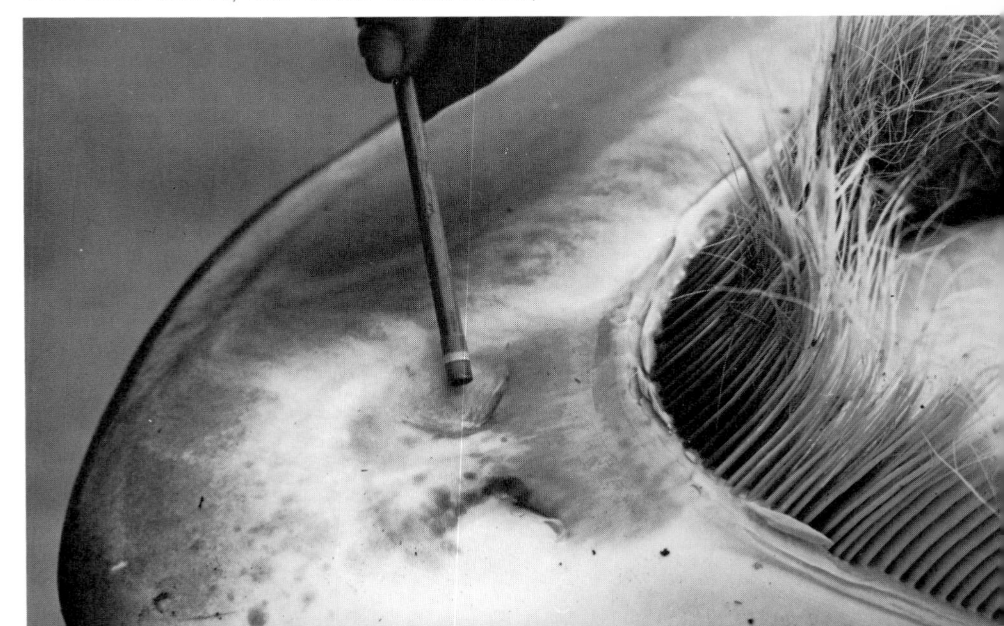

Only recently, after quick profit seekers had done untold damage, was the whale given any effective protection. In 1944 an international convention imposed limits on whaling fleets in the Antarctic, establishing a maximum annual kill of 600 "blue whale units," an internationally accepted standard of measurement for whale catches. The "blue whale unit" is equivalent to one blue whale, or two fin whales, or two and one-half humpback whales, or six sei whales. Under an agreement signed by those countries still interested in whaling (Japan, Norway, the Netherlands, South Africa, the United Kingdom, and Russia), the catch figure is controlled on the whaling grounds by a floating headquarters which tallies day-to-day catches and announces the end of the season as soon as the limit is reached.

Events leading to the present critical situation began hundreds of years ago. Since 1600 one species of whale after another has been wiped out by unrestrained exploitation. In the beginning, when whales were still abundant around coastal waters, they were killed from the land. The Basques hunted them along their shores for nearly 10 centuries, and as recently as 100 years ago whales could still be taken off the Scottish coast. Gradually, they retreated to more distant waters.

Whaling became an organized industry. Whaling ports were thriving communities throughout the world. One of the largest ports was New Bedford, Massachusetts, from which boats set sail on expeditions lasting as long as 3 or 4 years. These voyages were dangerous and primitive compared with those of the floating factories that sail the seas today. At that time whales still had a chance to survive. With the mechanization of whaling, however, which began with the invention of the explosive harpoon in 1856 and culminated in the first Norwegian "floating fleet" which sailed into the Antarctic in 1904, they were in great danger.

By 1912, 141 factory ships were stationed in whaling waters throughout the world. Each ship was capable of processing a giant blue whale carcass in 1 hour. Most of the ships in this fleet, 37 of which were Norwegian, were stationed in the Antarctic. Whales were abundant and prices were high. The average annual yield was 564,000 tons of oil, which sold at a price of $270 per ton. Whaling became internationally recognized as one of the most lucrative enterprises in the world.

The size of the whaling fleets continued to grow until 1939, the beginning of World War II. From 1939 to 1947 whales were left unmolested, but they were unable to recover from the ravages of the preceding years. In 1947, when the first fully equipped whaling fleet was again in operation, 31,000 whales were taken, giving a yield of only 302,000 tons of oil. Whalers have whaled themselves out of business. In 1965 there were only 15 factory ships left in operation. During the 1964–1965 season whalers could find only 20 blue whales to kill; during the 1966–1967 season they could find only four. Today there are fewer than 3000 blue whales left, perhaps as few as 500 in the home waters of the Antarctic, perhaps too few in too great an expanse of ocean to allow them to find mates. The fate of other species will no doubt be the same.

The Bald Eagle

bald eagle

For almost 170 years the bald eagle (Fig. 10-11) has been given the place of honor as our national symbol. Once ranging throughout most of North America, it has disappeared from many states and is becoming scarce even in Florida, where it was abundant until only a few years ago. Except in a few protected refuges, notably the Cape Romain National Wildlife Refuge, the sight of a bald eagle is extremely rare. It has been relentlessly persecuted everywhere. It was not only shot by "sportsmen" without the protection of laws but was also pursued by egg collectors. In many parts of the country, it had a price on its head, and these bounties, paid for each pair of talons brought in, resulted in an estimated 115,000 dead eagles between the years 1917 and 1952. In Alaska these bounties, offered by the territorial government itself in retaliation against the bird accused of robbing the country of its fish and game, resulted in wholesale slaughter. The bounty paid was 50 cents for each bird killed. It was only after the eagle population began to decline that investigation proved, contrary to popular belief, that the bald eagle was a poor fisherman; while it occasionally managed to catch a salmon as it approached or left spawning grounds, it fed primarily on those that died along the way.

Fortunately, the bounties are now on the heads of the real predators. Anyone found harming a bald eagle, or its nest, is subject to a fine of up to $500. This protection, however, may have come too late. There are only a few nesting areas left. Eagles nest only in very tall trees where they can raise their broods in comparative safety. Nests, used year after year by the same pair, are renewed each year with a new layer of branches; thus they become very large. One nest near St. Petersburg, Florida, probably in use for 30 or 40 years, is reported to have been 20 feet deep, 9½ feet in diameter, and to weigh 2 tons. The size of this nest is not

Fig. 10-11. American bald eagle. One of the endangered species. *(Courtesy U.S. Department of the Interior, Bureau of Sport Fisheries and Wildlife.)*

surprising, since eagles mate for life and are often long-lived. Some reach an age of 100 years.

The male and female usually take turns incubating the eggs (often only one per nest, maybe up to three) and then share the burden of caring for the young for a period of about 4 months. The eagle is fully matured after 4 or 5 years. The gleaming white feathers of his head and tail stand out so sharply against the brown-black coloring of the wings and body that he appears to be bald and tailless when seen from a distance. Eagles weigh about 12 pounds and can carry a weight of 7 to 8 pounds. While they are lazy fishermen, they are swift, powerful, expert fliers, and are not above robbing a more capable fisherman, the osprey, of its catch while in flight.

Unfortunately, bald eagles are not only endangered by the logging of forests and the vast real estate developments in Florida, but also by the pollution of rivers and streams, particularly by pesticides such as DDT which become concentrated in the fish on which they feed. Without food and nesting sites, they must seek refuge elsewhere, probably in our national parks. In 1961 the National Audubon Society reported a bald eagle population of 3576 in the United States. This number decreased to 600 breeding pairs in the contiguous United States by 1966. They remain in fairly large numbers in Alaska.

Plants

plants

Plant species are threatened by many of the same factors that menace animal species. The introduction of competitors, herbivores and parasites, the practice of "selective hunting" by lumber and pharmaceutical companies and ornamental plant collectors, pollution, agricultural land improvement, and many other influences can lead to the diminution of particular plant species. Plants are seldom deliberately exterminated and, if they are, their lack of movement and their unusually high reproductive potential make it likely that some are overlooked. Nevertheless, plant species can become endangered and eventually become extinct. There is little doubt that some have become extinct during recent times; but unless a plant is showy and its geographic distribution sharply delineated, its passing may go unnoticed (Fig. 10-12).

As a group, plants of the world's grasslands may have suffered the heaviest casualties, but other environments may lose plant species too. In the eastern part of the United States, several of the most beautiful native flowering plants have become extremely rare. The fringed gentian, an annual with tubular, violet flowers having petals fringed at the ends, is an inhabitant of bogs and marshes. As the reckless and excessive drainage of wetlands continues, this plant is becoming nearly impossible to find. The situation is worsened by its being an annual; when an entire clump of flowers is picked, it does not reappear in that location the next year. The trailing arbutus or mayflower is a low-growing, short-stemmed plant with

Fig. 10-12. *Psilotum* (wisk fern). This primitive fern grows in the Big Cypress Swamp of southern Florida. In this area it is rarely found other than in knotholes, fire-damaged areas, or crevices of cypress. The vertical distribution of *Psilotum* on the trees ranges from ground level to about 3 feet. This distribution may depend on water fluctuation. As the water recedes, floating spores collect in knotholes and crevices of the tree trunks. Draining of the Everglades could eliminate this "living fossil."

fragrant pink or white flowers. It grows in light shade and is a late successional or climax species. The trailing arbutus is very intolerant of soil disturbance and disappears rapidly in any area where man's activities are not strictly curtailed. In addition, because of its shape, it is difficult to pick the flowers without damaging the plant or uprooting it entirely. Another endangered species, the pinxter flower, or purple honeysuckle, is a branched shrub which grows to a height of 6 to 8 feet. It has pink-white flowers. It blooms in early spring before it and the surrounding deciduous trees have put out their leaves and is thus easily spotted from a great distance by ornamental shrub hunters. Once extremely common, similar to the fringed gentian and trailing arbutus, it is now encountered much less frequently in its native habitat.

Selective lumbering can threaten tree species, especially in tropical rainforests where artificial reseeding is difficult or impossible and where high species diversity and thin soils are unfavorable to the establishment of tree farms. In the North Patagonian Andes, for example, some of the conifers, which live for 1000 years or more, have become endangered even in the absence of bulldozers and chain saws by a few men working every summer with simple tools and selecting the species and particular trees most suitable for their needs.

Andes

As is the case with animal species, plants that have restricted ranges are vulnerable to a variety of disturbing factors. On several of the Galapagos Islands, some floral species may be lost because of excessive grazing by introduced goats. Unfortunately, the difficulty of preparing plant lists that are both up-to-date and taxonomically correct may prevent us from ever knowing which plants have become extinct because of our activities. There is no world list of endangered plants corresponding to that for vanishing mammals and birds issued by the International Union for the Conservation of Nature.

Fig. 10-13. Attwater's prairie chicken. This variety of the prairie chicken formerly ranged throughout most of the open coastal prairie from southwestern Louisiana westward almost to Mexico. The 1937 population of about 8700 dwindled to an estimated 750 by 1965. *(Courtesy U.S. Department of the Interior, Bureau of Sport Fisheries and Wildlife.)*

SOLUTIONS

Most extinctions caused by man's activities need not have occurred. With the exception of a very few species such as the passenger pigeon, which seemed unable to adjust to accommodate man to any degree, most species that have become extinct might be living today had there been a little care and foresight at the right time. Attempts are now being made to save some species from imminent extinction. No doubt others will benefit from these efforts in the future. American biologists and legislators have employed numerous techniques for restoring, maintaining, and increasing wildlife populations. They include protective laws, wildlife refuges, predator control, breeding and reintroduction of species and, most recently, habitat development. Some, such as Attwater's prairie chicken (Fig. 10-13) may be beyond saving. These techniques are discussed in the sections that follow.

Game Laws

game laws

Throughout human history there have been a few foresighted individuals aware of the importance of wildlife to man's happiness and welfare and of the ease with which this resource can be depleted, if not exhausted, by the unrestrained human "predator." Although game was generally available during early colonial times, the constant hunting of a few species such as the white-tailed deer prompted some states to enact protective laws. Thus in 1664 Rhode Island established the first closed season on deer, and by 1694 Massachusetts also protected this popular game species. In 1708 New York afforded protection to upland game such as the ruffed grouse, wild turkey, and heath hen. Rhode Island was the first state to enact legislation barring spring shooting of migratory waterfowl. In 1874 Congress was prodded into passing a protective law for buffalo; unfortunately, it was vetoed by President Grant. In 1894, just 3 years before the last wild bison was shot, a law was finally passed protecting it. Similar attempts in the 1890s to protect the passenger pigeon failed.

Most protective game laws were poorly enforced during the nineteenth and early twentieth centuries. Few officials had the courage to punish the numerous violators. The first real step toward a solution to this problem came with the employment of game wardens, by New Hampshire and California in 1878, charged with the responsibility of law enforcement. Thirty-one states had employed wardens by 1900.

With a more alert public demanding protection for our rare and endangered species, the enactment of new laws and the enforcement of these laws can preserve much of what we are now in danger of losing. If international agreements are made, this can occur throughout the world.

Wildlife Refuges

refuges

The federal wildlife refuge program had its beginning with the establishment of the Pelican Island Refuge on the east coast of Florida in 1903. Since then the program has expanded rapidly; it now includes 321 units embracing 28.6 million acres. Refuges range in size from the small Mille Lacs Refuge, consisting of two small islands in a Minnesota lake, to the large 8.9-million-acre Arctic National Wildlife Refuge in Alaska. Federal refuges are of three types: those designed primarily for waterfowl; those designed to serve big-game animals such as goats, antelope, and deer; and those designed to save endangered species from threatened extinction. In all instances the refuges provide favorable habitats for the species, and at the same time protect them from predators, including man. The Idaho bighorn sheep (Fig. 10-14) are in need of such protection.

Fig. 10-14. The first European settlers in Idaho found the bighorn sheep in all parts of the state. Now they occur in Idaho only in the Salmon River drainage area. The steady decline of the bighorn in Idaho and elsewhere in the West is attributed directly to pneumonia caused by lungworm infection. However, sheep that find adequate forage on the low winter ranges can resist the lungworm. When cattle grazing during the summer destroys the grass on the sheep's winter range, the bighorns suffer from malnutrition, contract pneumonia, and die. The state of Idaho now restricts cattle from certain of the bighorn winter ranges. (*U.S. Forest Service Photo.*)

Predator Control

The control of natural predators assumed a significant role early in wildlife management. It was only natural for a tired and disappointed hunter to vent his frustration, from not finding quail or cottontails, on a hawk or a fox. If a fox consumes 30 rabbits a year, the hunter assumed that one rifle bullet might mean 30 additional cottontails for hunters. In the past pressure has been exerted on state legislators to enact bounty laws, resulting in the expenditure of large sums of money which could have been spent more wisely in the development of wildlife habitats. Amounts paid by states for bounties have ranged from $2 for a fox to $35 for a timber wolf. This system is now under severe criticism.

crow

barn owl

Many accusations against predators are ill-founded. The crow has been accused by duck hunters of destroying nests and making a meal of eggs and ducklings. An analysis of the contents of crow stomachs in Michigan, however, revealed that two-thirds of their diet was composed of insects, all potential crop destroyers. The barn owl has frequently been persecuted for raiding chicken yards or for destroying quail and rabbits. A 3-year investigation by Michigan state biologists has revealed that the barn owl is not the culprit it is accused of being. An examination of 2200 barn owl pellets (regurgitated, undigestible bones and fur) showed no trace of poultry or game birds. Further, over 90 percent of the mammals represented in the pellets were mice, primarily meadow mice, a species capable of inflicting serious crop damage.

Most ecologists consider predation an essential part of the ecosystem. In many cases predators may actually promote the welfare of the prey species by culling aged, crippled, and diseased individuals from a population. Moreover, predators may serve a useful role in keeping the prey population within the limits imposed by the carrying capacity of the habitat (Fig. 10-15). A lack of predatory pressure might lead to a population explosion which could result in habitat destruction and culminate in massive death caused by starvation and disease.

Fig. 10-15. The coyote and other predators occupy a vital niche in keeping populations within the carrying capacity of the area. *(Courtesy Michigan Department of Conservation.)*

Breeding and Reintroduction of Species

Many species fail to breed successfully in captivity; a few do, and some have survived only because of captivity. Both the ginkgo and dawn redwood trees, once thought to be extinct, were rediscovered growing in the temple gardens of China. Today they are planted throughout the world. Another Chinese species, Père David's deer, was rediscovered in Peking's Imperial Garden in 1861, long after all wild individuals had disappeared. Today there are several hundred in zoos and private collections throughout the world.

whooping crane

A captive breeding program to save the whooping crane was initiated in 1967. Each season several eggs are taken from nests in Wood Buffalo Park in Canada and flown to the Federal Endangered Wildlife Research Station in Maryland where they are incubated. The captive birds are later bred to produce larger numbers for possible reintroduction into favorable habitats. The Hawaiian goose, or nene (Fig. 10-16), has been saved in a similar fashion. Its total population had been reduced to less than 50 when a captive breeding program was established in England. Within a few years a number sufficient to restock their native habitat in Hawaii resulted. Today the population contains over 500 individuals. A similar program is in progress for the Aleutian geese (Fig. 10-17).

Fig. 10-16. The nene goose on the nene goose range on Hawaii Island. The 1959 total population numbered fewer than 50. Through a successful captive breeding program in England and Hawaii, 200 birds were released and the population is now about 5000. *(Courtesy U.S. Department of the Interior, Bureau of Sport Fisheries and Wildlife.)*

Fig. 10-17. Aleutian geese. One of the rare and endangered species. These geese were reared at Patuxent, Maryland, after being taken from Buldir Island in the Aleutians. *(Courtesy U.S. Department of the Interior, Bureau of Sport Fisheries and Wildlife.)*

antelope

Several examples of the successful reintroduction of species are available. The Russian antelope or saiga was almost extinct by 1900. A careful breeding program increased the number to 1000 individuals. With more information about their habits, food, and nutrition, the saiga was reintroduced to its former range. By 1960 the population had increased to 2.5 million, and a controlled harvest was allowed for the first time. Since then, over one-quarter million saiga have been captured annually and are an important source of meat, hides, and industrial fat. This sustained productivity has been achieved on land unsuitable for cattle grazing.

Habitat Development

The most effective approach to the preservation of species and the conservation of wildlife populations is to increase the variety, number, and quality of suitable habitats (Fig. 10-18). Even with protective laws, predation control, breeding programs, and the reintroduction of species, populations will continue in jeopardy if, at the same time, their habitats are usurped, destroyed, or permitted to deteriorate.

Habitat acquisition and development programs are receiving high priority among governmental agencies responsible for wildlife conservation. These attempts have been effective and should become even more so in the future. Real contributions toward habitat development can be made by private landowners. Over 85 percent of the hunting lands in the United States are privately owned or controlled. With proper management of areas not in use for agriculture, private citizens can provide the variety of undisturbed habitats necessary to insure the survival of all our species.

Fig. 10-18. Reestablishing suitable habitats for wildlife is vital to the survival of many species. (*Courtesy Michigan Department of Conservation.*)

More and more people the world over are recognizing the dangers in what man is doing to the life-sustaining resources and are speaking out on the theme of man's responsibility to his natural environment. People are beginning to realize that the only valid option today is to try to live in harmony with nature, instead of trying to conquer and subdue it. Man cannot live without contact with nature. It is essential to his happiness and well-being.

We may think it sentimental to be concerned about the possible extinction of a bird, a fish, or a monkey. It is probably true that our pattern of life will not change if a particular species disappears forever. Our pattern of life will change, however, if this kind of thing goes on, and indeed if unchecked it could lead to the extinction of man himself.

Suggested Readings

Ehrenfeld, D. W. *Biological Conservation*. New York: Holt, 1970.
Fisher, J., N. Simon, and J. Vincent. *Wildlife in Danger*. New York: Viking, 1969.
Laycock, G. *America's Endangered Wildlife*. New York: Norton, 1969.
McClung, R. M. *Lost Wild America*. New York: Morrow, 1969.
Owen, O. S. *Natural Resource Conservation*. New York: Macmillan, 1971.
Pinney, R. *Vanishing Wildlife*. New York: Dodd, Mead, 1963.

The scimitar-horned oryx, identified by its curved horns, once ranged widely around the fringes of northern Africa's deserts. It has been eliminated from most of its northern range and now exists in a narrow strip along the southern edge of the Sahara Desert. Two factors, uncontrolled hunting and overgrazing by domesticated livestock, have decimated the former large herds of 1000 or more animals. Tribes such as the Haddad, whose survival depends on the oryx, need to learn the concept of a sustained yield. *(Courtesy Lion Country Safari, Laguna Hills, California.)*

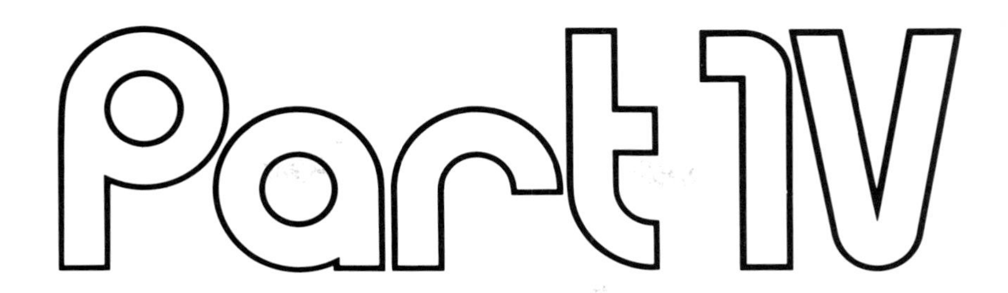

ASSESSING THE POPULATION PROBLEM

An osprey approaching its nest. Maintaining population numbers depends on successful reproduction that balances mortality. *(Courtesy Michigan Department of Natural Resources.)*

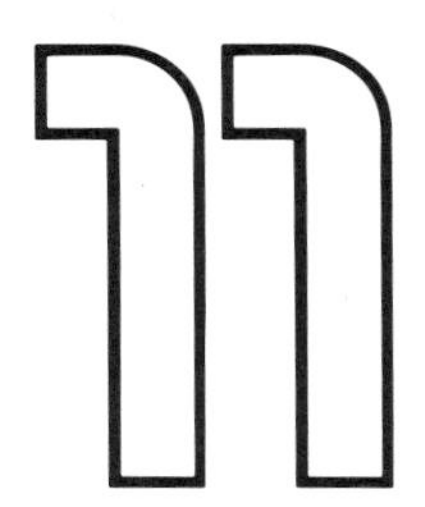

Population Dynamics

gene pool

A population is generally defined as a group of organisms of the same species that lives at a given time in a particular area. Genetically, the members of a population share in a common gene pool. Each population has characteristics which are a function of the group and not of the individuals. These characteristics include population density, birthrate, death rate, age distribution, reproductive potential, population pressures, and population cycles.

The interactions between communities and the physical environments they occupy tend to produce communities which become stabilized both in the kinds of organisms and in the numbers of organisms. For the number of organisms in a particular population to remain stable (without emigration or immigration), the birthrate and the mortality rate, although they may fluctuate from season to season or from year to year, must remain in balance. The fact that one species may differ from another within the same community in number of individuals, in birthrate, and in life-span indicates the variety of forces that are at work in the system of checks and balances that maintains a stable population.

The changes that occur in population size and the factors responsible for these changes constitute the area generally known as *population dynamics*. This area, as it relates to natural ecosystems, is one that justifies special consideration since the human population growth rate during the last few years has become a real concern to man in terms of his possible survival as a species. Studies of regulatory mechanisms among nonhuman species may very well suggest some reasonable options for controlling our own population size.

Although the factors that influence populations must be examined separately, it is important to remember that in any one instance several factors are at work simultaneously. This chapter describes some of the characteristics of populations in general, particularly as they relate to the regulation of population size. Chapter 12 discusses some of these same principles as they apply specifically to human population problems.

POPULATION DENSITY

density

biomass

The density of a population can be expressed as the number of individuals per unit area or volume (for example, 50 pine trees per acre), or in terms of biomass per unit area or volume (for example, 100 bushels of corn per acre). The number of individuals is a satisfactory measure of *population density* when the size of the individuals in the population is relatively uniform. When size is variable, as is frequently the case among both plants and animals, *biomass* gives a more accurate idea of the population's importance as it relates to the flow of energy and materials through an ecosystem. Regardless of the measure used, population density is usually variable, even though there are definite upper limits. This theoretical upper limit is determined by the interaction of several factors, including the total energy flow within the ecosystem, the trophic level to which the species in question belongs, and the size and metabolic rate of the individuals. Energy is always an ultimate limiting factor for any population. There is only a limited amount of energy available at a particular trophic level in any given ecosystem; there can be no more biomass than this amount of energy can support. Real population densities often fluctuate at levels well below the theoretical maximum. A study of the factors that regulate the densities of real populations in nature is a significant area of ecology.

Birthrate, Death Rate, and Survival

survival

The immediate determiners of population size are the numbers of births and deaths per unit time, and the average survival time. If the *birthrate* increases while at the same time the death rate decreases, or remains the same, or increases more slowly than the birthrate, population size will increase. Conversely, if the *death rate* increases while the birthrate decreases, or remains the same, or increases more slowly than the death rate, population size will decrease. The significant question involves how the birthrate, the death rate, and the average *survival time* are regulated. That they are regulated in part by environmental influences is clear from the fact that the observed values for these three population parameters are usually very different from the values that are theoretically possible.

The standard customarily used in evaluating observed birthrates is the maximum potential birthrate, or the theoretical maximum number of individuals that can be produced under ideal conditions. This number can be estimated by determining the birthrate per reproductive-age female (sometimes expressed as the number of births per 1000 individuals without regard to sex or age) under the best possible environmental conditions. Observed birthrates usually fall well below the maximum, an indication that environmental factors can influence population size by causing changes in the rate of production of new individuals.

mortality

The theoretical minimum mortality rate for a population is the number of deaths expected if all deaths are the result of old age; that is, if all individuals live to the end of their potential life-span. Of course, the average longevity in any real population is far below the potential physiological longevity, an indication once again that environmental factors influence population densities. Since minimum mortality is not usually found in nature, it is often useful to determine the mortality rate for the various age groups within a population. Such data indicate which stages in the life cycle are most susceptible to environmental controls and allow us to compute the percentage of individuals that will be alive at the end of each age interval. The results may be graphed as a survivorship curve (Fig. 11-1).

The curves in Fig. 11-1 illustrates several different *survival patterns*. Curve D approaches the pattern expected if all the individuals in a population realize average physiologically possible longevity. This curve indicates full survival through all the early ages (as shown by the horizontal portion of the curve), followed by death for all individuals more or less at the same time; the curve then falls suddenly. Curve A approaches the other extreme, where mortality is high among the very young but where individuals surviv-

Fig. 11-1. Survivorship curves of four different types. All populations are of similar-aged individuals, and the graphs represent the numbers surviving after given intervals of time. (A) An oyster population; (B) a population of fledged birds; (C) a population of mountain sheep; (D) a population of starved fruit flies. The figures on the vertical axis show the number of individuals surviving from an original population of 1000. The age scale intervals on the horizontal axis vary with the life-span of the particular species. (After A. S. Boughey, *Fundamental Ecology,* New York: Intext, 1971.)

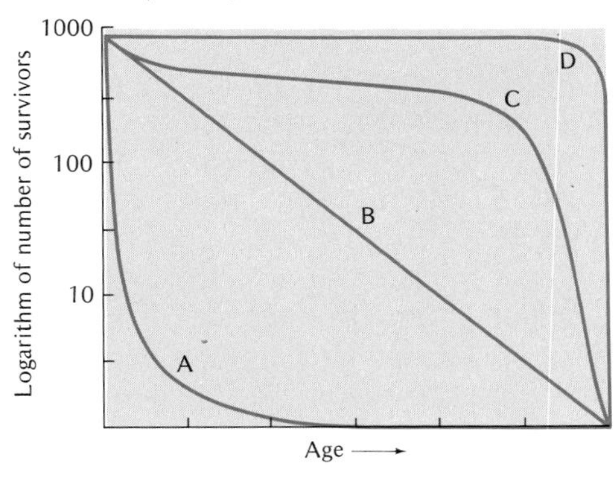

survival patterns

ing through the earlier stages have a good chance of surviving for a long time thereafter. Between these two extremes is the condition represented by curve B where the mortality rate at all ages is constant. The survivorship curve for most wild animal populations is probably intermediate between curves A and B, and the curve for most plant populations is probably near the extreme of curve A. In other words, high mortality among the young is the general rule in nature.

Changes in environmental conditions may alter the shape of the survivorship curve for any given population, and the altered mortality rates in turn may have profound effects upon the dynamics of the population and upon its future size. For example, one of the chief causes of the enormous increase in the human population has been a great reduction in mortality during the early years as a result of improvements in sanitation, nutrition, and medical care. These improvements have caused a shift in the human survivorship curve from one intermediate between curves A and B in primitive societies to one approaching D in the most advanced societies. There is no real evidence that the shift is due to an increase in the human birthrate per reproductive-age female (in fact, the birthrate has decreased).

Biotic Potential and Environmental Resistance

The maximum possible birthrate for a population almost always exceeds the actual ecological birthrate, and the potential physiological longevity is usually greater than the average ecological longevity. In other words, under natural conditions the number of births is lower and the number of early deaths is higher than would be expected under ideal conditions. One of the approaches to understanding the dynamics of a real population is to determine the potential population size under ideal conditions and then attempt to determine how actual conditions modify this expected pattern. Let us assume, then, that we can study a population that is stationary, one that has a stable age distribution, and one that exists in an unlimited environment (that is, one in which the amount of food, water, space, light, and other resources is plentiful and in which there is no competition, predation, or parasitism by other organisms). The maximum population growth rate under such ideal conditions is called the *biotic potential*. It is generally defined as the inherent property of an organism to reproduce and to survive; that is, to increase in numbers.

biotic potential

Every species, whether plant or animal, unicellular or multicellular, has a very large biotic potential. Numerous calculations are available that demonstrate what would happen if a pair of organisms were to produce a full complement of offspring, and if all these offspring survived to produce a full complement of offspring in turn, and so on, for several generations. For example, if a single pair of houseflies started to breed in April, they could produce 191 billion billion descendants by August if all eggs hatched and if all the resulting young survived to reproduce. A similar calculation has been made for 100 starfish found living along a small sector of the Pacific coast just north of San Francisco. If it is assumed that half of these starfish are females and that each produces one million eggs (a modest estimate), the population in the next year would be about 50 million. This would include about 25 million females, all of which would again produce about one million eggs each. It is obvious that if this rate of reproduction were to continue for even a few generations, with 100 percent survival of all offspring, soon the starfish would fill the seas and be pushed out across the land by sheer pressure of reproduction. At this rate, it would take only 15 generations for the number of starfish to exceed the estimated number of electrons in the visible universe (10^{79}).

Calculations similar to those above played an important role in leading Charles Darwin to formulate his theory of natural selection. Darwin made the following statement concerning reproductive potential.

elephant

> The elephant is reckoned the slowest breeder of all known animals, and I have taken some pains to estimate its probable minimum rate of natural increase. It would be safest to assume that it begins breeding when 30 years old and goes on breeding until 90 years old, bringing six young in the interval and surviving until 100 years old. If this be so, after a period of some 740–750 years there would be nearly nineteen million elephants alive, descended from the first pair.

Darwin's calculations have been extended to show that in 100,000 years one pair of elephants would have so many living descendants that they would fill the visible universe.

That population explosions such as those outlined above do not in fact occur is evidence enough that there are always limiting factors on any actual population, which prevent it from realizing its full biotic potential. The difference between the biotic potential and the actual rate of increase is a measure of environmental resistance, the sum total of the environmental factors acting on a population to limit its size.

If a population were not subjected to any limiting factors and could thus realize its full biotic potential, its growth would accelerate in an exponential fashion. Its growth curve when graphed would rapidly approach the vertical (Fig. 11-2). Real growth curves are usually exponential in early stages,

Fig. 11-2. An exponential growth curve.

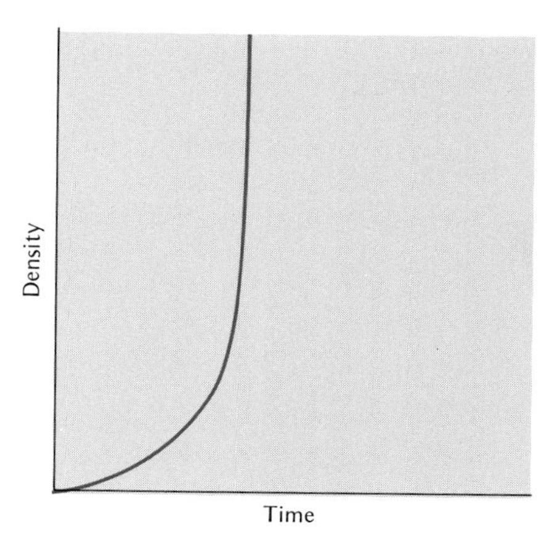

but *exponential growth* cannot continue for long because environmental limiting factors soon come into play. The limiting factors may cause an abrupt halt in the exponential growth. In this instance the limiting factors do not become effective in controlling population size until the number of individuals becomes relatively large; the factors then suddenly become very effective, usually causing a rapid decline in population density. Such a population growth pattern is characteristic of some small insect groups with short life cycles, among which the population size increases rapidly during a period of favorable weather and then decreases when the weather becomes unfavorable. Weather, which serves as the limiting factor in this example, is independent of population density. A change in the weather is not caused by an increase in population size, and its limiting effect would be as severe on a small population as on a large one. Figure 11-3 illustrates a different type of growth curve on which the increase is slow at first, then becomes very rapid, and finally slows down again as the population size approaches an equilibrium position. The curve is thus S-shaped (*sigmoid*). In mathematical terms, the growth accelerates logarithmically at first, then departs more and more from a logarithmic base until it reaches a point where the acceleration becomes negative, and then approaches a point (asymptote) that represents the limit in size of the population. This limited size represents the *carrying capacity* of the environment (Fig. 11-4); that is,

sigmoid curve

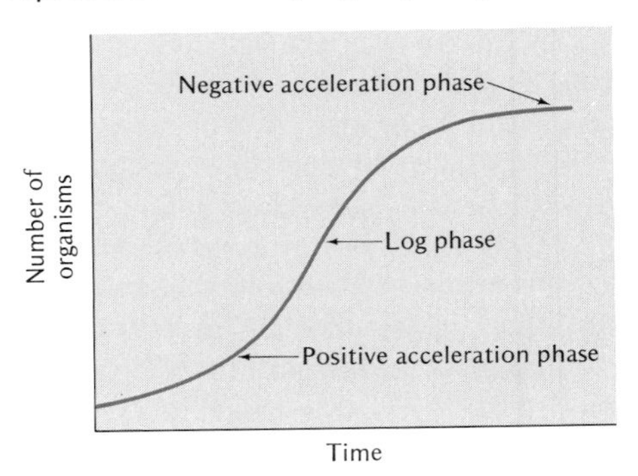

Fig. 11-3. An example of population growth; a sigmoid curve.

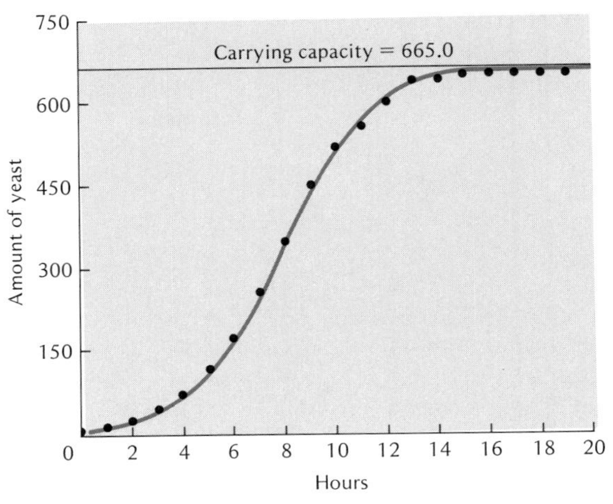

Fig. 11-4. Growth curve of a laboratory population of yeast cells. (Modified from W. C. Allee et al., *Principles of Animal Ecology*, Philadelphia: Saunders, 1949. After Keeton.)

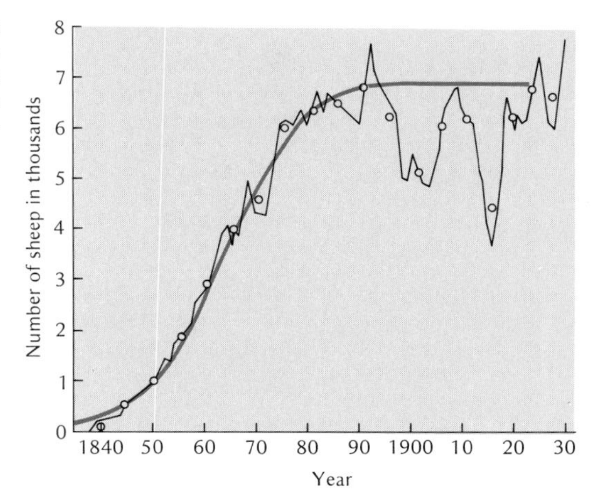

Fig. 11-5. Growth curve of the sheep population in South Australia. The smooth curve is a hypothetical one around which the real curve fluctuates. (Modified from J. Davidson, *Transactions of the Royal Society of South Australia,* Volume 62, 1938).

the maximum population that the environment can support indefinitely. Growth curves seldom trace a smooth curve (Fig. 11-5). Fluctuations are the rule; in fact, these fluctuations result from corresponding fluctuations in the physical factors that increase or decrease the carrying capacity of the environment. Density fluctuations also occur under laboratory conditions in which environmental factors are kept as constant as possible. Thus real growth patterns only approximate a sigmoid curve.

REGULATION OF POPULATION DENSITY

density-dependent factors

A sigmoid growth curve is produced when the limiting factors become increasingly effective as the density of the population increases; that is, when the limiting factors are at least partially density-dependent. The distinction between *density-dependent* and *density-independent* limiting factors is not always obvious, but the concepts are nevertheless useful in describing the kinds of environmental influences that help to regulate population density. Strictly speaking, a density-independent factor is one that exerts a constant influence regardless of population density, but it is also customary to consider factors that exert an influence of constant percentage as density-independent. For example, if a hunter kills five rabbits every year regardless of the density of the rabbit population, he is a density-independent limiting factor in the strict sense, but if he kills 5 percent of the local rabbit population he is also considered a density-independent factor. A density-dependent limiting factor, by contrast, is a hunter who kills 5 percent of the rabbits when the density is low, 20 percent when the density is intermediate, and 50 percent when the density is high. In other words, a density-dependent limiting factor is one whose intensity is at least partly determined by the density of the population it helps to regulate.

Climatic factors and other physical events (for example, the alternation of light and dark and the annual changes in day length) are ordinarily density-independent factors, whereas biotic factors (for example, parasitism, predation, competition, and endocrinological and behavioral changes) are usually density-dependent factors. There are exceptions in both instances, and there are factors that are either density-dependent or density-independent according to the particulars of the situation.

Food Supply

food

Food supply is a primary, density-dependent force in controlling population size. The number of animals in a population tends to increase and decrease in direct proportion to the food supply. At any given time the amount of food available to an individual depends directly upon the number of individuals in the population. If there are too many individuals for the food supply to support, some will starve and the population will decrease until an equilibrium is reached. If the supply increases, the population will usually increase proportionately.

Food Supply in Parasite Control

Regulation of the food supply is one of the methods man has used in an effort to control the size of animal populations. The tsetse fly, which serves as the intermediate host for the protozoan that causes African sleeping sickness, feeds chiefly on the blood of ungulates—large, hoofed, herbivorous animals. It feeds frequently, especially during hot weather, but because it does not stay on or near its host after taking a meal, each meal must be preceded by an independent search for a host. Following an epidemic of sleeping sickness in northwest Tanganyika (now part of Tanzania), wildlife experts suggested that the flies could be controlled if the animals on which they feed were reduced in number. Hunters were hired to shoot large mammals within a 10-square-mile area around the community in which the disease had been most severe. About 9000 animals were sacrificed, and the tsetse fly population was at least temporarily reduced. There was still plenty of food since one antelope can support large numbers of tsetse flies. However, the chance that a fly could find food every time it was needed was sufficiently reduced, according to reports, to exterminate the fly population and eradicate the disease from this area.

The reports of the success of tsetse fly control in Tanganyika may have been premature. An experience in Rhodesia indicates that sleeping sickness cannot be controlled by massive slaughter of ungulates. The theory behind the Rhodesian killing of an estimated one-half million antelope, gazelles, and zebras involved the source of the parasite carried by the tsetse fly. The assumption that the large herbivores were the only animals from which the tsetse fly receives trypanosomes (protozoans that are the cause of African sleeping sickness) was incorrect. Small mammals and birds also act as hosts. The killing of large herbivores in Rhodesia had little effect on the tsetse fly population and none on the infection of flies with trypanosomes.

Food supply represents only one of the environmental resources that are in limited supply. Other limited resources include *water, space,* and *light.* As population density increases, competition for these resources becomes more intense and the deleterious effects of competition become more and more effective in limiting population size. For example, if flowers are planted too close together in a flower bed, the plants will be weak and spindly and will produce almost no flowers. They will grow and do well only if they are thinned, either artificially or by the natural death of weaker individuals. The same kind of competition for space, light, water, and nutrients operates in a forest to control the number of trees that survive and continue to grow.

Parasitism and Predation as Limiting Factors

Predation, including *parasitism,* is a force that often operates to maintain a population size within limits that will insure survival. This is an association in which one organism (*prey* or host) is eaten or destroyed by another (*predator*). As the population of the host or prey increases, a higher percentage is usually victimized by the predator. Individuals that have been forced into less favorable situations, or have become weaker because of a greater drain on available resources, are easier to locate and attack. Furthermore, there is evidence that as the population density of prey is reduced, predators that attack a variety of prey species tend to alter their hunting patterns and concentrate on the most common species.

predators

As the density of a prey species increases, the density of the predators feeding on it usually increases. This increase in predators, together with their increased concentration on the particular prey species, may be one factor that causes the density of the prey to fall again. As the density of the prey decreases, there is usually a corresponding, but slightly later, fall in the density of the predator. The result is often a series of density fluctuations similar to those shown in Fig. 11-6 in which the fluctuations of the predator

Fig. 11-6. Fluctuations in the abundance of lynx and snowshoe hares over a period of 90 years. Index of abundance is based on the number of pelts received by the Hudson's Bay Company. (Redrawn from D. A. MacLulich, *University of Toronto Studies, Biology Series,* No. 43, 1938. After Keeton.)

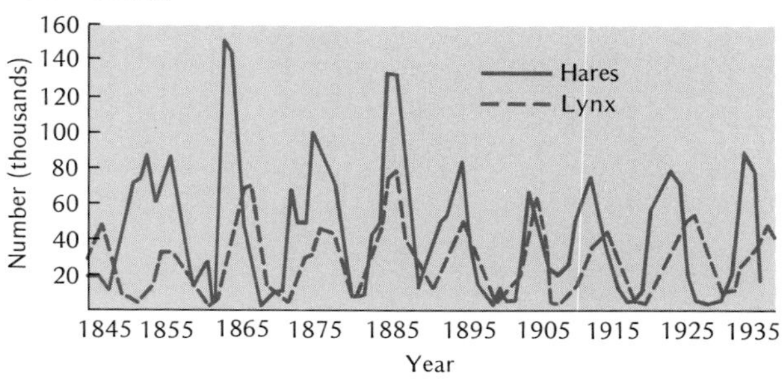

Fig. 11-7. Lynx. *(Courtesy Michigan Department of Natural Resources.)*

lynx

(the lynx, Fig. 11-7) closely follow those of the prey (the hare, Fig. 11-8), but with a characteristic time lag. Such linked fluctuations of predator and prey seem to indicate the major limiting factor for the predator to be the availability of food and that predation is probably one important limiting factor for the prey. The length of the time lag between the change in one population and the response in the other is one factor in determining the extent of density fluctuations.

In stable predator–prey associations, predation is often beneficial to the prey population even though it is destructive to individuals. This cardinal fact of ecology is frequently overlooked by well-meaning but misguided individuals who consider parasites and predators evil (forgetting or ignoring the fact that man himself is the most destructive predator alive). Wishing to

Fig. 11-8. Snowshoe hare. *(Courtesy Michigan Department of Natural Resources.)*

Fig. 11-9. Mule deer. *(Courtesy U.S. Department of Agriculture, Office of Information.)*

protect the prey from its enemies, they kill the predator. The results are often very different from those expected. A classic example involves the mule deer population of the northern rim of the Arizona Grand Canyon (Kaibab Plateau) (Fig. 11-9). The northern rim is a large plateau of approximately 10,000 square miles. This plateau is bounded on all sides by deep canyons which effectively isolate the animal population. In 1906 when the Grand Canyon was declared a national park, special attention was given to preservation of the existing deer herd (a population of about 4000). Professional hunters were brought in to trap and kill the natural predators. By 1931 over 9000 cougars, coyotes, bobcats, and wolves had been removed. The virtual absence of these natural predators, along with habitat alteration as a result of fire and grazing, allowed the deer to flourish; the size of the herd was reported to be increasing by about 20 percent each year. Although the Kaibab Plateau was rich in meadows, aspen groves, and ponderosa pine forests, it became apparent by 1920 that the area was overgrazed. The peak population was variously estimated at 100,000, 70,000, 60,000, 50,000, and 30,000 for the year 1924. On the basis of a 1926 estimate of 40,000 two severe winters are assumed to have reduced the deer herd by 60 percent if we accept the 100,000 figure. Other estimates of a drop from 50,000 to 20,000 give the same percentage loss. Forest supervisors thought that there was essentially no change in the size of the herd from 1923 to 1929, however. Apparently, there was a decline after 1929. These conflicting data are mentioned because of the tendency of so many authors to accept a maximum size estimate as correct.

In other areas, such as Isle Royale in Lake Superior, the effects of predators have been studied more accurately. The moose herd on the island numbers about 600. Prior to the introduction of wolves, overbrowsing threatened the herd and the vegetation. Wolves were imported and brought the herd into equilibrium with the available food supply.

Predation alone is sometimes sufficient to exclude an organism from an otherwise favorable environment. For many years native silkworms have been raised on wild cherry trees, each tree being enclosed in a protective net. Survival from egg to adult is as high as 80 percent under these conditions. Each year some trees are left unprotected; not one silkworm has survived through the larval stage on these unprotected trees. They are eaten by birds. The moth exists at a very low density in a nearby woods, but among cherry trees and open fields it cannot survive.

Hormonal Mechanisms as Limiting Factors

hormone action

Within recent years ecologists have come to recognize that within broad limits certain density-dependent mechanisms have evolved within animals themselves, which regulate population growth and curtail it short of suicidal destruction of the environment. In laboratory experiments with mice, increasing population density is accompanied by hypertrophy of the adrenal cortex and by degeneration of the thymus. Somatic growth is suppressed, sexual maturity is delayed (or even totally inhibited at very high densities), and reproduction by mature mice is diminished. This effect on reproduction includes delayed spermatogenesis in males and a prolonged estrus cycle in females, accompanied by reduced rates of implantation and inadequate lactation. There is also evidence of increased intrauterine mortality of embryos. Furthermore, it has been observed that the crowding of female mice before pregnancy sometimes results in permanent behavioral disturbances in the young they produce. This evidence points to an *endocrine feedback mechanism* which helps to regulate and limit population size by altering the reproduction rate. Presumably, as the density increases and as aggressive behavior increases, endocrine disturbances result and the reproduction rate decreases. Conversely, as density and aggressive behavior decrease, the reproduction rate increases.

Very dense populations often experience severe disease epidemics. Evidence now demonstrates that this proneness to epidemics is not due solely to easier spreading of the pathogen; a density-induced change in host resistance is apparently also involved. Numerous experiments have shown that increased population density is accompanied by a marked depression of the immune mechanism (inflammatory responses and antibody formation), and a resultant increase in susceptibility to infectious diseases or parasites. There is also an increased susceptibility to various toxic substances.

The theory that endocrine changes of the kind mentioned above act as major density-dependent limiting factors in nature has been severely criticized by some observers on the grounds that it is largely based on laboratory work with densities much higher than those that actually occur in nature.

Nevertheless, endocrine changes have been found in many wild animal populations under stress conditions. One of the best known examples of physiological and behavioral changes induced by crowding in natural populations is that of the solitary and migratory phases observed in several species of locust, particularly *Locusta migratoria* in Eurasia. Individuals in the migratory phase have longer wings, a higher fat content, a lower water content, and a darker color than solitary-phase individuals; they are more gregarious and more readily stimulated to march and fly by the presence of other individuals. The solitary phase is characteristic of low-density populations; the migratory phase is characteristic of high-density populations. As the density of a given population increases, the proportion of individuals entering the migratory phase also increases. The sight and smell of other locusts seem to play a significant role in triggering the development of migratory-phase characteristics. When the proportion of these individuals increases sufficiently, large swarms emigrate from the crowded area. They often consume all the vegetation in their path, completely devastating agricultural crops.

locust

Another example of a natural population that shows behavioral changes due to increased densities is the lemming (a mouselike rodent from the Scandinavian countries). Lemmings (Fig. 11-10) experience a peculiar cyclic increase and decrease in population size. In most lemming groups density fluctuations occur every 3 to 4 years. An adequate food supply and increased litter size apparently account for increased population densities. During the peak of the population cycle, large numbers of these animals leave their home range in mass migration. For many years these suicidal marches (so called because many drown in streams and in the sea) were be-

lemming

Fig. 11-10. Lemmings starting into the water during the mass migration in the autumn of 1963. The beaches became lined with the floating bodies of drowned lemmings. *(Reproduced with permission of Göran Hansson, Naturfotograferna, Österbybruk, Sweden.)*

lieved to be simply a quest for food, but most population biologists now believe that there are other causes of these migrations. As the population density and social stresses increase, the fighting and avoidance reactions that result seem to trigger the migrations. The migration reaction appears to be an innate mechanism which prevents the lemmings from eating all the available food in an area, thus destroying the entire species through starvation.

Behavior Patterns as Regulatory Mechanisms

behavior

A variety of defense mechanisms and behavioral patterns has evolved among various animal groups, which help to insure the survival of a population and which exert a regulating influence on population size. These include concealment and camouflage, mimicry, courtship behavior, parental care of young, societies, methods of communication, territoriality, and social hierarchies. The latter two are considered as examples in the discussion that follows.

territory

Populations of most wild animals show distinctive behavioral patterns which relate directly to the ownership of particular areas. This *territoriality* has been noted among both vertebrate and invertebrate animals. It is now well established that the claiming and defense of territory plays an important role in animal populations. The territoriality exhibited by wolves, rabbits, mice, sea lions, birds, and other animals demonstrates that some wild populations can and do regulate themselves in this manner.

rabbits

Territories are staked out by animals on both permanent and seasonal bases. Rabbits, for example, construct permanent burrows with established trails and warning signs surrounding the areas. Many of the signs utilized to mark and protect their territories are now known. Wild Australian rabbits possess two glands which produce marking secretions, an anal gland and a chin gland. The anal gland coats fecal pellets with a secretion easily recognized by other rabbits. When a burrow is constructed, the male rabbits excrete pellets in mounds and urinate around the circumference, thereby warning any intruder that the area is occupied. Chin gland secretions are utilized in marking twigs and grass to achieve the same result.

Songbirds establish territories early in the spring for breeding and nesting. The male flies from tree to tree within his area, singing, repelling intruding males, and attempting to attract a mate. The bird that sings the loudest and most often commonly occupies the largest territory. In general, an intruder respects the line drawn by the occupant. If the intruder is larger and stronger than the defender, the defender usually retreats and gives up part of his territory. It is assumed that defense of a territory by an individual, or by a group of animals, has the effect of limiting the population density of an area and conserving the available food and shelter. If an invader crosses into a territory that is too large for an individual or group to defend, he may remain. If, however, the population density is high in a territory, the invader will be forced to try another area. Territory size varies from species to species. It is usually large among animals that are only mildly social and small among animals that are highly social.

Even though most animals frequently travel beyond the borders of their defended territory, particularly in search of food, they tend to remain within a fairly limited area for long periods of time, sometimes for their entire lives. Such an area is called a *home range*. For example, the home range of the white-tailed deer is about $1/2$ to 2 square miles in size; that of the cotton-tailed rabbit averages about 14 acres; and that of the meadow mouse averages about $1/15$ acre.

peck order

Animal groups may be organized in a variety of ways. They are often led or directed by an older male or female. A type of social dominance which has been studied in detail is the *peck order* among chickens. A peck order is established whenever a flock of hens is kept together over a period of time. In any one flock one hen usually dominates all the others. She can peck any other hen without being pecked in return. A second hen can peck all hens but the first one, a third all hens but the first two, and so on through the flock, down to the unfortunate one that is pecked by all and can peck none in return. The hens that rank high in the peck order have special privileges at the food trough, the roost, and the nest boxes. As a consequence, they usually exhibit confidence and a sleek appearance. Low-ranking hens tend to look ruffled and unpreened and to hover timidly on the fringes of the group. A hen moved from flock to flock may have different ranks in different flocks, but in a stable flock the same rank is usually retained over a long period of time. Any new arrival usually ranks low and must peck her way up. During the time a peck order is being established, frequent fighting ensues, but once rank is established in a group, a mere raising or lowering of the head is sufficient to acknowledge dominance or submission of one hen in relation to another. Life then proceeds in harmony. If a flock is disrupted (by the addition of new members), the entire peck order must be reestablished.

Cocks do not normally peck hens, but they have their own peck order. A breeding flock therefore has two hierarchies, one for each sex. The dominant cock is the most successful sexually; the one lowest in the peck order is the least successful and may be an ineffective breeder. Chickens do not form permanent mating pairs. Among animals that do form permanent mating relationships and which also have a peck order, the female gives up her previous rank among those of her sex and assumes the dominance order of her mate.

Dominance hierarchies have been observed in other classes of vertebrates, but they have not been studied as extensively as in birds. Regardless of the group, however, a social hierarchy once established tends to give order and stability to a group. There is less tension, and there is less fighting and disturbance.

leadership

Leadership behavior, which represents a type of social dominance, appears to serve a more useful function than the peck order described above. Members of most social groups tend to follow a certain member within their group. The leader is usually an older female (as among sheep,

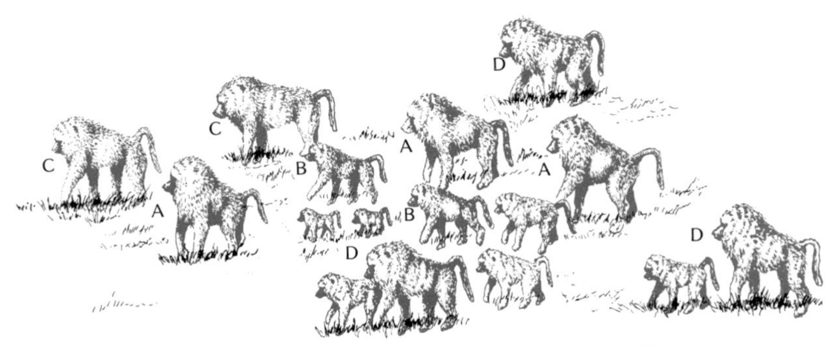

Fig. 11-11. Social organization of an African chacma baboon troop illustrating the arrangement of dominance hierarchies. Dominant males (A) protect females with young (B). Younger males (C) lead the troop; higher-dominance males escort females in estrus or form the vanguard (D).

lions, elephants, and red deer) who can stabilize the group and hold it together while on the move. She has usually earned the right to leadership by producing large numbers of offspring which acquired the habit of following her when they were young. Males, except during the breeding season, are usually away from the group, whereas females are always together. In some species, such as the gorilla, the males become leaders. Figure 11-11 illustrates social rank in a colony of baboons.

Social hierarchies and territories limit reproduction and also insure that individuals eliminated by food shortage or predators are more likely to be those already determined to be surplus and usually inferior, reproductively speaking. Both male and female rats of a lower dominance order are socially ostracized in crowded populations, while those of higher rank may continue to breed freely. The inferior rats show an absence of normal sexual interest. If they are removed from the restrictive environment, they will mate promptly and reproduce normally.

The breeding grounds of most animals usually afford better protection than surrounding areas. The muskrat, in its burrow, for example, is relatively invulnerable. Studies of muskrat carcasses found in the stomachs of minks, the chief predators of the muskrat, showed that 60 to 75 percent of those that had been eaten were diseased or otherwise disabled. Animals lower in the dominance order starve while those of higher rank eat normally.

Most organisms, if left undisturbed in their natural habitats, survive as a species and maintain a relatively stable population size. Various limiting factors, both density-dependent and density-independent, interact to balance the birthrate against the death rate. If this balance is upset, other factors, such as emigration or immigration, will begin to exert additional controlling influences upon population size.

Man is no longer part of a natural ecosystem. He has created an artificial ecosystem for himself in which he is the dominant species. His population size is no longer controlled by the kinds of factors that control populations in natural ecosystems. If he is to stabilize his population size and insure his survival as a species, he must devise some artificial means whereby he can balance birthrate against death rate to maintain a population size that his world is capable of supporting. Chapter 15 discusses some of the possibilities for man in terms of his survival in a world that can meet his needs as a human being.

Suggested Readings

Boughey, A. S. *Fundamental Ecology.* New York: Intext Educational Publishers, 1971.

Ehrenfeld, D. W. *Biological Conservation.* New York: Holt, 1970.

Errington, P. L. "The Phenomenon of Predation," *American Scientist,* 51(2): 180–192, 1963.

Guhl, A. M. "The Social Order of Chickens," *Scientific American,* 194(2):42–46, 1956.

Hairston, N. G., F. E. Smith, and L. B. Slobdokin. "Community Structure, Population Control, and Competition," *American Naturalist,* 94:421–425, 1960.

Hazen, W. E., Ed. *Readings in Population and Community Ecology.* Philadelphia: Saunders, 1964.

Kormondy, E. J., Ed. *Readings in Ecology.* Englewood Cliffs, N. J.: Prentice-Hall, 1965.

Slobodkin, L. B. *Growth and Regulation of Animal Populations.* New York: Holt, 1961.

Williams, C. M. "Third Generation Pesticides," *Scientific American,* 217(1):13–17, 1967.

Wynne-Edwards, V. C. "Population Control in Animals," *Scientific American,* 211(2):68–74, 1964.

The continuing expansion of housing projects symbolizes man's population problem. *(Courtesy U.S. Department of the Interior, Bureau of Reclamation.)*

12

The Human Population

Essentially all animals, including man, are obsessed with a breeding urge. They are provided with the biological capacity to overproduce, thereby insuring survival of their species. Since man is now able to influence many of the natural factors that once controlled his population size, he must learn to control his innate desire to reproduce excessively. All available evidence indicates that man has not yet learned this lesson.

THE CURRENT STATUS

The population of the world is now growing at an unparalleled rate of 2 percent per annum. Translated into a head count, this means that 132 persons are added per minute to the present world population. As time passes, this figure will increase.

It took man over 1 million years, from his emergence from a primate stock to 1830, to reach a population of one billion individuals. In 1930, 100 years later, the number had increased to two billion, and by 1960, only 30 years later, a third billion had been added. The Population Reference Bureau established the world population as of January 1, 1968, at 3.44 billion individuals. The 4.5-billion mark should be reached by 1976 if the present rate of increase continues.

The current increase of 2 percent per year may not appear to be an unusual rate of growth; however, a few calculations can demonstrate what this means. If the 2-percent growth rate had existed from the time of Christ until the present time, the increase would have been about 70 million billion. There would be over 20 million individuals for each person now living, or 100 persons for each square foot of surface area. At our present rate of 2 percent per year there would be over 150 billion people within two centuries.

Man is now in the logarithmic phase of a typical growth curve following a long lag period (Fig. 12-1). In nature no population has ever maintained such a logarithmic growth phase for a long period of time. The major factors that normally slow this rate of growth include exhaustion of food supply, accumulation of toxic products, decimation through disease, and the effects of some outside lethal agent which kills a high proportion of the population. Any one or all of these factors eventually forces a population into a no-growth phase or even a decline. Man must decide whether the future growth of human populations will be governed by such factors, or whether he will use his intelligence to bring about a stable population size which can be comfortably supported with the natural resources available. If population growth is not controlled, hundreds of millions of people may eventually die prematurely each year.

It is sometimes maintained that man is too recent in his origin and evolution to have evolved any mechanisms for regulating his population size. As a species, however, man is at least as old as the Pleistocene epoch and is therefore contemporary with the majority of animals and many of the plant species of the modern world, all of which, when investigated, are found to have regulatory mechanisms. Moreover, there is some evidence that man at one time possessed a regulatory mechanism in the form of the density-dependent carrying capacity of his environment. Starvation was, and under some circumstances still is, a density-dependent factor; so is disease. What seems to be missing is a regulatory mechanism to impose an increased environmental resistance before the food resource

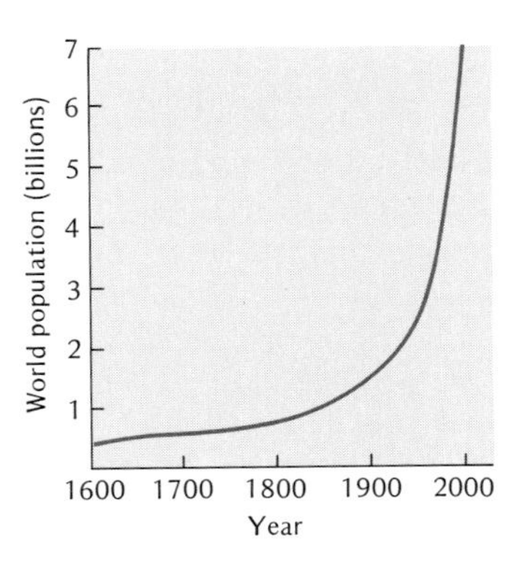

Fig. 12-1. Growth of the world's population, 1600–2000, based on publications of the United Nations. Data prior to 1800 are estimates. Those subsequent to 1965 are projections based on current trends.

carrying capacity is reached. If we can educate our society to the problems inherent in the absence of such a regulatory mechanism, we may very well find ourselves willing enough to take whatever steps appear necessary to offset the missing mechanism. This chapter represents an attempt to provide some information which will be helpful in arriving at this conclusion.

HUMAN POPULATION DYNAMICS

There are two ways in which a population can acquire additional members. One is by immigration, the movement of individuals into a population from other populations; the other is by increasing the birthrate. Likewise, there are two ways in which a population can lose members. One is by emigration; the other is by an increase in the death rate. This section describes the interactions between birthrates and death rates as factors that determine the growth characteristics of human populations.

Birthrates, Death Rates, and Population Increase

Birthrate is usually expressed as the total number of births per year divided by the total population at the midpoint of that year (July 1), multiplied by 1000. This calculation gives a figure that represents the average number of births per 1000 individuals within a population.

TABLE 12-1
Contemporary "Crude" Birthrate Statistics from a Selected Series of Countries[a, b]

Dahomey	54	Chile	35
Niger	52	Uruguay	24
Cameroon	50	Japan	19
Nigeria	50	United States	17.6
Guinea	49	Italy	17.6
Ghana	47	United Kingdom	17.1
Madagascar	46	Belgium	14.8
Costa Rica	45	East Germany	14.3
Mexico	44	Sweden	14.3
Uganda	43	Luxembourg	14.2

[a] Data from the 1970 world population data sheet, reproduced with the permission of the Population Reference Bureau.
[b] Figures indicate the number of live births per 1000 individuals.

Birthrates are variable from one country to another. Table 12-1 gives the actual birthrates for some contemporary populations. Dahomey has the highest figure, 54, followed closely by other West African nations such

birthrates

as Niger with 52. The lowest birthrates recorded for countries of any size are those for Sweden and East Germany with 14.3 each, only slightly lower than those of the United Kingdom and the United States, 17.1 and 17.6, respectively. These statistics suggest that birthrates seem to fall either in the 40s or 50s, or between 14 and the low 20s. This pattern is attributable to the fact that some nations have adopted population control methods on an extensive scale to reduce their birthrates; others have not yet done so. In the latter cases birthrates remain in the 40s and 50s. Demographers (individuals who study human population dynamics) refer to this cultural adjustment by a society from a high birth–high death rate to a lower birth–lower death rate as a *demographic transition*.

death rate

Death rate or mortality is calculated in a manner similar to birthrate. The total number of deaths occurring within a population during a year is divided by the total population at the midpoint of that year and multiplied by 1000. The result is expressed as deaths per 1000 individuals within the population. Table 12-2 lists mortality statistics for selected countries. West Africa heads the list with the highest mortality rate, Guinea and Cameroon having death rates of 26 per 1000. Hong Kong at 5 and Singapore at 6 have the lowest death rates recorded. Hong Kong and Singapore have lower death rates than more advanced industrial societies, such as the United States, the United Kingdom, and Belgium (which have death rates of 9.6, 11.9, and 12.8, respectively) because of differences in population structure, a concept that is examined in a later section of this chapter.

TABLE 12-2
Contemporary "Crude" Mortality Statistics from a Selected Series of Countries[a, b]

Guinea	26	Luxembourg	12.3
Cameroon	26	United Kingdom	11.9
Dahomey	26	Chile	11
Nigeria	25	United States	9.6
Niger	25	Netherlands	8.2
Uganda	18	Sri Lanka[c]	8
Pakistan	18	Costa Rica	8
India	17	Japan	7
East Germany	14.3	Singapore	6
Belgium	12.8	Hong Kong	5

[a] Data from the 1970 world population data sheet, reproduced with permission of the Population Reference Bureau.
[b] Figures indicate the number of deaths per 1000 individuals.
[c] Formerly Ceylon.

The remarkable reduction in death rates in such countries as India, Pakistan, Chile, and Sri Lanka, which is apparent in Table 12-2, occurred as recently as the 1950s. This was almost 100 years later than the parallel reduction that occurred in the industrialized societies. This reduction is apparently due to the large-scale introduction of two new medical procedures which have been highly effective in preventing widescale epidemics of previously fatal diseases. One has been the use of DDT and

DDT

other pesticides to control insect vectors of diseases such as malaria and typhus. The extensive use of these pesticides has unfortunately had effects other than those intended, as discussed in other chapters of this book. The discovery of drugs with antibiotic effects, first the sulfa series followed quickly by penicillin and others, coupled with the introduction of prophylactic serums, has provided protection against the fatal effects of such diseases as cholera, dysentery, polio, pneumonia, and yellow fever. This was the second major medical innovation that drastically reduced mortality in underdeveloped nations following World War II.

net gain

The *rate of natural increase* is the statistic generally used to express population growth. This figure is obtained by subtracting the mortality rate from the birthrate. For example, the mortality rate for Guinea (26) subtracted from Guinea's birthrate (49) gives a figure of 23. Instead of continuing to express this as an increase of 23 individuals per 1000 persons in the population, it is usually expressed as a percentage; the rate of natural increase for Guinea therefore becomes 2.3 percent. Table 12-3 lists the rates of natural increase for a selected list of countries. East Germany, which has a low birthrate and a fairly low death rate, has a natural increase of only 0.3 percent. This is a close approximation to a stable population; one that is neither increasing nor decreasing in size.

TABLE 12-3
Rate of Natural Increase Statistics from a Selected Series of Countries[a, b]

Country	Rate	Country	Rate
Costa Rica	3.8	Sri Lanka	2.4
Mexico	3.4	Guinea	2.3
Pakistan	3.3	Cameroon	2.2
Niger	2.9	Luxembourg	1.2
Madagascar	2.7	Uruguay	1.2
India	2.6	Japan	1.1
Nigeria	2.6	United States	1.0
Dahomey	2.6	United Kingdom	0.5
Hong Kong	2.5	Belgium	0.4
Singapore	2.4	East Germany	0.3

[a] Data from the 1970 world population data sheet, reproduced with permission of the Population Reference Bureau.
[b] These do not always coincide with the figures calculated from crude birth and death rates. The Bureau has sometimes adjusted them to allow for migratory movements, or what it believes to be a misleading figure for the birth or death rate.

West African countries such as Dahomey and Niger, which have the highest birthrates, do not have the highest rates of natural increase because their mortality rates are also high. Some Latin American countries that have high birthrates also have high rates of natural increase because of relatively low death rates. Costa Rica leads with a rate of 3.8 percent. One statistic, that of Kuwait, is not included in Table 12-3 because it has a very small population and because migratory movements cause it to be very misleading. Kuwait has actually been recorded, however, as having a birthrate of 47, a death rate of 6, and a natural increase of 8.3 percent, over twice that of Costa Rica.

In order to provide more readily comprehensible statistics for population increase, demographers sometimes convert the percentage rate of population increase to *population doubling time*. This is the length of time it would take for the size of the population to double if the existing rate of natural increase continued unchanged. The predictable doubling times for a selection of countries based on current rates of natural increase are provided in Table 12-4. East Germany, as already noted, has almost stabilized its population size. It will take 233 years for East Germany to double its population at its present rate of natural increase. Costa Rica, which has the highest rate of natural increase, will require only 19 years to double its population size. This is almost quadrupling the size of the population in every human generation. In the exceptional case of Kuwait, a doubling time of 9 years is indicated, allowing for an eightfold increase during every generation.

doubling time

TABLE 12-4
Population Doubling Time for a Selected Series of Countries[a,b]

Costa Rica	19	Sri Lanka	29
Mexico	21	Guinea	31
Pakistan	21	Cameroon	32
Niger	24	Luxembourg	58
Madagascar	26	Uruguay	58
India	27	Japan	63
Nigeria	27	United States	70
Dahomey	27	United Kingdom	140
Hong Kong	28	Belgium	175
Singapore	29	East Germany	233

[a] Data from the 1970 world population data sheet, reproduced with the permission of the Population Reference Bureau.
[b] Values are given in years.

Population Structure and Survivorship

Age distribution is an important population characteristic which influences both the birthrate and death rate. The ratio of the various age groups in a population determines the current reproductive status of the population and indicates what can be expected in the future. Usually a rapidly expanding population contains a large proportion of young individuals; a stationary population, a more even distribution of age classes; and a declining population, a large proportion of older individuals.

A convenient way to picture age distribution in a population is to arrange the data in the form of an age pyramid, the number of individuals or the percentage in the different age classes being shown by the relative widths of successive horizontal bars.

age pyramid

Figure 12-2 represents the structure of a young, growing population. In this theoretical example it is assumed that half the children produced are male and half female and that all survive to an age of between 60

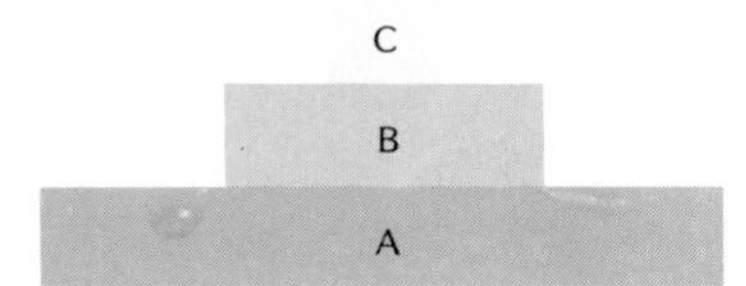

Fig. 12-2. Population structure of a theoretically unstable population. The population is divided into three categories: (A) infants and juveniles, (B) reproductive adults, (C) postreproductive adults. The numbers in the first class far exceed those in the second; unless a very high mortality rate occurs there, the population must be unbalanced and experiencing a high rate of natural increase. (Reprinted with permission of The Macmillan Company from *Man and the Environment* by A. S. Boughey. Copyright 1971 by Arthur S. Boughey.)

and 90 years. With these assumptions it becomes possible to construct the diagram in such a way as to indicate the number of males and females in the three categories: children (0 to 13 years), adults (14 to 44 years), older people (45 to 90 years). If one-third of the older people in this group were to live to age 90 to 120, and as a consequence there was a reduction of one-third in the birthrate for a generation, the pyramid would have the form of a declining population as shown in Fig. 12-3.

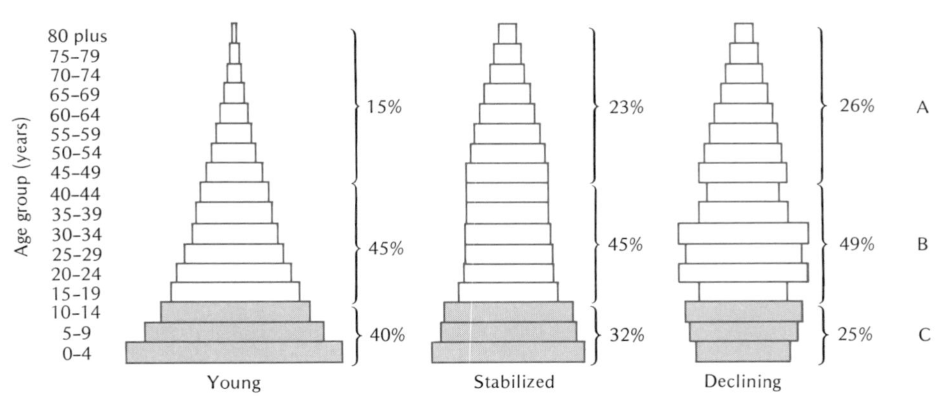

Fig. 12-3. Comparison of young, stabilized, and declining human populations. Although the size of the reproductive segment (B) is not widely contrasted in these three conditions, those of the prereproductive (C) and postreproductive (A) segments are. Any factor that markedly reduces the mortality rate in the prereproductive element of a young population therefore produces a population explosion a few years later unless the fertility rate is reduced. (After A. S. Boughey, *Fundamental Ecology*, New York: Intext Educational Publishers, 1971.)

Whether population growth is recorded as the rate of natural increase per 1000, or whether it is expressed as a percentage rate of natural increase, it is a measure of the difference between birthrate and death rate within a population. The addition of members to a population by birth is a relatively simple factor to express. Mortality, however, may be con-

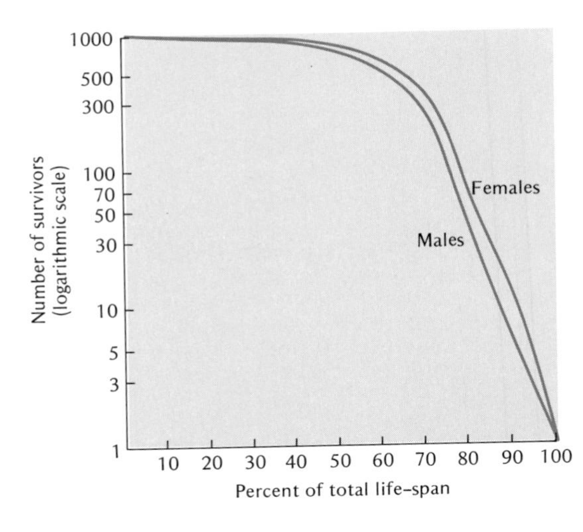

Fig. 12-4. Survivorship curve for a human population in an industrialized country. Lowering infant and juvenile mortalities produces a pattern which with improving health services conforms increasingly to that shown above. Mortality is light until individuals approach the average life-span of approximately 70 years. Female mortality is slightly lower than male at most ages. (After A. S. Boughey, *Fundamental Ecology,* New York: Intext Educational Publishers, 1971.)

ceived of in several different ways. Because ecologists are interested primarily in the organisms that survive rather than those that die, mortality is commonly expressed in the form of a survivorship curve (Fig. 12-4).

survivorship curve

The survivorship curve for the human population in an industrialized society approaches a right-angle. This is because mortality losses among infants and juveniles have been significantly reduced in such societies, as have losses among individuals in their fifties and early sixties. In spite of this decrease in the death rate for individuals in their fifties and sixties, the actual life-span for man has not been increased substantially, a factor that tends to concentrate the ages at which people die in their late sixties, seventies, and early eighties.

Within a given type of survivorship curve, there are differences arising from several sources. One of the most consistent of these, in heterosexual species, is the difference in the shape of the survivorship curve for males and females of the same species. This sex difference in human populations is illustrated in Fig. 12-4. Other differences in the shape of survivorship curves may result from variations among different races of the same species. The shape of the survivorship curve is also partially determined by the initial density of the population, which if higher than optimal may reduce individual chances of survival.

In a young population mortality rates are much lower than in a declining population, for there is a higher proportion of younger people whose mortality rate is lower. This is the reason why populations such as that of Hong Kong, as illustrated in Table 12-2, attain a mortality rate of only 5. It also explains why it is not sufficient in such populations, if the objective is the attainment of zero population growth, to restrict the number of offspring of married couples to two. Until the population age structure assumes a stable form, zero population growth may be attained only by keeping the number of offspring lower than two, as is illustrated in Table 12-5.

TABLE 12-5
Fecundity and Stable Populations[a, b]

When mortality rate reaches this figure	*Birthrate to maintain zero population growth must be*	*Fecundity (number of offspring per female) to provide this birthrate is*
45	45	6.7
35	35	5.2
25	25	3.7
15	15	2.2
10	10	1.5
5	5	0.7

[a] After A. S. Boughey, *Fundamental Ecology,* New York: Intext Educational Publishers, 1971.
[b] By making several assumptions regarding the structure of a human population, it is possible to derive this table. It shows the fecundity rate that will produce zero population growth for given levels of mortality and birthrates. This table illustrates the point that for societies that have achieved a mortality rate lower than 15 the fecundity rate must be less than 2.0. In other words, a stable population in existing industrialized countries, and in areas such as Hong Kong and Singapore with a high percentage of young people, can be attained only by keeping the average number of children per female to *less than two.* Indeed, in the last two instances it has to be *less than one.* In the light of such figures, proposals such as that for an "international childless year" appear less unrealistic.

TABLE 12-6
Fecundity Rates—Variations in the Number of Live Births per Female in Various Populations[a, b]

Population	*Live births per female (to nearest integer)*
Eskimos (hunters)	3
North American Indians (hunters)	3–4
Australian aborigines (hunter-gatherers)	5
Central Africa (agriculturalists)	ca. 6
India (agriculturalists)	6–8
United States (midtwentieth century industrialists)	3
United States (early nineteenth century preindustrial stage)	7

[a] After A. S. Boughey, *Fundamental Ecology,* New York: Intext Educational Publishers, 1971.
[b] These fecundity figures vary considerably according to whether the population has undergone a demographic transition. Underdeveloped countries have invariably high fecundity rates but not necessarily high rates of natural increase. Whether they do so or not depends on their mortality rate.

age

Age structure also affects the birthrate in a similar manner, for the number of births per 1000 in a population will be much higher if the population contains a larger than usual number of females or if it contains a large proportion of young people. For this reason demographers frequently use instead the *fertility rate,* the number of births per 1000 women in the population between the ages of 15 and 44. The fertility rate in the United States is illustrated in Fig. 12-5.

The actual amount of reproduction by individual female members of the population is another statistic that should be examined. In human populations that is commonly expressed as the *fecundity rate,* the average number of live offspring born to an individual married woman (Table 12-6).

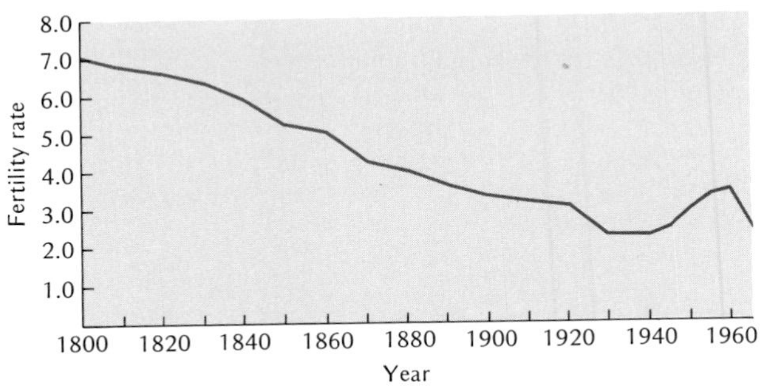

Fig. 12-5. Fertility rate for the United States population—mean rate for whites 1800 to 1970. The rate has steadily declined from the beginning of the nineteenth century to the depression years of the 1930–1940 decade. The subsequent slight upward trend since then was reversed in the last decade. (After A. S. Boughey, *Fundamental Ecology,* New York: Intext Educational Publishers, 1971.)

Whether we are dealing with human populations or with those of other kinds of organisms, and whatever parameters we use to measure such features as birthrate, death rate, and natural increase, in ecosystems that have a steady-state balance, population growth has to approximate that of a stable population. The great majority of the populations encountered in contemporary communities are therefore stable. Considerable attention has been given to determining the factors responsible for the regulation of population numbers in order to achieve this steady state.

Other Factors

The available statistics on birthrates, death rates, rates of natural increases, doubling times, and population structures might appear sufficient to make valid comparisons between different populations, always assuming that immigration and emigration effects can be discounted. This is not the case. Many other kinds of variations exist. Populations vary with respect to marriage age, life expectancy, and differential age group mortalities and birthrates. Any comparisons and predictions based on demographic statistics must make appropriate allowances for these kinds of variations.

The average human female is capable of becoming pregnant and giving birth from age 13 to age 45. As gestation takes 9 months and since conception can occur within 2 or 3 months following parturition, there is a theoretical possibility of an individual female producing 32 offspring. In actuality the maximum number of single live births recorded for a human female is 24. Several factors account not only for the reduction of the theoretical expectation to this value, but also for the usual occurrence of an even lower figure. The nursing of an infant for a period of up to 3 years is associated with failure to resume ovulation following parturition. This factor alone reduces the reproductive potential from 32 offspring to 8 or 9. An average life expectancy of 30 years would halve this value, providing a maximum number of four or five children per female. Child mortality losses of as high as 50 percent are not unusual, even in the contemporary world. This could bring the maximum number of children per female down to two or three. The birthrate would then barely suffice to maintain existing numbers.

All these factors have operated in the past to reduce population growth. The general practice of a prolonged nursing period in many tropical African societies probably represents the persistence of a very ancient practice. Supposedly, it accounts in part for the fact that the average number of live births per female never seems to exceed eight in any contemporary society. Estimates of life-span at various periods of man's history suggest that prior to the twentieth century many women died before they had completed the menopause. This is still the case in some contemporary societies.

infant mortality

Figures for infant mortality (Fig. 12-6) show that some contemporary societies still lose large numbers of children before they reach the age of 12 months in spite of modern hygiene and the new medical techniques and materials available. Two of these three factors affecting the rate of natural increase, namely, a reduction in infant mortality and a longer life expectancy, are determined largely by the extent to which health care is available. Adequate health care is dispensed most readily in an urban

Fig. 12-6. Infant mortality is expressed as the number of deaths of infants under 12 months of age per 1000 live births. These figures from a selection of countries illustrate that, although the various disease control measures introduced following World War II reduced the rate in all countries, the effect was proportionally greatest in those countries with an already low mortality rate. (Reprinted with permission of The Macmillan Company from *Man and the Environment* by A. S. Boughey. Copyright 1971 by Arthur S. Boughey.)

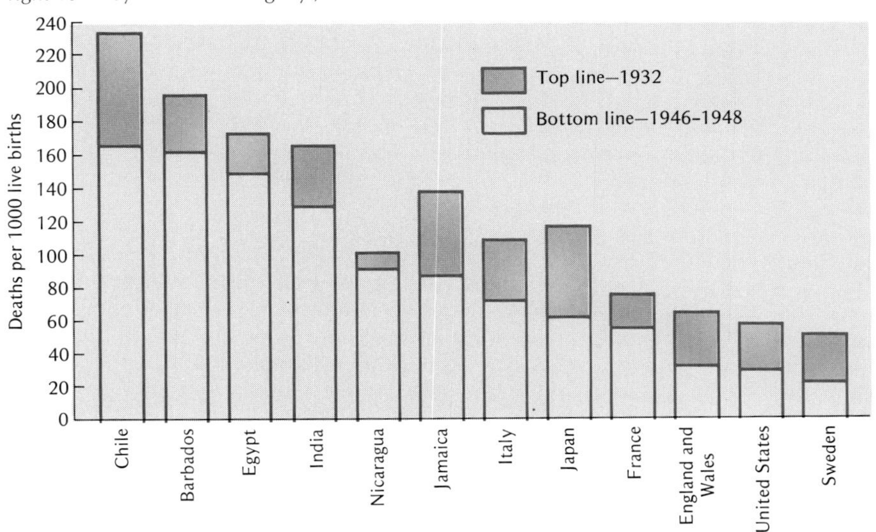

environment. This encourages early weaning since it offers food substitutes for the infant and provides employment opportunities for the mother. It is not surprising therefore that the massive reductions in infant mortality, the lengthening of the life expectancy to cover all the nubile years, and the shortening of the nursing period all coincide with the industrial revolution and the greatly accelerated concentration of populations in urban centers.

Western nations were able to undergo industrialization and show rapid population growth a century ago without producing effects comparable to those that resulted between 1940 and 1960 for two reasons. First, the frontiers of the world were still open and the excess population could be accommodated in sparsely settled lands. Second, the actual proportion of the total human population of the world that these western nations represented was relatively low. The utilization of additional resources a century ago was not so extensive that the increased demands could not be met simply by the exploitation of existing supplies. Populations were increasing by the millions at that time, not, as now, by the billions.

TABLE 12-7
Population Growth in India[a, b]

Census year	Total population (millions)	Birthrate	Death rate	Rate of natural increase (%)
1901	236	46	44	0.2
1911	249	51	43	0.8
1921	248	49	48	0.1
1931	276	46	36	1.0
1941	313	45	31	1.4
1951	357	39	27	1.2
1961	439	40	21	1.9
1965	490	41	17	2.4
1969	537	43	18	2.5

[a] Reprinted with permission of The Macmillan Company from *Man and the Environment* by A. S. Boughey. Copyright 1971 by Arthur S. Boughey.
[b] During this century, the birthrate has remained constant, fluctuating about a mean, while the death rate has fallen almost two-thirds. The population growth has increased over 2000 percent since the beginning of this century.

reduction in death rate

One example from the many statistics available illustrates the dramatic effects of the post-World War II reduction in mortality on the rate of natural increase. Table 12-7 shows some population statistics for India at intervals during this century. The rate of natural increase has continued to rise since the beginning of the century. The recent upward trend in the mortality rate, if real, will reduce this in the future. The actual family size India must attain if it is to achieve zero population growth at present death rates is estimated at about 2.5 children per female. This statistic, known as the

fecundity rate

fecundity rate, is somewhat confusing at first sight since a fecundity rate of 2 normally maintains a zero population growth. It does so, however, only in a population with a mature structure. Because of the dramatic reduction in mortality rates over the past 30 years, the population in many underdeveloped nations contains a disproportionate number of young people. A birthrate based on a figure per 1000 individuals is not offset as much as it should be by a given death rate because, being younger, many do not die when they should according to mortality statistics. Thus India, with a death rate of 17 or 18, should not permit a fecundity rate higher than 2.5. Hong Kong and Singapore, with death rates of 5, will have to attain a fecundity rate of 0.7 to stabilize their populations at zero growth. This is tantamount to having 3 girls in 10 remaining single while the rest marry and have only one child each. No modern society can discourage child bearing to this extent.

REGULATION OF HUMAN POPULATION SIZE

Man has recently turned to a more intensive search for a possible regulatory mechanism in his own species. Two contrary schools of thought exist. One is based on the work of Thomas Malthus (1790), which assumes that human populations are always maintained at a maximum size as determined by such density-dependent limiting factors as war, famine, and pestilence. The other school of thought, founded by A. M. Carr-Saunders in 1926, assumes that there is an optimum human population size for each ecosystem. This optimum size is maintained by various limiting, but equally density-dependent, cultural factors. These factors regulate population size so that it never reaches a maximum at which more extreme limiting factors would intervene.

regulation

Most human societies have escaped from the regulatory mechanisms that might maintain an optimum population density. Many of the environmental crises with which we are now confronted arise from the seeming inevitability that war, pestilence, and famine, or some combination of these, are the only regulatory mechanisms remaining to check continued population growth within some national boundaries. In order to reestablish cultural regulatory mechanisms, nations must go through a demographic transition. As mortality rates are reduced through the introduction of modern medicine and hygiene, birthrates must also be lowered by the adoption of population control techniques.

Man has always been and always will be concerned with preserving his personal freedom. The only way to do this is to control population growth. The means whereby this can be accomplished are already available, but we must first overcome our superstitions, religious objections, political barriers, and ignorance. When all the peoples of the world fully understand the consequences of unrestricted growth, present solutions can be effective and better ones can be developed. If this approach fails, the other alternative is to impose government restrictions that will limit birthrates. Surely people can be motivated by moral, economic, and social means to preserve the freedom that allows them to determine the number of children they produce.

family planning

Family planning is the real hope for successfully controlling population size. This idea is not synonymous with birth control. The purpose of family planning is to plan for the size of a family; help is given to couples who have problems conceiving children as well as those who wish to prevent pregnancy. Many underdeveloped countries have established family planning programs, but the major emphasis has been on birth control. Some religious groups have been unrelenting in their opposition to these programs; however, the National Council of Churches sanctioned the practice of birth control as early as 1961. Attitudes toward birth control are further complicated by some political groups who argue that such practices are applied toward political or even racist ends. As long as birth control information and devices are made available to everyone who wants them without coercion, such charges seem unwarranted.

In order to understand the principles involved in birth control, or contraception, a section describing the human reproductive process follows.

HUMAN REPRODUCTION

gametes

Reproduction in man, as in most animal forms, is accomplished by the union of two gametes—the ovum or egg produced by the female and the sperm produced by the male. Every part of the reproductive systems of both sexes, as well as the various physiological and psychological phenomena associated with sex, has a single purpose—to insure the successful union of the egg and sperm and the subsequent development of the fertilized egg into a new individual. The functioning of the male and female reproductive systems is discussed in the following sections.

The Male Reproductive System

male reproductive system

The *testes* are paired organs which develop within the abdominal cavity of the male. The testes descend shortly before or after birth into the scrotal sac, an outpocketing of the body wall. Each testis consists of several hundred highly coiled *seminiferous tubules.* Between the seminiferous tubules are located masses of interstitial cells which are responsible for producing the male sex hormone, *testosterone.* It is the cells within the *germinal epithelium* that have the potential to undergo meiosis and to produce sperm. While some of the cells undergo meiosis, others undergo regular mitotic divisions in order to replenish the cell supply within the germinal epithelium. The seminiferous tubules also contain nurse cells which nourish the developing sperm (Fig. 12-7).

At the end of each seminiferous tubule is a very small tubule called the *vas efferens* which empties into the *epididymis,* a single highly coiled

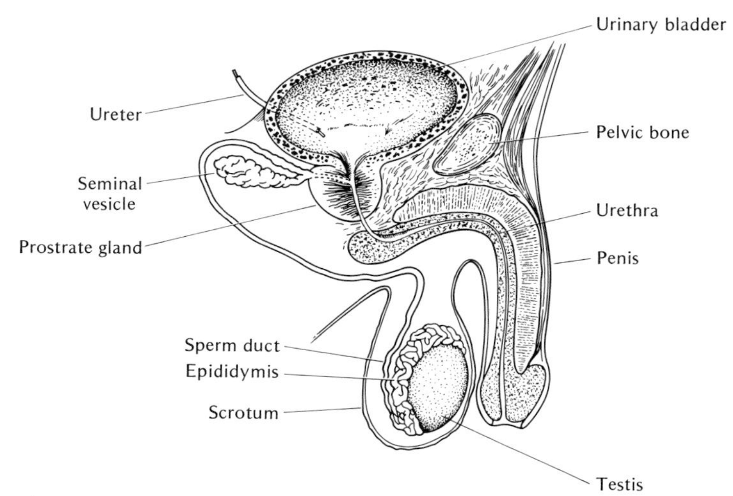

Fig. 12-7. Section through the pelvic region showing genital organs of the human male. Note the relationship to excretory organs.

tubule in which the sperm are stored temporarily. The epididymis is continuous into the *sperm duct,* or *vas deferens,* which passes into the abdominal cavity and eventually joins the urethra.

Spermatozoa (sperm) are suspended in a fluid called *semen.* This fluid is produced by three pairs of glands—the *seminal vesicles,* the *prostate,* and *Cowper's glands.* The semen contains buffering agents which protect the sperm against the acids normally present in the urethra and in the female reproductive tract. Glucose and fructose are present in small amounts to support the metabolism of the sperm, and substances for lubricating the passages through which the sperm must travel are also present.

The *urethra,* which serves as a passageway both for urine from the urinary bladder and for semen, leads to the outside of the male body through the *penis.* Semen is usually ejaculated only after the penis has been stimulated during sexual excitement. Ejaculation is the result of reflex waves of muscle contraction in the walls of the sperm duct, the urethra, and the associated organs.

The Female Reproductive System

female reproductive system

The *ovaries* produce both the *female hormones* that are primarily responsible for the reproductive cycle in the female and *ova.* These paired organs are about the size and shape of almonds. They are located on either side, within the lower portion of the abdominal cavity. The human ovaries normally produce one mature ovum during each *reproductive cycle.* The ovaries alternate in releasing their eggs, but the alternation is irregular and unpredictable. When the mature ovum is released into the abdominal cavity during *ovulation,* it normally passes immediately into

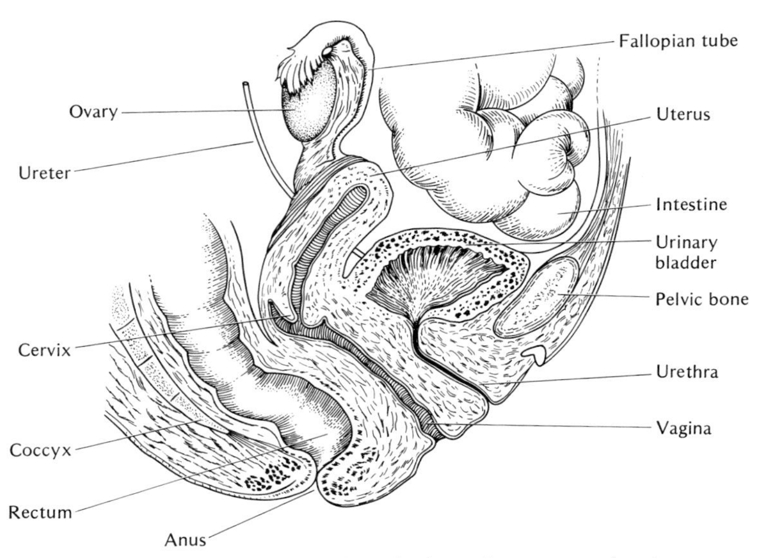

Fig. 12-8. Longitudinal section through the pelvic region showing genital organs of the human female.

one of the paired *oviducts* or *fallopian tubes,* the ends of which are funnel-shaped and ciliated to "catch" the ovum and conduct it along the length of the oviduct.

The two oviducts empty into the upper portion of a pear-shaped organ called the *uterus* or womb. This organ, which houses the developing embryo during the period of *gestation,* lies in the middle of the lower portion of the abdominal cavity between the urinary bladder and the rectum. It is normally about the size of a clenched fist. The uterus has a thick wall composed primarily of smooth muscle tissue, and a mucous lining richly supplied with blood vessels. Its lower end narrows into the *cervix* which projects a short distance into the posterior region of the *vagina,* the relatively large tubular organ which serves both as a receptacle for semen and as a birth canal (Fig. 12-8). Since the uterus is not divided it is termed a *simplex* uterus.

The external genital organs in the human female, collectively known as the *vulva,* consist of two sets of folds of skin (the outer *labia major* and the inner *labia minor*) which enclose the opening of the vagina. At the point where these folds meet in front of the vagina is located the *clitoris,* an organ homologous to the penis in the male. Immediately behind and posterior to the clitoris is the opening of the urethra, which in the female has only a urinary function.

The Female Reproductive Cycle

In most mammalian species the female experiences rhythmic variations in the intensity of her sex urge which are coordinated with the time of ovula-

female reproductive cycle

tion. The period when the sex urge is greatest, when the animal is said to be "in heat," is known as *estrus*. The females of most species accept the male in copulation only during the estrus period. The estrus cycle is accompanied by hormonal influences which cause changes in the lining of the vagina and uterus, particularly the uterus. The lining of the uterus becomes thick and soft, and its blood vessels and glands increase in number and size. These changes become maximal a short time after ovulation. If fertilization does not occur, the lining of the uterus gradually reverts to its original condition.

menstruation

The human female and the female anthropoid ape do not experience any distinct period of estrus. Instead, the reproductive cycle is marked by *menstruation* which occurs about every 28 days and lasts for 4 or 5 days. The menstrual flow results from a sloughing off of the lining of the uterus, accompanied by capillary bleeding. Immediately following menstruation, one or more follicles in the ovary, under the influence of follicle stimulating hormone (FSH) from the pituitary gland, are stimulated to develop. Each follicle contains a primary oocyte. The developing follicles themselves are stimulated to secrete *estradiol* (one of the *estrogens*), a hormone that causes the lining of the uterus to be replaced and to become thicker and more vascular.

ovulation

Ovulation under the stimulus of FSH and a luteinizing hormone (LH) also secreted by the pituitary, occurs about 15 days following the beginning of the previous menstrual period. A mature egg is released from the ovary by a rupturing of the follicle, whereupon the follicular cells are transformed into the *corpus luteum* (yellow body). The corpus luteum, under the stimulation of LH, secretes the second hormone involved in the menstrual cycle, *progesterone*. This hormone completes the development of the uterine lining, preparing it to receive the embryo. It also prevents the development of additional follicles within the ovary and stimulates the growth of mammary glands (Fig. 12-9).

Fig. 12-9. The reproductive cycle in the human female.

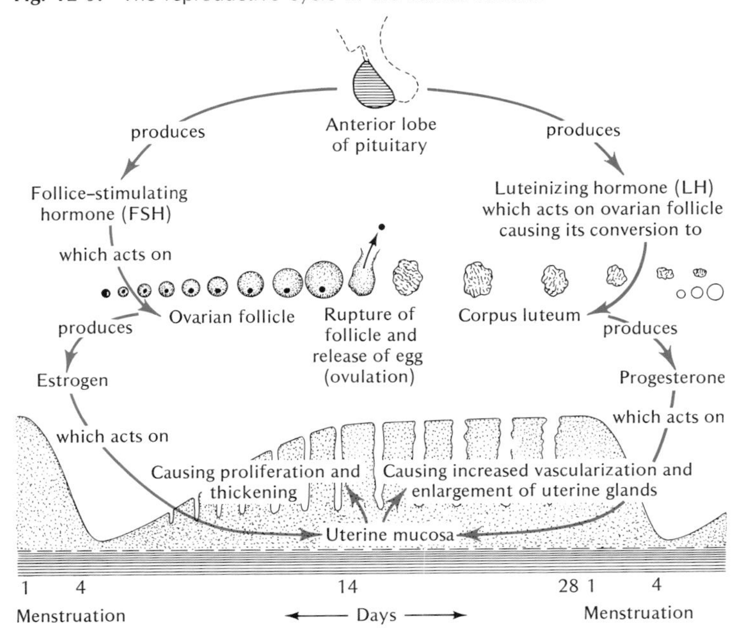

If the egg is not fertilized, the corpus luteum undergoes regression about the twenty-seventh day of the menstrual period, thus terminating the secretion of progesterone. Since progesterone is necessary for the maintenance of the uterine lining, the regression of the corpus luteum marks the beginning of menstruation. The absence of progesterone also permits the follicles within the ovaries to develop in preparation for the next cycle.

If pregnancy occurs, the corpus luteum remains and continues to secrete progesterone throughout the period of gestation. The continued secretion of this hormone is necessary, particularly during the early stages of pregnancy, for the maintenance of the uterine lining in a state that will sustain the developing embryo. This hormone also inhibits ovulation during pregnancy. If the corpus luteum is destroyed, the pregnancy is terminated immediately. Progesterone also stimulates the growth of the mammary glands during the later part of pregnancy and prepares them for the action of two pituitary hormones, *prolactin* and *oxytocin,* which cause the flow of milk.

Fertilization

fertilization

A female human fetus at 4 or 5 months of age contains several hundred thousand potential egg cells in its ovaries, apparently the only oocytes that the female will ever develop. Only about 1 in 1000 of these potential egg cells actually becomes a mature ovum ready for fertilization. Ovulation begins at *puberty* and lasts until the onset of *menopause,* a period of approximately 30 years. This means that a woman with an average 28-day reproductive cycle produces approximately 400 eggs during her lifetime.

During copulation the male ejaculates his semen into the vagina of the female. A single ejaculation usually involves 3 to 5 milliliters of fluid containing upward of 300 million sperm. The sperm travel through the vagina and into the uterus, partly under their own power but primarily by force of muscular contractions in the walls of the female organs. Most of the sperm expend their energy within a short distance, but a few find their way to the opening of the oviducts and swim up them. The same current (caused by the beating of cilia in the oviduct) that draws the egg from the abdominal cavity into the oviduct probably stimulates the sperm in their swimming movements. If ovulation has occurred shortly before or after copulation, the egg that passes into the oviduct has a good chance of being fertilized by one of the sperm, as fertilization occurs in the upper part of the oviduct. By the time the egg reaches the lower part of the oviduct it has lost its capacity to be fertilized. The time during which fertilization may occur is thus very short, probably no more than 24 to 48 hours. The ovum at the time of ovulation is surrounded by a layer of cells called the *corona radiata.* The cells forming this layer are held together by an organic substance known as *hyaluronic acid* which is acted upon by the enzyme *hyaluronidase.* Because each sperm contains only a very small amount of hyaluronidase, large numbers of sperm (probably hundreds of thousands) are required to break down the hyaluronic acid and separate some of the corona radiata cells in order that one sperm can enter and fertilize the egg. As soon as the egg has been penetrated by a sperm, it develops a *fertilization membrane* which prevents the entrance of other sperm. Sperm and unfertilized eggs die within the oviducts or the uterus and are phagocytized by white blood cells.

Available evidence indicates that human sperm remain alive and retain their ability to fertilize the ovum for only 24 to 48 hours at most after having been deposited in the female genital tract. There are several reasons for this. Sperm are extremely small. Because their cytoplasm contains very little food material, their energy supply is limited. They are very sensitive to heat, not being able to withstand even body temperature for long periods of time, and they are also sensitive to changes in pH within the genital tract. The leukocytes of the vaginal epithelium which engulf millions of sperm constitute another hazard.

implantation

Following fertilization the zygote begins to undergo *cleavage* as it passes down the oviduct to the uterus. From 8 to 10 days elapse before the human embryo is *implanted* in the wall of the uterus, during which time the developing embryo is nourished by secretions of the uterine glands. By the time of implantation, the embryo consists of hundreds of cells which have already begun to undergo some differentiation. During implantation the embryo sinks into the soft lining of the uterus and becomes firmly attached, usually in the upper part of the uterus on either the dorsal or ventral surface. The embryo becomes completely covered with the lining of the uterus. *Extraembryonic membranes* and the *placenta* develop shortly after implantation. The extraembryonic membranes serve to protect the embryo, and the placenta provides for the exchange of materials between the circulatory system of the embryo and the mother's bloodstream. The embryo or fetus is attached to the placenta by the *umbilical cord.*

Selected Readings

Balinski, B. I. *An Introduction to Embryology.* 3rd ed. Philadelphia: Saunders, 1970.

Berelson, B., and R. Freedman. "A Study in Fertility Control." *Scientific American,* 210(5):29–37, 1964.

Leisner, R. S. and E. J. Kormondy, Eds. *Population and Food.* Dubuque, Iowa: Brown, 1971.

Moog, F. "Up From the Embryo." *Scientific American,* 182(2):52–55, 1950.

Naylor, A. W. "The Control of Flowering." *Scientific American,* 186(5):49–56, 1952.

Patten, B. M. *Foundations of Embryology.* 2nd ed. New York: McGraw-Hill, 1964.

Pincus, G. "Fertilization in Mammals." *Scientific American,* 184(3):44–47, 1951.

Population Council. *Country Profiles* (a series). New York: The Population Council, 1969.

Salisbury, F. B. "The Flowering Process." *Scientific American,* 198(4):108–117, 1958.

Torrey, J. G. *Development in Flowering Plants.* New York: Macmillan, 1967.

Oral contraceptives come in a variety of colors and with different devices to keep track of the days.

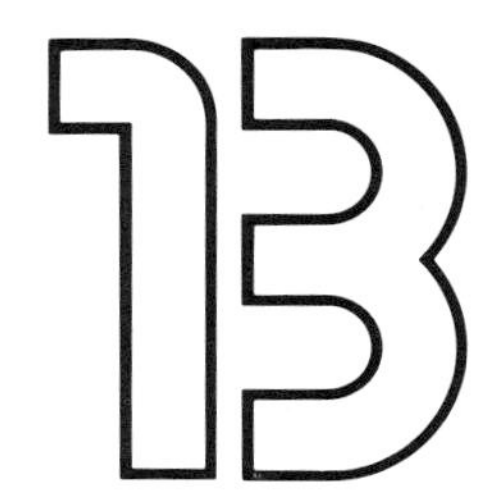

Contraceptive Technology[1]

METHODS

Introduction

Until recently, the scientific basis of most contraceptive methods was the realization that the ejaculate represents or contains the male factor responsible for fertilization. For centuries, mankind attempted to prevent pregnancy by the simple and direct procedure of withdrawing the penis prior to ejaculation; by mechanical devices such as the condom and, later, the diaphragm; by a variety of chemical spermicides introduced into the vagina; and retroactively by postcoital douching. The effectiveness, or lack of it, of these procedures depends on their success in preventing sperm from making their way to the arena of fertilization, the fallopian tubes, on the occasion of a particular coitus. Permanent blockage of sperm passage was achieved by surgical procedures on the male and female genital organs.

Contraceptive technology caught up with the twentieth century when scientists turned their attention to the ovulatory cycle in the female and the

[1] Pages 287–330 by Sheldon J. Segal, Ph.D., Vice-President and Director, Bio-Medical Division, The Population Council, and Christopher Tietze, M.D., Associate Director, Bio-Medical Division, The Population Council. Reprinted from A Report on Population and Family Planning, July, 1971, with permission of The Population Council.

hormonal control of reproduction in both sexes. The principle of periodic abstinence timed to avoid coitus near the day of ovulation was the first method of fertility regulation that had as its basis a modern scientific understanding of the reproductive process. That the rhythm method has never proven to be an effective contraceptive practice does not detract from its significance in focusing upon the ovulatory process as a key event for control of fertility. It was several decades before the necessary knowledge was marshaled to develop effective means to prevent ovulation, but when that moment came, the practice of contraception was revolutionized. The era of hormonal contraception was launched and, with it, the search for different ways to achieve the regulation of fertility by interfering with specific links in the reproductive chain of events.

Almost without exception, experimental efforts to inhibit fertility can be described as attempts to manipulate a key event in the endocrine control of reproduction. With the gradual elucidation of the normal hormonal requirements of the reproductive process, it becomes apparent that there are many steps in this sequence that are vulnerable to controlled interference.

This report attempts to review the pertinent studies in this area, to point out the types of work needed, and to project the kinds of control mechanisms that could evolve from our present knowledge of reproductive physiology.

Currently Available Contraceptive Methods

The term contraception, as used in this chapter, includes all temporary and permanent measures designed to prevent coitus from resulting in pregnancy. This objective may be achieved by preventing ovulation, fertilization, or implantation.[2] Interference with fetal survival following implantation is not included in the definition. The discussion covers the following 10 methods, which together account for virtually all contraceptive practice in the world today.

Folk methods:
1 Coitus interruptus
2 Postcoital douche
3 Prolonged lactation

Traditional methods:
4 Condom
5 Vaginal diaphragm
6 Spermicides
7 Rhythm

Modern methods:
8 Oral contraceptives
9 Intrauterine devices

Permanent method:
10 Surgical sterilization

The scope of the evaulation encompasses effectiveness, acceptability, and safety (side effects).

[2] For a discussion of abortion see page 330.

Coitus Interruptus

History

Withdrawal of the penis prior to ejaculation, or coitus interruptus, is probably the oldest contraceptive procedure known to man. It is referred to in the Old Testament and has been observed by anthropologists among diverse tribes throughout the world. In western and northern Europe during the Middle Ages and early modern times, where relatively late marriages co-existed with close and frequent contacts between unmarried adults and where pregnancy out of wedlock was strongly condemned, coitus interruptus appears to have been the principal method for averting the consequences of premarital intercourse. Transfer of the practice into married life on a scale sufficient to influence the trend of the national birth rate occurred in France toward the end of the eighteenth century and in other countries of the region during the nineteenth century.

Mode of Action

Coitus interruptus results in the deposition of semen outside the female genital tract.

Advantages

The method requires no supplies and no particular preparation; it costs nothing.

Disadvantages

The successful practice of coitus interruptus makes great demands on the self-control of the male; some men are physically or emotionally unable to use the method. Unless the woman reaches orgasm prior to withdrawal, additional manual stimulation may be necessary for sexual satisfaction.

Effectiveness

The contraceptive effectiveness of withdrawal has been traditionally underestimated by the medical profession. Careful studies in the United States and the United Kingdom have revealed pregnancy rates only slightly higher than those achieved by the same populations using mechanical and chemical contraceptives.

Reasons for Failure

Escape of semen prior to ejaculation, delayed withdrawal, or deposition of semen in the woman's external sexual organs may result in pregnancy.

Side Effects

A wide variety of gynecological, urological, neurological, and psychiatric ills have been attributed to the practice of coitus interruptus, but the cause-and-effect relationship has never been demonstrated. Many couples continue to use the method for years without apparent ill effect and with adequate sexual satisfaction for both partners.

Extent of Use

Coitus interruptus does not appear to be popular in the United States but apparently still occupies the first place in many countries of Europe and the Near East. In general, its use is inversely associated with socioeconomic status.

Postcoital Douche

History

Postcoital douches with plain water, with vinegar, and with various solutions advertised as "feminine hygiene" products have long been used for contraceptive purposes. The bidet, presumably designed for ablution of the genitalia, made its first appearance in France early in the eighteenth century. The douche was the principal method recommended by Charles Knowlton (1883), one of the earliest American writers on family planning, and was widely used until World War II.

Mode of Action

It is intended to provide for the mechanical removal of semen from the vagina. Addition of vinegar, and so on, is intended to produce a spermicidal effect.

Advantages

It is useful only as an emergency measure, for example, if a condom breaks.

Disadvantages

It is ineffective and inconvenient.

Effectiveness

While the postcoital douche reduces the chance of conception to a limited degree, it is the least effective of the methods currently in use.

Reasons for Failure

Spermatozoa have been demonstrated in the cervical mucus within 90 seconds after ejaculation.

Side Effects

Frequently douching and the use of strong solutions may damage the bacterial flora of the normal vaginal mucosa.

Extent of Use

The use of the douche has declined markedly over the past 30 years, at least in the United States, where its employment for contraceptive purposes is concentrated among the poor and uneducated.

Prolonged Lactation

History

Since time immemorial, women have known that they are less likely to conceive after a delivery if they breast-feed their babies than if they do not. This awareness led to the deliberate extension of lactation in order to delay conception.

Mode of Action
It delays the return of ovulation.

Advantages
It costs nothing and requires no special preparation.

Disadvantages
See "Effectiveness."

Effectiveness
Breast feeding prolongs postpartum amenorrhea but cannot do so indefinitely, nor is it possible to predict the extent of the prolongation. Ovulation may return at any time and may even precede the first postpartum menstrual period. If this happens, conception may occur while the woman is still amenorrheic.

Reasons for Failure
See "Effectiveness."

Side Effects
If nursing is prolonged beyond the appearance of incisors in the baby, trauma to the mother's nipple may result. Exclusive reliance on lactation beyond infancy may result in malnutrition of the child.

Extent of Use
Breast feeding over long periods is the rule in most of the developing countries, but it is not known to what extent this is done with the intention, or even the hope, of delaying conception.

Condom

History
The contraceptive sheath made its first appearance in England during the eighteenth century. Early condoms were made from the intestines of sheep and other animals. Since the latter part of the nineteenth century, these so-called skin condoms have gradually been replaced by the cheaper and more convenient rubber sheaths.

Mode of Action
The condom serves as a cover for the penis during intercourse and prevents the deposition of semen in the vagina.

Advantages
The condom offers reliable protection not only against pregnancy, but also against venereal infection. It can be used in almost any situation in which coitus is possible. The evidence immediately after intercourse of an intact contraceptive barrier adds reassurance.

Disadvantages
Foreplay must be interrupted to put on the condom. Some men and some women perceive the rubber membrane as an obstacle to sexual sensa-

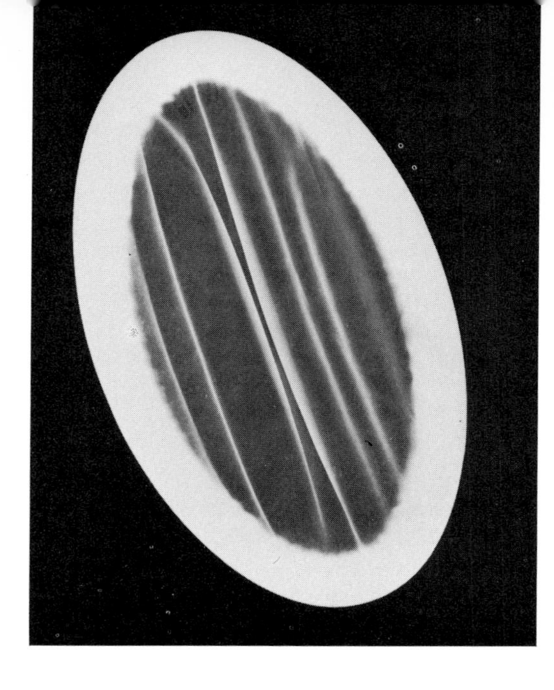

Fig. 13-1. A rolled condom.

tions; some men do not like to use a condom with their wives because of its association with prostitutes.

Effectiveness

The manufacture of condoms requires a high level of quality control. In the United States, where these devices are under the jurisdiction of the Food and Drug Administration, about 997 out of every 1000 condoms sold are free from defect demonstrable by current tests. Given a high-quality product and consistent use, the condom is one of the most effective means of contraception.

Reasons for Failure

Pregnancy may result from a break or tear—estimated in one study to occur once in 150 to 300 instances of use—or from the escape of semen at the open end of the condom if withdrawal is delayed until after detumescence. The most common cause of failure is, however, "taking a chance." The risk of pregnancy associated with a single unprotected coitus, enjoyed at random during the menstrual cycle, is on the order of 2 to 4 percent, which is probably more than the pregnancy rate during a full year of consistent use of the condom.

Side Effects

Side effects are extremely rare; an occasional individual may be sensitive to rubber or to the powder used for dusting the condom.

Extent of Use

Prior to the introduction of oral contraceptives, the condom was the most widely used method in the United States. According to a survey in 1955, about 27 percent of all couples practicing contraception reported the condom as the method used most recently; the corresponding figure for 1965 was 18 percent. Use of the sheath is also important in most countries of Europe, in Japan, and increasingly in developing countries where the method has been incorporated into national family planning programs.

Vaginal Diaphragm

History

The vaginal diaphragm, invented by Wilhelm P. J. Mensinga, a German physician, some time before 1882, is still in use with minor modifications. Prior to the introduction of oral contraceptives, the diaphragm was the contraceptive method most often recommended by physicians both in private practice and in birth control clinics in the United States, the United Kingdom, and, indeed, throughout the world. The diaphragm is always prescribed in combination with a vaginal jelly or cream.

Mode of Action

The diaphragm serves as a mechanical barrier to the entry of spermatozoa into the cervical canal. The jelly or cream acts as a spermicide and as a lubricant for inserting the diaphragm.

Advantages

The vaginal diaphragm is a reliable method for the woman's use and is virtually without side effects.

Disadvantages

Because the diaphragm must be fitted, it requires a pelvic examination by a physician or other trained health worker. For this reason, it is not suitable in situations in which such services are not available. Nor is it a suitable method if privacy is lacking, since it must be inserted either daily as a bedtime routine or before intercourse. The method makes great demands on purposeful behavior and is easily abandoned if poor motivation or neurotic behavior intervenes. Some women object to the genital manipulation associated with the insertion of the device.

Effectiveness

Used consistently, the diaphragm offers a high level of protection against unwanted pregnancy, reflected by a failure rate of two to three pregnancies per 100 women per year. However, perfect use without any omissions is rarely achieved. In clinical practice a pregnancy rate of 10 per 100 women per year is quite satisfactory and much higher rates (20 to 30) have been reported for populations not yet accustomed to the practice of contraception.

Reasons for Failure

Even a well-fitted diaphragm may be incorrectly inserted so that it fails to cover the cervix, or it may be displaced during the orgastic expansion of the inner two-thirds of the vaginal barrel.

Side Effects

Reactions to rubber or to one of the components of jelly or cream have been reported in rare cases.

Extent and Continuation of Use

Prior to the introduction of oral contraceptives, the diaphragm was used by about one-fourth of all couples in the United States who used any form of contraception. By 1965 the share of the diaphragm in total contraceptive practice had dropped to about one-tenth of all users. It has always been less popular in Europe and in other regions for which information is available.

In general, use of the diaphragm is directly associated with socioeconomic status. While middle-class women have found the method acceptable, rates of continuing use among the clientele of birth control clinics have been disappointingly low.

Spermicides

History

Jellies and creams and, more recently, vaginal foams, with highly spermicidal action, intended for use without a diaphragm, have been developed by the pharmaceutical industry in recent years. Other vehicles for the introduction of spermicidal substances are suppositories or wafers which melt in the vagina and tablets which crumble and dissolve on contact with moisture to release carbon dioxide, producing a dense foam.

Mode of Action

These materials immobilize sperm on contact with the ejaculate.

Advantages

Spermicides are relatively simple to use and do not require a pelvic examination; foam tablets, at least, are inexpensive.

Disadvantages

Many users complain of vaginal leakage ("messiness") and excessive lubrication. Suppositories and foam tablets require a waiting period of several minutes to allow for melting or disintegration.

Effectiveness

Spermicides used alone appear to be less effective than spermicides used in combination with the diaphragm. According to clinical trials in the United States, vaginal foams appear to be more effective than other types of spermicides.

Reasons for Failure

An inadequate quantity or quality of the spermicidal material is the most obvious reason for failure. Some spermicides dissolve or disintegrate slowly and are inadequately distributed throughout the vagina. Some couples fail to observe the waiting period.

Side Effects

Irritation or inflammatory changes of the mucous membrane have been reported in rare cases.

Extent of Use

The use of spermicides is of relatively minor importance in the United States, vaginal foam and suppositories being the leading types. They are apparently more popular in Europe. In spite of many failures, foam tablets are still an important method in the family planning programs of some developing countries.

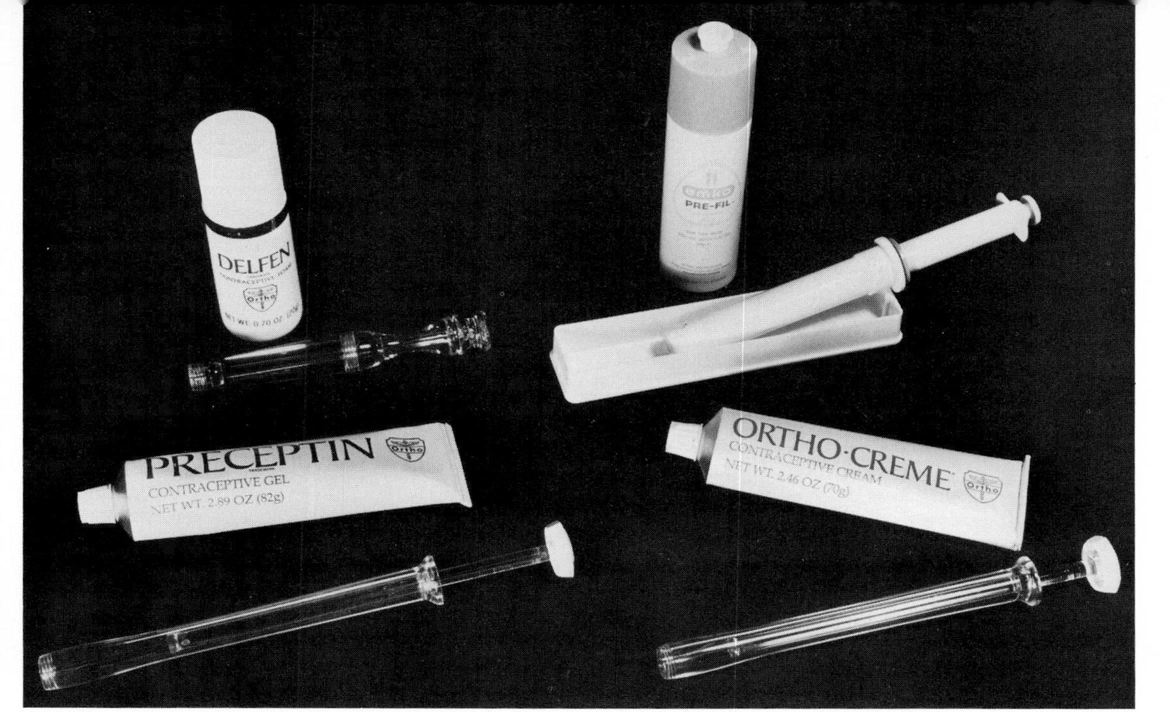

Fig. 13-2. Some of the many spermicides.

Rhythm Method

History

The notion that women are able to conceive during part of the menstrual cycle only is very old. However, early ideas of the fertile and sterile periods of the menstrual cycle were often the opposite of what is known today.

At the present time two varieties of the rhythm method are practiced: calendar rhythm and temperature rhythm. Calendar rhythm was developed independently in the 1920s by Ogino in Japan and Knaus in Austria. According to these investigators, the day of ovulation can be estimated by means of a formula based on the individual woman's menstrual history recorded over a number of months. Abstinence is prescribed for a few days before and after the estimated day of ovulation.

More recently, observation of the typical changes in basal body temperature during the menstrual cycle has been used to determine that ovulation has occurred and conception is no longer possible during the cycle in question. Intercourse is permitted during the postovulatory phase only.

Mode of Action

The rhythm method is based on the avoidance of coitus on the days it could result in the simultaneous presence of a fertilizable ovum and motile spermatozoa.

Advantages

Rhythm is the only contraceptive method currently sanctioned by the Roman Catholic church.

Disadvantages

Opportunity for coitus is greatly reduced, especially with temperature rhythm. It is unsuitable for women with grossly irregular menstrual cycles.

Effectiveness

The contraceptive effectiveness of the rhythm method has been the subject of much controversy. Correctly taught, correctly understood, and consistently practiced, the rhythm method may be quite effective, especially the temperature rhythm method. However, successful practice requires considerable self-control and an equally strong desire to control fertility. Self-taught rhythm, haphazardly practiced, is a very ineffectual method of contraception.

Reasons for Failure

Apart from "taking a chance" on a day known as "unsafe," the principal reasons for failure of calendar rhythm are errors in recording the menstrual history, errors in computation, the inherent variability of the menstrual pattern, and exceptionally long survival of sperm in the female genital tract. The principal reasons for failure of the temperature rhythm method are errors in reading the thermometer and errors in interpreting the temperature curve.

Side Effects

There are no side effects.

Extent of Use

The rhythm method is widely used among Roman Catholics to whom other methods are prohibited by their faith. According to a survey in 1955, rhythm accounted for about one-half of all contraceptive practice by Catholics in the United States. By 1965 the fraction had declined to about one-third. In general, use of the rhythm method among Catholics is directly proportional to socioeconomic status.

Oral Contraceptives (OCs)

History

Attempts to control human fertility through oral medicaments are as old as written history. Ancient manuscripts exist that contain prescriptions for potions and other medication for the prevention of pregnancy. The exotic content and the application of these ancient prescriptions have little or no rationale for the modern student.

In our own times the work done in the United States by such pioneers as John Rock and the late Gregory Pincus has brought oral contraceptives (OCs) into the realm of practical reality. Following intensive investigation in the laboratory and some clinical study in the early 1950s, field trials began in 1956. By 1960 the first OCs were approved by the U.S. Food and Drug Administration for general distribution in the United States, and since then their use has spread rapidly throughout the world.

Mode of Action

The OCs discussed in this chapter are synthetic steroid compounds, some of which are similar in structure to the natural hormones associated with the menstrual cycle and with pregnancy in the human female. In the dosages currently used, these compounds suppress ovulation. The nature and the importance of any subsidiary antifertility effects are still under investigation.

The OCs in general use today are prescribed according to two distinct regimens—combined and sequential. Under the combined regimen 20 or 21 identical tablets, containing one of several synthetic progestins as well as estrogen, are taken from the fifth to the twenty-fourth or twenty-fifth day of the cycle. Under the sequential regimen the tablets are divided into 15 or 16 containing estrogen only, followed by 5 containing progestin and estrogen. Under both regimens suspension of medication ordinarily results in withdrawal bleeding within a few days, with resumption of medication on the 5th day of the new cycle. An alternative procedure is "3 weeks on the pill, 1 week off," with inert tablets sometimes included for the "off" week to facilitate following the regimen.

A third regimen, involving daily medication with small doses of progestin, without estrogen, has been investigated but not approved for general use in the United States.

Advantages

The outstanding advantage of the OCs is their almost complete effectiveness, according to available clinical studies. Women who faithfully adhere to the regimen can be free of the fear of unwanted pregnancy. Another distinct advantage is that their use is not related to sexual activity.

Disadvantages

See "Side Effects."

Effectiveness

Taken according to the prescribed regimen, OCs of the combined type are almost 100 percent effective in preventing unwanted pregnancy. Major reports cover an aggregate of about 325,000 cycles of medication, with 17 pregnancies not associated with the omission of one or more tablets, according to the statements of the users, corresponding to a pregnancy rate of less than 0.1 per 100 women per year.

Sequential OCs are somewhat less effective and have a pregnancy rate of about 0.3 per 100 women per year, based on 116,000 cycles and 30 pregnancies, apparently resulting from method failures.

Reasons for Failure

The only important reason for failure is the omission of one or more tablets during the prescribed cycle of medication.

Side Effects

The OCs have wide-ranging metabolic effects extending far beyond their primary target organs, that is, the genital tract, the pituitary gland,

and the hypothalamus. Laboratory studies of carbohydrate metabolism, and of hepatic and thyroid function, to mention only a few, have revealed deviations from normal values in many women taking these compounds. Although they do not produce clinical symptoms, these abnormal findings are a cause for concern, since their significance for the health of women is not fully understood.

Clinically, the early use of OCs is frequently associated with symptoms similar to those occurring during early pregnancy, such as nausea, vomiting, or breast engorgement, which are primarily related to the estrogen content of the tablets. Other common complaints are breakthrough bleeding during medication, weight gain, headache, dizziness, and a brownish discoloration of the facial epidermis known as chloasma. Whereas most of these symptoms last only a few months, some users become discouraged and turn to other methods or even abandon their efforts at family planning.

In addition to the relatively minor symptoms associated with early use, a wide range of adverse experience has been observed, some of which has been attributed to the medication. The evaluation of these findings is extremely difficult because the conditions under consideration also occur among women who are not "on the pill" and may therefore be expected to occur among the millions of women taking OCs.

The condition for which a statistical association with OCs has been most firmly established is thromboembolic disease, including its sometimes fatal outcome pulmonary embolism. Cases of this type among users of OCs were reported as early as 1961, generating a lively and occasionally heated controversy. The first definitive statistical evidence associating the two phenomena was provided by three major investigations in the United Kingdom and one in the United States published in 1967–1969. Each of these studies was based on a retrospective comparison of "cases" (with thromboembolic disease) and "controls" (without thromboembolic disease); each yielded statistically significant results.

The first of the investigations was sponsored by the Royal College of General Practitioners. A selected group of 29 doctors reported on 97 women of reproductive age with pulmonary embolism or other forms of thromboembolic disease of the venous system not associated with pregnancy or the puerperium. For each patient two control cases, matched for age, marital status, and parity, were selected from the doctors' files. The patients and controls were interviewed by the doctors concerning their reproductive and contraceptive histories. OCs had been used prior to the relevant illness by 16 of the 97 women treated for thromboembolic disorder as compared with only 13 of the 194 controls for this group. The risk of thromboembolic disorder was estimated to be about three times higher among users than among nonusers.

The starting point of another British study was a group of married women, under 40 years of age and discharged from 19 large general hospitals in London during 1964–1967 with a diagnosis of deep vein thrombosis or pulmonary embolism without evidence of a predisposing cause. They were subsequently interviewed in their homes, at which time information was obtained about their medical, obstetric, and contraceptive histories. For comparison, married control patients admitted to the same hospitals with acute surgical or medical conditions, or for elective surgery, and matched with the affected patients for age, parity, date of admission, and absence of any predisposing cause of thromboembolic disease, were selected and similarly interviewed.

Of 84 patients with thromboembolic disorder, 42 had been using OCs during the month preceding the onset of the illness, whereas only 23 of the corresponding 168 controls had done so. From these data it was estimated that the risk of hospital admission for venous thromboembolism was six to seven times greater in women who use OCs than in those who do not.

The third British study was concerned with women of reproductive age who died in 1966. Inquiries were made about the use of OCs by these women, compared with controls selected from the practices of the general practitioners under whose care the deceased women had been. Of 36 women who died from pulmonary embolism or infarction or from cerebral thrombosis, without evidence of a predisposing condition, 21 or 58.3 percent had used OCs, compared with only 15.8 percent expected on the basis of the experience of control women of the same age and parity. The mortality from these two conditions was estimated to be about seven times as high among users as among nonusers.

In the United States an investigation of the association of thromboembolic disease with the use of OCs was completed in 1969. The subjects were 175 women aged 15–44 discharged from 43 hospitals in five cities, after initial attacks of idiopathic thrombophlebitis, pulmonary embolism, or cerebral thrombosis. They and 175 carefully matched controls, free of chronic conditions either associated with thromboembolism or constituting contraindications to pregnancy, and presumably fertile, were interviewed to provide information on their use of OCs prior to hospitalization. Sixty-seven women with thromboembolism and 23 controls had used OCs until within 1 month before they were hospitalized. From these data it was estimated that the risk of hospital admission for thromboembolism was four to five times greater for women taking OCs than for nonusers. It was found that medication that terminated over 1 month before hospitalization was not associated with thromboembolism.

In summary, then, the incidence of idiopathic thromboembolic disease of the venous system (pulmonary embolism or cerebral thrombosis) appears to be several times higher among women taking OCs than among women with comparable characteristics not using these compounds. Duration of use does not appear to affect the risk.

A recent retrospective study in the United Kingdom has demonstrated an association between postoperative deep-vein thromboembolism and the use of OCs. Of 30 patients with thromboembolism, 12 had taken OCs during the month preceding surgery, compared with only 9 of 60 matched controls, suggesting a three to four times greater risk for users than for nonusers.

The incidence of thromboembolic disease among women taking OCs in the United Kingdom, Sweden, and Denmark was found to be directly associated with the estrogen content of the tablets, according to another study published in 1970. The same study also established for the first time a statistically significant association of coronary thrombosis with OC use. On the basis of these findings, the regulatory agencies of several countries have taken steps to discourage, or even prohibit, the use of OCs with more than 50 micrograms of estrogen per tablet.

According to a cooperative investigation conducted in the United States, the United Kingdom, and Sweden, the risk of thromboembolic disease among women taking OCs is almost three times higher for those with blood groups A, B, or AB than for those with blood group O.

The annual rate of mortality from idiopathic thromboembolic disease in the United Kingdom, in the absence of predisposing conditions, per 100,000 women of specified age, has been estimated as 1.5 for OC users 20 to 34 years old, and as 3.9 for OC users 35 to 44 years of age. The corresponding estimates for nonusers are 0.2 and 0.5 per 100,000 women, respectively. The differences between these two sets of rates have been interpreted widely as measuring the excess mortality due to the taking of OCs. This interpretation overlooks two factors: (1) mortality from coronary thrombosis is excluded and (2) the use of OCs by the general population appears to have been substantially overestimated. If allowance is made for these two factors, the estimated rates of mortality among British women taking OCs rise to 3 per 100,000 in the younger age group, 20 to 34, containing the majority of users, and to 9 per 100,000 in the older group, 35 to 44 years of age.

The meaning of the excess mortality from thromboembolic disease should be interpreted against a current annual death rate from all causes in the United Kingdom of about 100 per 100,000 women of reproductive age, 15 to 44 years, and against a risk to maternal life resulting from or associated with pregnancy and delivery, exclusive of death from illegal abortion, of about 25 per 100,000 pregnancies. The comparable rates in the United States are slightly higher.

It should be noted that all epidemiological studies linking thromboembolic disease to the use of OCs have been carried out in Europe and the United States. It is commonly believed, without firm statistical evidence, that thromboembolic disease is far less common in other parts of the world and the relevance of the available data to these other populations is therefore unknown.

Although epidemiological evidence is lacking, clinical observations have established beyond reasonable doubt that the taking of OCs can be the cause of a wide range of diseases and pathological conditions in susceptible women. These conditions include jaundice, hypertension, obesity, depression, lack of libido, and occasional persistent amenorrhea after discontinuation of treatment. These changes are probably in most cases reversible. It is not known whether observed changes in carbohydrate metabolism can precipitate clinical diabetes.

Some biologists have expressed concern about a possible carcinogenic effect of OCs, especially of the estrogen component. Their views are supported by observations in several species of laboratory animals. Others have held that these findings are not applicable to humans.

In 1969 a survey of women attending the clinics of Planned Parenthood of New York City revealed a higher prevalence of epithelial abnormalities of the uterine cervix, diagnosed as carcinoma *in situ,* among women using OCs compared with those using the diaphragm. The diagnosis was made in each case on the basis of a biopsy, examined by two pathologists without knowledge of the contraceptive used.

Women who had used OCs for 1 year or more were individually matched against diaphragm users with respect to five variables: age, parity, age at first pregnancy, ethnic group, and family income. The prevalence of carcinoma *in situ* was about twice as high among those who used OCs as among those who used the diaphragm. The difference was significant at the 5-percent level.

For the interpretation of these findings, it would be important to know how diaphragm users differ from pill users with respect to age at first coitus, frequency of coitus, number of sexual partners, and previous screening, all of which may contribute to the prevalence of carcinoma *in situ.* One would assume that such factors are highly correlated with one or more of the variables used in matching, but important residual differences may remain.

In another study, in California, a significantly higher prevalence of dysplasia was found among women who had chosen OCs as their method of contraception but had not yet taken them, compared with women selecting other methods. The two groups of women were quite similar with regard to such variables as age at first intercourse, age at first pregnancy, and age at clinic attendance, number of children, ethnic group, and religion.

In 1970 several types of OCs were withdrawn from the market or from clinical investigation because of the development of nodules in the mammary glands of beagle bitches after long-term administration of the products according to the regimen and in the dosage per body weight used in women. The relevance of these findings to human beings is open to question. Several major studies of the possible association of breast cancer in women with the use of OCs are now in progress.

The latest available statistics on the incidence of, and mortality from, breast cancer do not show any changes in trend attributable to the increasing use of OCs. Continued evaluation of the possible carcinogenic action of OCs remains a matter of high priority. Since the effect of all known carcinogens in humans is delayed, with a latency period of about 1 decade, it will not be possible to make a definitive statement on this point until substantial numbers of women have used OCs for prolonged periods.

On the basis of current experience, fertility is unimpaired among most women who discontinue the use of OCs in order to plan a pregnancy.

Whereas no increase has been noted in the prevalence of gross congenital malformations among children born to women taking OCs, a highly significant excess of chromosomal abnormalities has been reported in spontaneously aborted embryos from women who conceived within 6 months following the discontinuation of OCs. It should be noted that the abnormalities in question (polyploidy) are always lethal early in pregnancy.

Extent and Continuation of Use

The high acceptability of the OCs in the United States is attested by the fact that, within 5 years of the approval of the first product by the U.S. Food and Drug Administration, current users reached about 3.8 million, accounting for close to one-quarter of all contraceptive practice in the United States. By early 1969, 20 brands of OCs were distributed in the United States at a rate of approximately 8.5 million cycles per month. The Senate hearings of 1970 resulted in at least a temporary reduction in the number of users, the long-term effect of which has not been evaluated.

Other countries together account for a slightly larger total of users than the United States, with the highest rates of use reported from Australia and New Zealand, Canada, and several countries in western Europe. The number of users in Latin America exceeds 2.5 million and appears to be growing rapidly.

Although in the United States OCs are most popular among younger women with better than average educations, clinicians generally agree that most women can be taught to take them with reasonable consistency. This method of birth control has also proved acceptable to many couples who had been unwilling to try the traditional methods or were unable to use them successfully.

A major measure of the acceptability of a contraceptive method is the continuation rate, indicating the proportion of couples still using the method at a given time after use was initiated. For OCs, data from a national survey in the United States suggest a continuation rate of 73 percent after 24 months, excluding women who discontinued use because they wanted to plan a pregnancy. The same survey also showed that the continuation rate was higher among younger women and those with a better education than among older and less-educated women, but the differences were not very large. According to preliminary evaluations, recent attempts to introduce OCs into the family planning programs of several developing countries have been characterized by lower continuation rates than had been expected on the basis of experience in the United States. Because the continuation rates are so low, failure rates based on all unplanned pregnancies, including those occurring after discontinuation of contraception, tend to be disappointingly high.

It is reasonable to assume that small women, especially those suffering from malnutrition, react with more severe vomiting and other gastrointestinal symptoms to amounts of estrogen that are well tolerated by heavier and well-fed women. Nevertheless, since OCs have proved themselves far superior to the traditional contraceptive methods among low-income groups in the more developed countries, they may be expected to be useful in the less developed countries as well.

Intrauterine Devices (IUDs)

History

Various types of intrauterine devices (IUDs) are known to have been in use since the nineteenth century. Although some physicians approved of these devices, the majority of gynecologists rejected them as abortifacients and also because their use, at least in untrained hands, was associated with inflammatory conditions of the pelvic organs which, before the introduction of antibiotics, were difficult to treat and not infrequently fatal.

In 1928 in Berlin, Ernst Gräfenberg reported on his experience with an intrauterine ring, which he first made of silkworm gut and, later, of silver wire. After a brief flurry of popularity, opposition to the new device, which became known as the Gräfenberg ring, developed rapidly and universally among gynecologists, the majority of whom had never had any experience with it but judged it by what they had been taught about its forerunners. For almost 30 years textbooks of gynecology, if they discussed contraception at all, mentioned the Gräfenberg ring only to condemn it. The rehabilitation of the IUD began in the 1960s when the Population Council inaugurated an intensive research program, which included experimentation in the laboratory, clinical and field trials, and a large-scale statistical evaluation, known as the Cooperative Statistical Program (CSP).

Today's IUDs are manufactured in a great variety of shapes from such materials as plastics, usually polyethylene, and stainless steel. Among the newer models a plastic "shield" deserves mention; another recent development is a T-shaped device of inert plastic supporting a spiral of metallic copper wire.

Mode of Action

In all species of mammals that have been studied, the presence of a foreign body in the uterus causes infiltration of the endometrium by polymorphonuclear leukocytes. The fluid from a uterus with a foreign body resembles an inflammatory exudate, containing large numbers of leukocytes and products released by these cells, which thus create an environment hostile to spermatozoa as well as to fertilized ova. It is not known whether the spermatoxic or the blastotoxic effect predominates in the human female. There is no evidence that the antifertility effect of the IUD involves interference with the implanted embryo.

Advantages

The IUD is a method of contraception particularly suitable for large-scale programs in which the motivation of the individual members of the target population may vary considerably. For the majority of couples, the IUD is effective, safe, and acceptable. Furthermore, the fact that its use is disassociated from the sexual act appeals to couples at all social levels. The IUD is economical, and national programs for mass application of its use are relatively easy to develop.

Disadvantages
See "Side Effects."

Effectiveness
The largest body of clinical data on the effectiveness of IUDs has been assembled under the CSP. The most recent comprehensive report, as of 1968, covers more than 400,000 woman-months of use experienced by almost 24,000 women. This project represents the first attempt in the history of fertility regulation to evaluate a new method from its inception by the systematic analysis of pooled data, using uniform procedures and a sophisticated statistical approach. Other sources of data are independent investigations in Chile, Taiwan, South Korea, Pakistan, and elsewhere.

Analysis of the CSP data has revealed failure rates for the most widely used IUDs on the order of three pregnancies per 100 women during the first year and lower rates during subsequent years. Somewhat higher pregnancy rates have been observed under the conditions of public health programs, since the IUD may be expelled without its being noticed by the wearer. The frequent checkups, customary in clinical studies, increase the chance of discovering an unnoticed expulsion before pregnancy occurs.

The contraceptive effectiveness of the IUD is largely independent of the patient's psychology and social background. Well-educated couples emotionally adjusted to vaginal contraception should not expect a higher level of protection from today's IUD than from the consistent use of traditional methods, such as the diaphragm or the condom, and they can certainly achieve lower accidental pregnancy rates with OCs than with IUDs. The higher effectiveness of the IUD is most apparent in populations not accustomed to the sustained use of any form of preventive medicine and among women who have been unsuccessful in the use of birth control methods that require consistent practice.

Reasons for Failure
Failures with IUDs are for the most part method failures, since the majority of the few accidental pregnancies that do occur are observed in women with the device *in situ*. Some undesired pregnancies could be avoided if all IUD wearers checked periodically for the presence of the device.

Side Effects
Since the antifertility effect of the IUD appears to be local rather than systemic, side effects are also primarily of a local character. Among women fitted with IUDs, the most common complaints are bleeding or spotting and pain, including cramps, backache, and similar discomforts. These symptoms most often occur soon after insertion and tend to disappear within a few months. In some cases, however, the bleeding or pain is sufficiently severe to require removal of the device.

A more important adverse experience associated with the use of IUDs is pelvic inflammatory disease (PID), reported in the CSP during the first year after insertion for 2 to 3 percent of the women using the major types

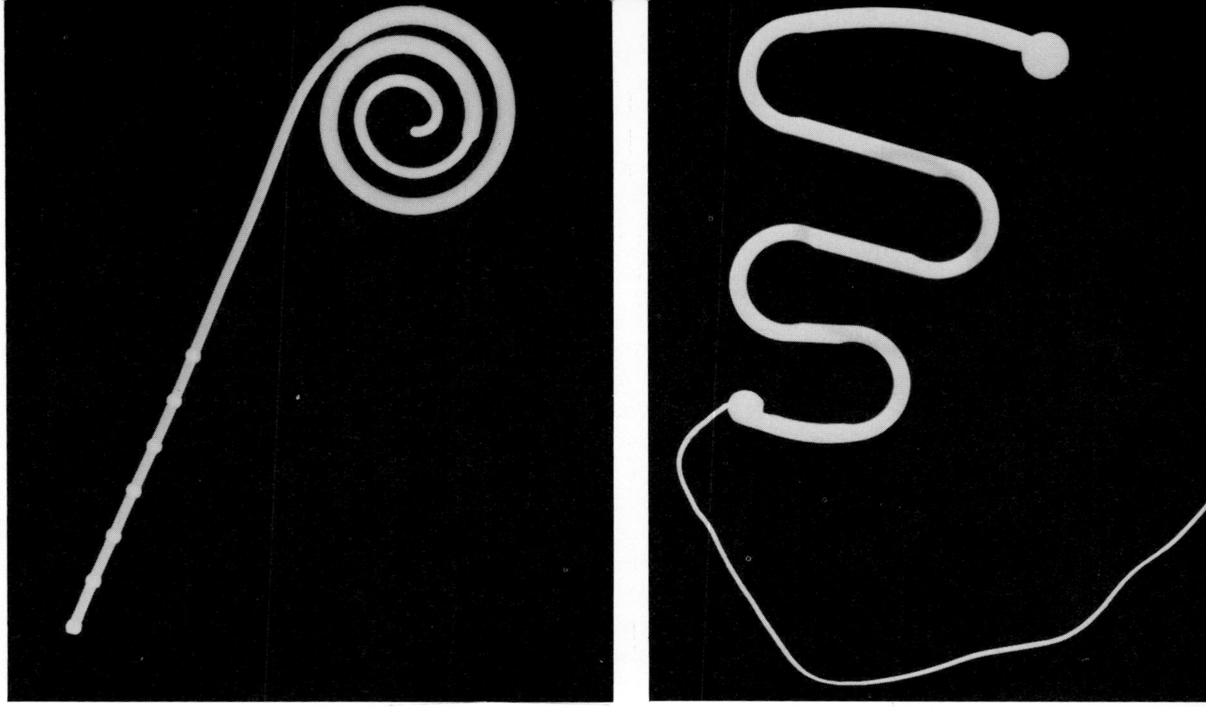

Fig. 13-3. Intrauterine devices. (Above, left) Margulies spiral: expulsion rate, 20 percent; pregnancy rate, 1.5 percent (in any one year, 15 of 1000 women using this IUD become pregnant). (Above, right) Lippes loop: expulsion rate, 10 percent; pregnancy rate 2.7 percent. (Below, left) Double loop: expulsion rate 16 percent; pregnancy rate, 2.8 percent. (Below, right) Dalkon shield: expulsion rate, 2.3 percent; pregnancy rate, 1.1 percent.

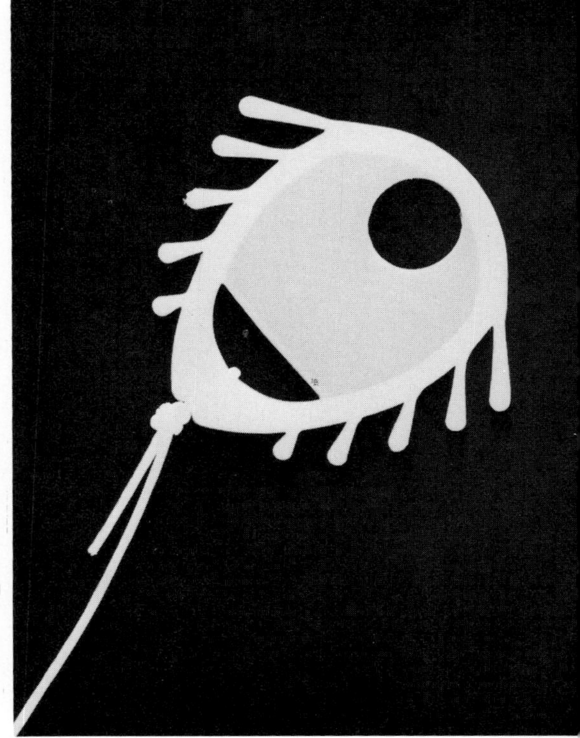

of IUDs. The incidence was much higher during the first 2 weeks following insertion than later. Comparable data on the incidence of PID in the general population are not available, but at least one study suggests that in a population with a high rate of PID the incidence is even higher among women using IUDs.

To a considerable extent episodes of PID associated with IUDs use have been interpreted as reactivations of preexisting chronic or subchronic conditions, brought on by the insertion procedure, rather than by new infection. Whether the insertion of IUDs in women with healthy pelvic organs can produce PID is not known.

The majority of cases of PID among women using IUDs are relatively mild and can be treated successfully with antibiotics and without removing the device. Some patients with PID, however, with or without an IUD, develop serious complications and a few still die, even with adequate medical care. Ten deaths attributed to PID and associated with IUD use are known to have occurred in the United States prior to mid-1967. The case histories of these patients suggest that in four instances the insertion of the IUD may have been the precipitating cause of the fatal illness.

Perforation of the uterus with translocation of the IUD into the abdominal cavity is an infrequent complication. Most perforations of the uterus are entirely asymptomatic. The incidence of this accident with the most widely used device was 1:2500 insertions, according to the CSP A much higher rate (1:150) has been reported from Singapore for the same type of IUD. Closed IUDs, that is, devices with a fixed aperture large enough to accommodate a segment of the intestine, are known to have caused intestinal obstruction following perforation. It is now accepted as good medical practice to remove closed devices without delay when a perforation is discovered. In the absence of symptoms, removal of a translocated open IUD is less urgent.

There is no evidence that the IUDs cause cancer in women or accelerate the transition from dysplasia to cervical carcinoma. However, since the effect of all known carcinogens in humans is delayed, with a latency period of about a decade, it will not be possible to make a definitive statement on this point until substantial numbers of women have used IUDs for prolonged periods.

On the basis of current experience, fertility is unimpaired among women who discontinue the use of IUDs in order to plan a pregnancy. While no increase has been noted in the prevalence of malformations among the children born to women wearing IUDs, the frequency of spontaneous abortion appears to be increased.

Extent and Continuation of Use

The number of IUD insertions in the United States is not known, but available data on manufacturers' sales suggest between 1 and 2 million as the order of magnitude. Abroad, IUDs have been widely used in several national family planning programs, especially in Asia, with the largest numbers of insertions reported from India and Pakistan, and the highest percentages of current users in South Korea and Taiwan. The total number of women in the world currently wearing one of the modern IUDs is probably about 6 million.

Continued use of IUDs is primarily determined by the incidence of expulsions and of side effects that necessitate removal of the device. The incidence of expulsion varies markedly among the various IUDs, with about 10 percent during the first year after insertion reported for the most widely used model. The expulsion rate tends to be higher among young women of low parity than among older women of higher parity, with age being the more important factor. Most expulsions occur during the early months after insertion; usually, but not always, during the menstrual flow. Expulsion after the first year is uncommon. The risk of repeated expulsion after reinsertion is much higher than the risk of primary expulsion. The most important medical reasons for the removal of an IUD are bleeding and pain, including cramps, backache, and other types of discomfort. Because the monthly rates of removal do not decline so rapidly as do the corresponding rates of expulsion, removal is the major cause of discontinuation.

According to observations under the CSP, which reflect mainly clinical practice in the United States, 70 to 80 percent of the acceptors were still using an IUD 1 year after the first insertion, and 60 to 70 percent continued use 2 years after the first insertion. These figures include women who used an IUD after one or more reinsertions. Continuation rates in national family planning programs have been significantly lower than in the CSP, with typical 2-year rates on the order of 50 percent. Preliminary data for the T-shaped device with metallic copper justify the hope of a continuation rate of 85 percent after 1 year of use.

Surgical Sterilization

History

Surgical sterilization was originally used to protect women whose lives or health was threatened by pregnancy or delivery. James Blundell of London is credited with having first suggested the procedure in 1823. Effective techniques were developed in the latter part of the nineteenth century when aseptic surgery and anesthesia became available. At about the same time, sterilizing operations began to be used on males, mainly in connection with surgery of the prostate gland.

Growing confidence in the efficacy and safety of surgical sterilization led to its use for eugenic purposes, that is, to prevent persons suffering from hereditary disabilities from having offspring. In recent years discussion has centered on the legality and the propriety of voluntary sterilization as a method of family limitation and on the use of sterilization in countries where high birthrates and rapid population growth threaten to produce serious economic and social difficulties.

Mode of Action

Cutting, ligation, and removal of a portion of the fallopian tube in the female (tubal sterilization) or the spermatic duct in the male (vasectomy) prevent fertilization of the ovum.

Advantages

The operation provides maximum protection. No further action is needed at any time.

Disadvantages

While the chances of restoring fertility by a second operation have probably improved in recent years, this cannot be counted on and the decision to undergo sterilization should in each case be considered definitive.

Effectiveness

In the hands of a well-trained surgeon, the operation is virtually 100 percent effective.

Reasons for Failure

Inadequate surgery and rare anatomical aberrations.

Side Effects

The risks of surgical misadventure and complications are small, especially in the male, but they exist nevertheless. Adverse emotional reactions occur in predisposed individuals.

Extent of Use

The number of voluntary sterilizations in the United States during the late 1950s is estimated at 110,000 annually, including 65,000 operations on women and 45,000 vasectomies. In recent years the numbers have dramatically increased, with vasectomies alone estimated at 750,000 in a recent nationwide survey. Sterilization of women has been popular in Puerto Rico since the 1940s and has gained wide acceptance in Japan. In India, Pakistan, South Korea, and other Asian countries, vasectomy is encouraged by the government as a method of birth control.

Developments at a Stage of Clinical Evaluation

Surgical Sterilization of the Female

Present methods of surgical sterilization carry the prospect of irreversibility and, particularly with female sterilization, the requirement of abdominal surgery and hospitalization. Consequently, efforts are being made to develop new and simple means to interrupt or occlude a portion of the fallopian tube. Several investigators have achieved success in the use of culdoscopic or peritoneoscopic instruments to interrupt the human oviduct. By eliminating open surgery, these procedures obviate the need for postoperative hospitalization.

Another approach to surgical sterilization that eliminates the need for hospitalization is the transcervical instillation of cytotoxic or sclerosing chemicals to cause occlusion of the tubal openings into the uterine cavity. A solution of atabrine has been used successfully for this purpose. Success in 11 out of 12 cases has been reported following monthly instillations into the uterine cavity repeated three times. The investigator is now trying to evaluate the permanency of the chemically induced tubal closure and eliminate the need for multiple instillation. Other investigators are using the transcervical approach to achieve closure of the tubes by applying a sclerosing chemical or physical treatment directly to the tubal openings. Since the tubal ostia are not readily identified, this research includes the development of new instruments required for visualization and delivery of the treatment. The fallopian tube has a remarkable capacity to recanalize following physical trauma, and perhaps the same tendency to recover from chemical insult will be revealed. It will therefore require a significant period of observation before the level and duration of effectiveness of these experimental procedures can be established.

Temporary occlusion of the fallopian tube has been attempted by the injection of a liquid silicone polymer (polydimethylsiloxane) which vulcanizes at body temperature to form a pliable plug. So far, this experimental procedure has been attempted in only a few women scheduled for hysterosalpingectomy so that the contraceptive efficacy has never been tested in human subjects. Animal work suggests that such plugs are readily dislodged by normal tubal motility and can, in fact, be extruded intraabdominally.

Surgical Sterilization of the Male

Several foreign materials have been used to cause a partial or complete occlusion of the vas deferens in the male. The insertion of a prosthetic device made of silk thread, which partially occludes the vas deferens and is anchored to its exterior surface, causes in the human male, as in animal species, either azospermic ejaculates or ejaculates containing a high percentage of disrupted spermatozoa (head-tail separation). The effect in man, however, is short-lived, so that after several months restoration of high sperm counts is not infrequent, thus rendering the procedure unacceptable for contraception. Intravas devices designed to plug the vas completely have been made but have not been sufficiently tested to permit meaningful evaluation. These are made of polyethylene, polypropylene, silicone polymers, or various metals, in shapes including cylindrical, conical, dumbbell-shaped, or round. The use of a microvalve, which could be operated either manually or by magnetic alignment, has been mentioned frequently. All these possibilities share the problem of requiring that the devices be anchored with sufficient pressure to close the lumen of the vas, without causing a pressure necrosis that would result in damage to the organ, thus defeating the purpose of the procedure—easy reversal.

The application of removable clips or sutures to constrict the vas and prevent sperm passage has the same liability and has not proven acceptable.

Other modifications of the vasoocclusion procedure include crushing a segment of the vas with a surgical instrument or the intravas injection of a necrotizing fluid (formalin), in each case without opening the scrotum and visualizing the vas. The risk involved in these procedures is not negligible since it is difficult to isolate the vas deferens from the adjacent spermatic artery.

Vaginal Contraception

A plastic condom has been developed but not yet subjected to evaluation of clinical effectiveness or acceptability. Its major advantages are ease of manufacture and unlimited shelf life. Cost is comparable to conventional latex condoms, but could be lower in circumstances of mass quantity manufacture.

Some interest has developed in the use of a dissolving film for applying a spermicidal agent intravaginally, but again there are no data to establish effectiveness. This procedure was tested on a limited basis about 20 years ago, then abandoned. Now there are at least two versions of a film 2 inches square, which can be either inserted manually into the vagina prior to intromission, or introduced atop the glans penis. It remains to be established whether the intravaginal distribution of spermicide is sufficiently uniform and complete to assure a high level of contraceptive effectiveness.

Intrauterine Contraception

Several modifications in intrauterine device design, size, or material have been suggested. Two experimental devices are made of stainless steel and employ a spring principle in an attempt to reduce the likelihood of spontaneous expulsion. However, the problems encountered in removing these devices, when it is required, have discouraged physicians from using them for their patients.

Other innovations include devices that contain a magnetic material which can deflect a galvanometer needle, a miniaturized radio transmitter, or a supply of a chemical or hormonal antifertility agent which can be released gradually into the uterine lumen. Attempts are underway to replace the medium-density polyethylene now used in the manufacture of intrauterine loops with other synthetic materials, including silicone polymers; this is intended to reduce the incidence of bleeding and pain associated with the loop. Most of these are experimental designs that have been either used in only a few women, or used more generally without adequate statistical analysis of efficacy. Evaluation of the double S-shaped loop (Lippes loop) made of a stiffer polymer than the original has revealed improvement in performance.

Studies in animals have shown that the intrauterine placement of certain metals, such as copper, reduces fertility by the prevention of implantation. This interesting finding has been incorporated in the design of a T-shaped plastic device which carries a small quantity of copper. The small size of this device, compared with others in use, facilitates the insertion procedure and reduces the risk of uterine perforation or endometrial trauma. Now under clinical investigation, this device appears to have considerable promise. Continuation rates near 90 percent for the first 6 months to 1 year of use have been reported in independent studies in Chile and the United States. Pregnancy rates of 0.5 and 1 per 100 women per year have been reported from these two studies. Significantly, the device can be used successfully in women who have never been pregnant. At least one other copper-carrying device is being studied. This, and others that are likely to emerge, will have to be evaluated independently to establish performance statistics.

Hormonal Suppression of Ovulation

Oral Preparations

Pharmaceutical firms continue to investigate different combinations or doses of synthetic progestins and synthetic estrogens in an effort to reduce side effects of antiovulant therapy while maintaining a high degree of effectiveness. The progestogenic dose has gradually been reduced from over 9 milligrams in the original preparations to products with 0.5 milligram. The dose of estrogen, ethinyl estradiol or its 3-methyl ether, has remained constant around 0.1 milligram until recently when products with 0.05 milligram appeared. There are insufficient data to evaluate the comparative performance of the various dosage combinations with respect to either efficacy or minor side effects, but it has been established that the risk of thromboembolic disease is less with preparations containing 50 micrograms of estrogen. A product of growing popularity contains 500 micrograms of a racemic mixture of the synthetic progestin, norgestrel, and 50 micrograms of ethinyl estradiol. Since only one enantiomorph has identifiable biological activity, this combination is probably giving antiovulatory therapy at an effective dose of 250 micrograms of the progestin and 50 micrograms of the estrogen.

A series of steroids which are stored in adipose tissue after absorption from the gastrointestinal tract is being investigated for possible one-pill-a-month contraceptive therapy. The investigation is aimed at calibrating the oral dose of the combination that will result in a month-long release of steroid from the adipose tissue at a level that will suppress ovulation while maintaining an acceptable endometrial bleeding pattern. So far, the problem of unpredictable endometrial bleeding has not been resolved.

Parenteral Preparations

Intramuscular injections of steroids can give a depot effect, adjusted to last a single month or for many months. A widely tested monthly injection regimen has been based on the use of an injectable estrogen-progestin combination which "wears off" approximately 30 days after administration. For some clinical situations in which the physician wants control of drug therapy not left to the patient's responsibility, this procedure has distinct advantages, but more studies are required to establish the total pattern of safety, efficacy, side effects, and reversibility.

The most widely studied compound for injectable hormonal contraception is 6-alpha-methyl-17-alpha-hydroxyprogesterone acetate; several thousands of women have been included in studies in many countries. The regimen investigated most completely is 150 milligrams injected every 90 days, although studies are also in progress with semiannual injections of 500 milligrams. With this procedure ovulation is generally suppressed through an interference with midcycle elevation in the production of LH. Ovarian follicle development appears, nevertheless, to proceed so that endogenous estrogen production may not be completely obliterated. The endometrial pattern, however, reveals that the established estrogen-progestin balance is far from normal. As a result, uterine bleeding is totally unpredictable for women on this regimen. There is considerable patient variation, but by the end of a year the majority of women have atrophic endometria and are amenorrheic. An extremely low pregnancy rate has been obtained with this procedure. There is, however, considerable delay in the restoration of ovulatory cycles at will. Delays in ovulation from 12 months to 21 months are not uncommon, and the time required for the posttreatment establishment of a regular ovulatory pattern is still not certain. In order to regularize the pattern of endometrial bleeding, some clinicians have employed the periodic administration of estrogen either orally or by injection as an adjunct to injected progestin. Since this requires monthly return visits, or self-administered monthly courses of oral estrogen, it detracts considerably from the method's simplicity.

The same compound has been tested for antiovulatory activity following absorption through the vaginal mucosa. A novel mode of administration, in the form of a vaginal ring, has been designed for this purpose. The compound is homogenized with a nonvulcanized form of a silicone polymer, and the mixture is molded in the form of a ring, similar to the rim of a diaphragm. The physico-chemical properties of the polymer are such that the hormonal steroid diffuses from the ring at a relatively constant rate and can be absorbed through the vaginal mucosa. The daily release rate is sufficiently high to provide systemic levels of the hormone that cause pituitary suppression and the inhibition of ovulation. The vaginal ring can be positioned by the woman and left in place for approximately 1 month. After removal of the ring, endometrial sloughing and bleeding occur. Subsequently, a new ring can be used for the next month. Thus a simulated anovulatory menstrual cycle can be induced similar to that achieved with antiovulatory therapy using oral hormones.

Hormonal Contraception without Suppressing Ovulation

Daily Oral Progestins

A new type of hormonal contraception is now being investigated in several countries. At least one product is commercially available in some countries. This innovation in hormonal contraception is continuous low-dose progestin therapy, which imparts an antifertility effect without added estrogen and without necessarily inhibiting ovulation. The total basis of the antifertility action remains an uncertainty.

During the investigation of oral progestin-estrogen contraceptives, several investigators concluded that ovulation was occurring in a significant percentage of cycles, even though the antifertility effect was almost absolute. In 1965 work was reported which implied that ovulation suppression could be dispensed with entirely while still retaining a potent antifertility effect. Approximately two-thirds of the patients had some cycle irregularity during the study period. Subsequently, at least six other synthetic progestins have been placed in clinical investigations at doses intended to replicate the low-dose effect. Experience is sufficient with three of these compounds to indicate confirmatory results. With one of these, antifertility activity can be achieved with the extremely low oral dose of 40 micrograms daily. A preparation containing 350 micrograms of norethindrone is now marketed in some countries. It is now evident that the low pregnancy rate reported in the initial study is not achieved generally with low-dose progestin therapy. Pregnancy rates ranging from two to seven per 100 women per year have been reported, and the extent of irregular bleeding is considerably greater than that first reported.

Although the mechanism by which the uninterrupted daily administration of these progestational agents creates a state of infertility without suppressing the pituitary and inhibiting ovulation remains uncertain, the possibilities can be narrowed down considerably. That the therapy does not necessarily interfere with ovulation suggests that the mode of action may be on sperm or ovum transport, the fertilization process itself, transport of the zygote, or the preparation of the endometrium for nidation. Histological evidence from biopsy material suggests that endometrial changes are not responsible for the antifertility effect. Sperm transport could be affected at the level of passage of sperm through the cervical mucus, or higher in the female tract. Although the preliminary reports tended to emphasize changes in cervical mucus that could create a barrier hostile to spermatozoa, it now appears that these changes are not necessarily correlated with the antifertility effect. Future investigations will be required to establish the effect of microdoses of progestins on such key factors as tubal transport rates of gametes and fertilization itself. The effect of progestins in preventing sperm capacitation in other mammals may very well reveal the mechanism behind the antifertility effect observed in clinical usage.

Subdermal Progestin Implants

The discovery of the antifertility action of low-dose progestins, based on uninterrupted administration, opens for the first time the possibility of single-administration, long-term, reversible control of fertility by hormonal means in a manner that would allow for maintenance of ovarian function and menstrual cycles. A possible application of this principle was suggested by experiments demonstrating that steroid hormones may be released at low and constant rates from capsules made of various silicone polymers. One such material, polydimethylsiloxane, is already used widely in surgery and is found to be nonreactive when implanted subdermally in human subjects. Capsules containing the synthetic progestin, megestrol acetate, have been inserted subdermally in female rats, rabbits, and monkeys, and biological evidence of slow and constant release of the hormone has been obtained.

A capsule small enough to be inserted under the skin through a hypodermic needle can be filled with an adequate supply of this progestin to last for more than 3 years. Capsules that could provide continuous low-dose progestin release for 5, 10, 15 years, or longer may prove to be within practical limits. Contraceptive implants of this type could be removed at will and there is no reason to believe that subsequent fertility would be impaired. This form of low-dose progestin contraception, reversible and simple, may very well be among the next developments in contraceptive methodology. The method is now being evaluated clinically. Its effectiveness for a full year after capsule insertion has been established.

Progestin-Releasing Intrauterine Devices

Local delivery of antifertility agents to the endometrium has been suggested as a means of improving intrauterine contraception. The concept has been tested clinically with the use of progestational steroids released from a carrier of polydimethylsiloxane attached to a conventional contraceptive device. The preliminary results are too limited in scope to determine whether this procedure changes significantly the performance of the intrauterine device itself.

Weekly Oral Preparation

Efforts to alter the endometrium sufficiently to prevent nidation but not so drastically as to cause unacceptable patterns of irregular bleeding include intermittent as well as daily use of oral steroids. The weekly method now under investigation in some countries involves the use of a steroid with minimal hormonal activity (13-ethyl-17-alpha-ethinyl-17-beta-hydroxy-gona-4,9,11-trien-3-one). If the compound does not have a prolonged biological half-life, an issue that has not yet been studied, its antifertility effect could be interpreted in terms of the influence exerted on the affected end organ during the specific days of intermittent administration. Although more studies are required, the end organ involved is most likely the endometrium. The intrusion of the exogenous steroid once a week appears to disrupt the normal endometrial preparation for nidation in response to the endogenous cycle of ovarian hormones. With this compound the titration of an antifertility effect that does not cause irregular endometrial bleeding is in progress. Currently, studies involve a total of several hundred women in the various regimens tested.

Menses Inducers, Oral Use

Specific inhibition of progesterone synthesis by the corpus luteum has long been recognized as a possible means to prevent or interrupt nidation in all eutherian mammals, including the human female. In laboratory animals this has been achieved by the use of natural luteolytic substances extracted from uteri, by immunization with protein enzymes involved in the biosynthesis of progesterone, and by various hormonal interventions that inhibit the maintenance of luteotropic stimulation by pituitary gonadotropic hormones during the period that corpus luteum function is required for the maintenance of pregnancy. This concept has now been extended to human studies. A fall in postovulatory serum progestin levels has been observed with steroidal compounds administered during the luteal phase of the cycle. The obvious contraceptive implication of this observation is now being investigated in the United States and Sweden on a small series of women with two different compounds, norethindrone and oxymetholone. In another clinical trial still in a preliminary phase, a compound that interferes with a key enzymic step in progesterone synthesis by the corpus luteum has been used for menses induction. The compound, aminoglutethamide, is a known adrenal suppressant. Investigations with this and related compounds will have to focus on relative effects on the corpus luteum and the adrenal cortex.

Postcoital Estrogens

The postcoital administration of estrogens hastens egg transport in several species (see below for a more complete discussion). This type of activity may account for the reported antifertility effect in women of the postcoital administration of known estrogens or synthetic compounds with weak estrogenic activity. One such compound, F-6103, was under limited clinical evaluation as a postcoital contraceptive, but the results have shown that it is ineffective in preventing pregnancy in humans.

Prostaglandins

The prostaglandins are a class of C_{20} long-chain hydroxy fatty acids found in many tissues and are known to have a wide spectrum of biological activities. One of the first to be described was the ability to cause uterine contractions. Recently, progress in isolation and purification of prostaglandins has revived interest in this specific biological activity and has led to clinical study of the use of two of them (prostaglandins E_2 and $F_{2\alpha}$) for induction of labor or the induction of abortion. Although there is evidence in lower animals that prostaglandins may be luteolytic under certain conditions, there is no evidence that the imputed abortifacient activity in women is based on any action other than the initiation of myometrial contractions. There is an observation from studies with women that at least one of the prostaglandins increases the level of activity of the musculature of the fallopian tubes. Thus three types of contragestational action have been suggested, with the hope that a specific effect on the reproductive system will be separable from the general pharmacological activity of the compound. These are medical induction of abortion, menses induction in the case of a missed (or late) period, and postcoital use to interrupt normal zygote transport and thereby prevent pregnancy. Since the prostaglandins so far isolated or synthesized are not potent by the oral route of administration, other means of administration have to be used for their clinical application. The intravenous route has significance only for a hospital or clinic situation, that is, for the induction of abortion, not for monthly or postcoital use. Yet published experience with intravenous administration of prostaglandins for induction of abortion reveals failure to evacuate the uterus completely in a significant number of cases. This shortcoming, coupled with a high incidence of side effects, makes it unlikely that this procedure will replace surgical methods now in use. An alternative route of administration under study is absorption by the vaginal mucosa. This drug delivery system would be possible for any of the three approaches to contragestational activity mentioned above; it is being tested clinically to induce abortion and menses. The limited data so far available reveal the problems of incomplete evacuation and failure to disassociate the action on the myometrium from action on other smooth muscles. This route of administration requires further study, but it is already clear that for menses induction and abortion a success rate of considerably less than 100 percent would have to be accepted with the use of the current experimental procedures. Administration of prostaglandins by the transcervical intrauterine or transabdominal intraamniotic routes is being studied for the termination of pregnancy. The advantages, if any, of these procedures over those currently in use remain to be established. In general, the subject requires a considerable effort in fundamental research before the pharmacological applications of the prostaglandins, for any therapeutic purpose, can be established. Fortunately, the current revival of interest in their potentiality for the control of fertility has stimulated the pace of investigation.

Methods for Use by Men

Attempts to develop methods of contraception for use by men are limited by the small number of steps in the male reproductive sequence that are amenable to controlled interference. There are three: sperm production, sperm maturation in the epididymis, and sperm transport. Under-skin capsules that release minute amounts of testosterone are being evaluated to determine if a dose can be established that inhibits spermatogenesis without eliciting the more general metabolic effects of an androgen. Compounds that interfere with sperm maturation in animals are also being tested in men by the use of under-skin capsules. For interfering with sperm transport, the surgical procedure of vasoligation is simple and effective, but it carries the disadvantage of irreversibility in many cases. In an attempt to overcome this problem, various procedures to achieve reversible vasoocclusion are being evaluated (see above). Other potential methods of male contraception are still at the stage of infrahuman animal experimentation.

Laboratory Research Related to Fertility Regulation

Ovulation Suppression

Although the original hormonal contraceptives (progestin plus estrogen) probably have other biological effects that contribute to their contraceptive effectiveness, they are basically ovulation inhibitors. The effect on ovulation is subsequent to suppression of pituitary gonadotropin release that in turn is the result of an action by the administered steroids on the hypothalamus or higher brain center. Ovulation suppression by means of a primary action at the level of the central nervous system (CNS) can be achieved experimentally by a number of other pharmacological agents, including tranquilizers, narcotics, and some cardiovascular drugs. Morphine, for example, has been shown to inhibit ovulation in women. A practical application of these observations for the purpose of controlling ovulation seems unlikely, however, since there is no evidence that the antiovulatory effect can be isolated from the general pharmacological effect of these compounds.

Another possibility for the interference with ovulation through an effect at the CNS stems from our growing understanding of how the CNS-pituitary link operates. Humoral substances from the brain that regulate the release by the pituitary of gonadotropins have been identified. They are relatively simple molecules, and their precise chemistry is now being established. The decapeptide that controls the production and release of LH has been isolated, described, and synthesized. Whether it is dissimilar from the substance controlling the release of FSH remains to be established. The possibility of using these releasing factors or analogs that may act as competitive antagonists as a basis for fertility control is very real. Of greater potential applicability perhaps is the recent finding that the hypothalamus may produce, in addition to gonadotropin-releasing factors, inhibitory substances that provide a physiologically normal means to suppress gonadotropin production. The inhibitor has been found in the brains of infants and prepubertal children, suggesting that it may play a role in holding the pituitary-gonadal circuit in check until puberty. The role of the pineal gland in this regard is under investigation and may prove to be of considerable importance.

Direct suppression of gonadotropin production at the pituitary level, or interference with action of circulating gonadotropins, can be achieved by immunological means. Antibodies to gonadotropins can be induced in experimental animals by immunization. Immunized animals, either males or females, have typical manifestations of gonadotropin deficiency. In the male spermatogenesis is impaired; in the female ovum maturation or ovulation is prevented. There remain, however, several important issues to be resolved; a practical application of these experimental findings in human subjects is not imminent. We cannot, for example, envisage now a means of imparting controlled reversibility to a method of fertility inhibition based on active immunization with gonadotropins. Also, our present inability to purify completely LH and FSH makes it difficult to separate an immunologically induced interference with the gamete-producing function of the gonad from undesirable interference with the gland's hormone-producing function.

Another approach to the inactivation of circulating gonadotropins has been the study of natural plant products. Several plants, one a North American prairie grass, have been reported to have this activity, but years of study have failed to reveal an active and stabile constituent that is devoid of undesirable side effects. In general, the evaluation of plant extracts for antifertility action by means of gonadotropin inactivation or any other route of activity has been discouraging and unrewarding. From time to time a plant product is described that has a clear antifertility effect in laboratory rodents. Almost invariably these results can be ascribed to a mild estrogenic activity common in many legumes and other plants, an activity that has no practical significance for contraceptive purposes.

It appears therefore that control of fertility based on ovulation suppression will, in the near future, continue to depend on the action of synthetic hormones similar to those now in use as constituents of the widely used OC agents. It is possible to envisage, however, application of our growing understanding of the preovulation surge of LH as a basis for controlled interference. Premature ovulation by the use of exogenously administered releasing factors, or asynchronous LH peaking by the use of estrogens early in the cycle, are distinct possibilities for future study.

Remarkably little is known of the various steps involved in the ovulatory process itself. It is as yet not clear what factors cause one antral follicle to mature and ovulate whereas adjacent follicles of the same size and the same histological and cytological appearance become atretic. The specific actions of the gonadotropic and steroid hormones in the gross development of follicles and at the cellular level can be described only in the most general terms. We need to know what initiates the first changes in the follicular apparatus that lead to its growth and development.

The phenomenon of follicular atresia can be recognized if it is reasonably well advanced, but the basic biochemical or endocrinological aspects of its origin are unknown for any mammal. A great variety of endocrine insults have been shown to induce atresia at all levels of follicular development and there may be other factors that need to be explored in detail. There apparently is a definite interrelationship established very early in development between the egg, the corona, and the cumulus cells. Increasing evidence indicates that if the egg is not completely surrounded by follicular cells the meiotic process proceeds to diakinesis, and follicular atresia results. A better understanding of atresia could very well contribute information as to the factors involved in normal follicular development and ovulation and their control.

Tubal Transport of Ova

The zygote or newly fertilized egg normally spends several days passing through the female reproductive tract before it begins the process of implantation in the uterus. This is a carefully timed sequence which must proceed in phase with preparatory changes occurring in the endometrium. If the zygote arrives too early from the fallopian tube, the uterus is inadequately developed and the zygote will degenerate.

Ovarian steroid hormones have a major regulatory influence on the tubal transport of ova. Estrogens increase the rate of secretion of tubal fluid, stimulate ciliary activity at the upper portion of the tube, and increase the peristaltic activity of the tubal musculature. Progesterone generally has the opposite effect on each parameter. Upsetting the proper sequence of hormonal influences can therefore, disturb the normal passage of cleaving ova in the fallopian tubes. Indeed, this has been demonstrated by many experiments, but no simple unifying concept can be synthesized from the reported observations. Nevertheless, the apparent lability of the regulatory mechanism for normal tubal transport of ova provides an attractive basis for controlled interference with fertility. Indeed, this site of action has been postulated as the basis for the antifertility effect of intrauterine devices.

Studies of tubal transport in ova in rhesus monkeys have shown that the antifertility action of an intrauterine device may occur not only in the uterus but at the tubal level as well. This finding is compatible with the extensive clinical data on intrauterine contraception, which establish that such devices prevent both uterine and ectopic tubal pregnancies. Rapid expulsion of tubal ova occurs in gonadotropin-stimulated monkeys with intrauterine devices, whereas gonadotropin treatment itself does not have this effect. Ova cannot be found in the tubes of monkeys with these devices within 24 hours after ovulation. Since gonadotropin-treated animals tend to have multiple ovulations, an elevated level of ovarian steroids may contribute to the acceleration of tubal ova. That this phenomenon in itself does not account for the rapid passage of ova is evidenced by the recovery of ova from gonadotropin-treated animals without the devices. In subsequent experiments ova were recovered from the tubes of normally ovulating monkeys, although the precise rate of tubal passage in this situation has not yet been established. Whereas the additive effect of gonadotropin stimulation and the presence of an intrauterine device may be required to demonstrate a dramatic tube-flushing effect within 24 hours, each stimulus alone may influence the tubal transport rate to some lesser extent. Careful timing of tubal transport rates in monkeys with the devices and under different conditions of ovarian function is required to elucidate this issue. Investigations with human subjects have established that sperm may reach the fallopian tubes, but whether fertilization occurs in women using intrauterine devices effectively has not been established, and the recovery of several hundred human ova will be required before this important point can be clarified.

An influence on tubal transport of ova or zygotes may be involved in studies on the effectiveness of postcoitally administered hormonal agents in the prevention of nidation in infrahuman primates. Pregnancy in the rhesus monkey can be prevented by the administration of an estrogen during the 4-day period after mating when the fertilized egg is traversing the fallopian tube. Estradiol, stilbestrol, ethinyl estradiol, mestranol, or an experimental compound that is both antiestrogenic and estrogenic were administered orally following over 300 matings in a colony of rhesus monkeys that normally achieves a 70-percent pregnancy rate after mating. When the estrogen was administered postcoitally, not a single pregnancy ensued.

In these experiments there were no direct attempts to ascertain the cause for the antifertility action. Accelerated tubal transport is implicated because, in other species, the relationship has been established between postcoital estrogen treatment and acceleration of tubal transport of ova.

Estrogen-induced premature expulsion of the ova from the fallopian tubes has been demonstrated in rats, rabbits, and guinea pigs. The effect has now been reported with a variety of compounds, both steroidal and nonsteroidal, but it becomes evident that the common feature of all is their estrogenicity. The data available on the application of this principle to human subjects are too few to permit an evaluation, although it has been reported that no pregnancies occurred in a limited number of rape cases and volunteer subjects treated with stilbestrol or ethinyl estradiol for several days immediately after midcycle insemination. A systematic analysis of the potential antifertility action of postcoital estrogen treatment in women is required. On the assumption that the activity in animals carries over to the human, an interesting method of fertility control is suggested. Pills taken orally for 1 or 2 days after intercourse would prevent pregnancy even if fertilization had occurred. There would be no disruption of the natural ovarian cycle and no manifestation whatever of the pregnant state. Indeed, there is no physiological criterion by which any stage of the process could be classified as a pregnancy.

Biology of the Ovum

The blastocyst remains free in the cornual or uterine lumen for several days before attachment and implantation begin. Although information is not available for lower primates and man, comparative studies of other mammalian species reveal that the zona pellucida remains intact until shortly before embryo attachment begins. It is further known that embryos escape from the zona pellucida by a variety of mechanisms, depending upon the species. Such escape is effected only if the ovum is fertilized. Studies in mice reveal that the expanding blastocysts mechanically rupture the zona pellucida. In the rat the zona pellucida is shed during the afternoon of the fifth day after mating or just before embryo attachment. Rupture of the zona in this species is effected also by embryo expansion. In one experiment a macaque blastocyst, washed out of the uterus 8 days after ovulation, still had an intact zona pellucida. Likewise a human blastocyst recovered from the uterine lumen about 4 days after ovulation was enclosed in the zona. There is no information as to the mechanism of blastocyst escape from the zona pellucida in primates. Whether the trophoblast participates in depolymerizing the zona so as to effect escape is unknown. If so, as is the case for other mammals, then knowledge of methods of inhibiting depolymerization and thereby preventing the escape of the blastocyst may be a promising target in pregnancy control.

Corpus Luteum Function

The uterine environment is not essential for blastocyst survival, implantation, and development, since in some species ova can be fertilized and cultured *in vitro,* and a human ectopic pregnancy is quite independent of the uterine environment. Nevertheless, in all species studied a successful intrauterine pregnancy requires adequate progestational preparation and maintenance of the endometrium. It seems likely that the same situation prevails in the human female.

On this assumption several investigators have sought steroid inhibitors of implantation by examining a variety of compounds for their contraprogestational effect, and a few compounds have emerged which appear to have implantation-inhibiting activity while being virtually devoid of estrogenic activity. In the *A*-norandrostene series of steroids, at least one compound with an excellent record in biological assays has been carried to preliminary clinical trial and evidence for contraprogestational activity obtained. Further trials in infrahuman primates will be required before the antifertility potential of this and related compounds can be established.

In addition, several laboratories have reported contraprogestational or antiimplantation activity for several synthetic di- or triphenyl hydrocarbons which have been available for many years and several of which are used clinically as synthetic estrogens. One (clomiphene) that has excellent antifertility activity in laboratory rats has proved to be remarkably effective in the induction of ovulation in cases of human infertility and is now prescribed for that purpose. Another was in fact tested as an antifertility agent in women, but without success. Still others, in spite of interesting laboratory findings, have not been evaluated in the human subject. These compounds have been variously described as antizygotic, blastotoxic, antinidational, weak estrogens, antiestrogens, or antiprogestational, depending on the assay system used. A systematic analysis of the potential usefulness and safety of this type of compound for human contraception should be undertaken.

Another approach to the elimination of the progesterone needed for nidation has been to interfere with the function of the corpus luteum by pharmacological means. Several amine oxidase inhibitors with phenylhydrazine structure, when studied in the rat, appear to have this effect either directly or through a depressing effect on the pituitary production of luteotrophin. The role of the biological amines, amine oxidases, and their inhibitors in the reproductive process is as yet inadequately investigated. However, this could be one of the more important pharmacological approaches to fertility regulation.

A similar luteolytic effect by inhibition of luteotrophic hormone release is believed to account for the antifertility activity of ergocornine, an antihistaminic of the ergot series, which prevents implantation in the rat or mouse when administered during a limited period of tubal transport of fertilized ova. Future development of this particular compound for contraception is not likely in view of apparent toxicity in clinical trials.

The present disappointments notwithstanding, antifertility action through a luteolytic contraprogestational effect is one of the more intriguing prospects on the research horizon. In theory an oral preparation active in such a manner could be taken by a woman either monthly, at the time of the expected menses, or only on the occasion of a suspected fertile cycle as evidenced by delay in the onset of menstruation. Efforts along these lines will be stimulated by the growing evidence for the existence, in many species, of a humoral luteolytic substance produced by the uterus and transmitted by tissue diffusion and common blood supply to the ovary.

It has been suggested that luteolysis may be a normal physiological role for prostaglandins. This class of ubiquitous C_{20} fatty acid derivatives can be extracted from many tissue sources including seminal fluid, umbilical cord, lung, iris, thymus, kidney, brain, and endometrium. Studies in laboratory rodents suggest that the compounds may be luteolytic, although prostaglandins may stimulate *in vitro* progesterone synthesis. Whether the prostaglandins can in fact suppress corpus luteum function in man remains to be established.

Myometrial Stimulation

The initiation of myometrial contractions heralds the onset of labor at or near term, or the beginning of a spontaneous or threatened abortion at earlier stages of pregnancy. Although the steroid hormones can influence uterine contractility and probably play a role in the endocrine control of the myometrium, other factors are undoubtedly involved, not all of which are understood. Oxytocin, for example, is an effective stimulant of the late gestational uterus, but is less effective when administered earlier in pregnancy. It has, however, been studied for nonsurgical induction of abortion. However, some of the prostaglandins can initiate myometrial activity in early or late pregnancy, or in the nongravid uterus. This activity led to the testing of prostaglandin $F_{2\alpha}$ or E_2 as abortifacients, from the eighth to twenty-second week of pregnancy. Since the compounds are not effective orally, they were used intravenously for the purposes of study. The initiation of myometrial contraction was usually followed by evacuation of the uterus. So far, it has not been established that an effect on the myometrium can be separated from effects on other smooth muscle. The percentage of cases with incomplete abortions is unacceptably high, and the prospects of serious side effects (respiratory or cardiovascular acute changes, profuse bleeding) are too great to consider the intravenous administration of prostaglandin E_2 or $F_{2\alpha}$ as useful replacement for surgical abortion. There are insufficient data published to evaluate the potential of intravaginal or intrauterine modes of administration as a means to avoid the problems encountered following intravenous administration.

Suppression of Sperm Production

The testis, similar to the ovary, depends upon stimulation by pituitary gonadotropic hormones in order to perform its normal function of producing sex hormones and sperm. Similar to the ovary, the testis can be secondarily suppressed by a procedure that stops the production of gonadotropins. The oral progestins, for example, could be effective agents for the inhibition of sperm production. Doses required, however, have the unwelcome effect of inhibiting the secretion of sex hormones by the testis; as a consequence libido and potency are reduced. There are, however, long-acting androgen esters now available that may provide long-term suppression of spermatogenesis while maintaining libido and general well-being.

Although testosterone and other androgens can suppress spermatogenesis in man, the feasibility of such treatment for contraception depends on establishment of a dosage and mode of administration that would provide the antispermatogenic effect of an androgen without causing the more general metabolic effects such as changes in blood chemistry or stimulation of the prostate, effects that would give cause for medical anxiety.

There are other orally active chemical compounds that stop sperm production without interfering with testicular hormone secretion. Several classes of chemical substances have been found to have this effect, but all have accompanying side effects that render them unsatisfactory for contraceptive purposes. One group of such compounds is the nitrofurans, which have had wide use as inhibitors of bacterial growth. They are very effective in stopping sperm production in man, but the doses required cause nausea and headache. In 1960 great promise was seen in another group of compounds, bisdichloroacetyl diamines, originally of interest for their usefulness in the treatment of intestinal amebas. The drugs were tested with volunteers among penitentiary prisoners; complete suppression of sperm count was demonstrable. When the drug was discontinued, sperm count returned to normal. With high hopes trials were expanded to include men in more usual social circumstances. The first unexpected observation was that the drug enhanced the vascular effects of imbibed alcohol. As surveillance of the original prisoner volunteers continued, it was suspected that the drug was associated with a frequent occurrence of hepatitis, so further investigation was halted. It is now uncertain whether or not the suspected hepatotoxicity was indeed drug related; thus reevaluation of the prospects of this class of drugs should be made.

Another promising compound, a dinitropyrrole, impaired spermatogenesis in the rat for as long as 4 weeks after a single oral dose. An infertile state could be maintained indefinitely by administering single doses at intervals of 4 weeks. Sperm production recovered fully when treatment was finally stopped. Subsequent toxicological findings resulted in the withdrawal of the compound from investigation, but it is possible that a related compound may be discovered that retains the antispermatogenic activity while being devoid of toxicity. The compound fluoracetamide may meet this specification. It suppresses spermatogenesis in rats with no evidence of toxicity. The compound is now being studied in infrahuman primates; but preliminary results suggest that a similar effect does not occur, even with doses considerably higher than those that are effective in the rat.

Testicular antigens that can be used for specific immunization to prevent spermatogenesis have been isolated. Indeed, even nonpurified, crude testicular extracts can cause aspermatogenesis in the guinea pig and in the rat. An attempt has been made to immunize human males with testicular extract believed to be purified for the aspermatogenic factor, but the results have not been notable. Immunization with tissue extracts for the purpose of inducing sterility in either the male or the female organism seems distant at this juncture. The basic problems of tissue cross-reactions, specificity of antigens, controlled reversibility of the immune response, and development of acceptable adjuvants still impede progress in this field.

Fertilizing Capacity of Spermatozoa

Mammalian spermatozoa may have complete morphological appearances of normalcy and normal motility without possessing the capacity to fertilize ova. This final maturation stage of sperm has been termed "capacitation." Evidence for capacitation has been obtained in several mammalian species, including rabbit, rat, hamster, sheep, and ferret. From the viewpoint of fertility control research, the intriguing extension of the capacitation concept is an understanding of the manner in which it can be inhibited. Sperm do not capacitate in the uterus of a rabbit injected with progesterone or in the uterus of a rabbit in the pseudopregnant state (progestational condition). In fact, it has been suggested that fully capacitated sperm can lose their fertilizing capacity by exposure to a female reproductive tract under progestin domination.

Although the process of capacitation has not been demonstrated as an essential element of sperm maturation in primates, the assumption that it does occur seems reasonable. With the greater availability of infrahuman primates for reproductive research, this issue should be clarified before long. In any event the evidence from other eutherian mammals suggests that interference with sperm capacitation may account for the contraceptive action of continuous low-dose progestin therapy that imparts an antifertility effect without added estrogen and without inhibiting ovulation.

Scientists have not been successful in finding a morphological or biochemical criterion that can be linked unequivocally with capacitation. Changes in oxygen consumption by sperm, surface-coating proteins, or detachment of the acrosome have been implicated, but these phenomena are not proven characteristics of the capacitation process. Ultimately, the proof lies in the actual fertilizing capacity of the sperm. Work on the process of *in vitro* fertilization is therefore directly related to this problem.

The role of progestin in influencing the ability of sperm to capacitate in the female tract raises the question of a possible influence of this class of hormones while sperm still reside in the male storage system. The effect of progestins in the male at doses below the threshold for inhibition of spermatogenesis has not been systematically analyzed in any species. Although the fertilizing capacity of spermatozoa is dependent foremost on an intracellular component, namely, the DNA of the sperm chromatin, changes on the surface coat of the sperm could perhaps control the ability of the sperm to fulfill its role in fertilization. Whether such changes could be brought about, by hormonal or other means, while the differentiated spermatozoa reside in the epididymis or ductule system, warrants careful study.

Human Seminal Fluid

A primary function of the ejaculate in all species is to act as a carrier of the spermatozoa from the male to the female reproductive tract. However, the chemical complexity of the seminal plasma suggests other and more subtle relationships to the function of spermatozoa. The chemical composition of the seminal fluid in man has been analyzed in considerable detail. It is known, for example, that the seminal fluid contains several trace metals, for example, iron, copper, zinc, and magnesium. Chelating agents at a concentration of a few parts per million can be toxic to ram, bull, rabbit, and human spermatozoa; and the addition of copper or zinc enhances the spermatotoxic activity of many chelating agents. Considerable work has been done on seminal carbohydrates, which appear to be an important source of energy for spermatozoa. In human semen sorbitol, which may be oxidized to fructose by the enzyme sorbitol dehydrogenase, is also present. Fructose is formed in the seminal vesicles and its formation is dependent upon the secretion of testosterone by the testis. It remains to be explored if mild antiandrogenic substances can influence seminal fluid chemistry at doses below the threshold for other, undesirable, antiandrogenic effects. Even so, the influence of such a change on the fertilizing capacity of spermatozoa is uncertain since the seminal fluid may not be essential to the fertilizing capacity of sperm, according to some studies.

Although such conclusions tend to minimize the significance of the seminal fluids, it remains a distinct possibility that adverse conditions of the fluid medium could influence the spermatozoal surface in a manner that would impair the ability of sperm to reach the fallopian tube or to penetrate an ovum. From experimental data now available, it seems that "an adverse condition" is more likely to be the addition of a deleterious constituent to the seminal fluid than the deletion of something normally present. For example, zinc, which is believed to be essential for normal sperm metabolism, is contributed chiefly by the rat's dorsolateral prostate, yet removal of this organ does not reduce the fertility of the male rat. The possibility of exogenous substances finding their way to the seminal fluid is real. Following the administration of estrogen to the male rat or rabbit, for example, estrogen can be found in the ejaculate. Ethanol and sulfonamides from exogenous sources have also been found in seminal fluid. A pertinent question therefore is whether spermicidal substances might be found which could reach seminal fluid after ingestion or parenteral administration but which would not have general toxicity.

Pheromones

The study of pheromones provides another approach to the possible development of new agents or methods for fertility control. The term pheromone designates a volatile substance transferred from one individual to another of the same species, which generally in minute amounts evokes some behavioral or developmental response in the recipient.

Investigations in this field have centered largely on insects, but there have been demonstrations of analogous phenomena in laboratory mammals. One example is the failure of implantation that occurs in a mated female mouse placed in the presence of a "strange" male within the first 4 days, but not later, after mating. The effect does not occur if the female is first rendered anosmic by removal of the olfactory lobes. Also, administration of prolactin or ectopic implantation of a pituitary gland overcomes the block to implantation. Another pheromone-induced phenomenon is the occurrence of pseudopregnancy when several female mice are caged together, but not when they are housed individually. Again the effect is overcome if the olfactory bulbs are removed. Excessive crowding of female mice results in anestrus even when contact between animals is prevented by compartmentation. Both anestrus and pseudopregnancy are overcome by the introduction of one or more males, direct contact being again unnecessary for this effect.

Among the insect pheromones those concerning the control of reproductive behavior and gametogenesis are of special interest. In the honeybee a pheromone normally derived from the queen has been isolated and identified which inhibits oogenesis in workers. The manner in which this substance acts to suppress ovulation is uncertain. Also, there have been only occasional, and rather incomplete, investigations of the possible action of these or of chemically related substances in mammals. From other information concerning the action of hormones of mammalian origin on lower organisms, including insects, it is not unreasonable to expect the reciprocal situation to pertain with respect to physiological processes that are basically similar.

Possible Means of Fertility Control — Distant or Near

The foregoing is a fairly comprehensive review of topics in reproductive biology that have some apparent relationship to fertility regulation. From these areas of research, as well as some others, it is possible to compile a list of potential methods to regulate fertility that have a realistic basis in terms of our present knowledge.

For Use by the Female

1. *Once-a-month antiovulant pill.* Now under limited clinical evaluation; this is a modification in the manner of using estrogen-progestin combinations. The steroids used are absorbed from the gastrointestinal tract, stored in adipose tissue, and released gradually over a month.
2. *Once-a-month antiovulant injection.* Now in clinical investigation; this is a modification in the manner of using estrogen-progestin combinations. Long-acting steroid esters are injected at a dose calibrated to last 1 month.
3. *Once-a-month vaginal ring.* Preliminary trials completed. This procedure is based on antiovulatory action of a synthetic progestin released from an elastomer and absorbed through the vaginal mucosa.
4. *Long-term antiovulant injection.* Extensively studied in clinical trials, this procedure is based on the antiovulatory action of a synthetic progestin without estrogen. Microcrystallized suspensions of the steroid are injected in doses that will last 3 or 6 months.

5 *Long-term antiovulant implant.* Not yet studied clinically, this procedure would provide chronic release of estrogen-progestin to suppress ovulation and menstruation.

6 *Continuous low-dose progestin.* This provides an antifertility effect without inhibiting ovulation or the normal endometrial cycle.

a Pill or oil-filled capsule taken orally. Several compounds are under investigation.

b Subdermal implant. Provides for continuous absorption from an elastomer at a constant rate. One version now proven to work for one full year. Studies continuing to improve and simplify the method.

c Removable vaginal ring. Provides for continuous absorption and may act locally on cervical mucus glands or systemically to give the low-dose progestin antifertility effect.

d Long-acting injection. Requires the development of a preparation that would provide a depot effect that gives constant, low absorption below level that will affect pituitary or endometrium.

e Skin-contact absorption. Requires the development of highly potent progestin, active at levels that could be absorbed through the skin—from a finger ring, for example or by being added to a cosmetic.

f IUD-released. Provides for continuous absorption and may act either locally in the uterus, or systemically to give the low-dose progestin antifertility effect.

7 *Long-term luteotrophin injection.* Requires a better understanding of the trophic control of the human corpus luteum. The purpose would be to lengthen the postovulatory phase of the cycle to perhaps 90 days so that a woman would have fewer ovulations per year.

8 *Corpus luteum maintenance by injection of LTH releaser.* Requires identification and purification of LT-RF active in human female. Objective same as in method (7).

9 *Monthly oral preparation to cause luteolysis.* At least two compounds, active orally, have been claimed as having luteolytic activity in animals, and possibly in humans. Taken regularly, the drug would bring on menstruation whether or not the cycle had been fertile.

10 *Monthly injection to cause luteolysis.* This would be an application of the proposed uterine luteolytic factor. If active in the human, it will probably require injection since preliminary work indicates a peptide structure. Objective would be as in method (9).

11 *Vaginal tampon to cause luteolysis or uterine contraction.* This would be a method to apply substances, such as prostaglandins, that are not orally active. Used regularly, it would bring on menstruation whether or not the cycle had been fertile.

12 *Nonregular use of methods (9) or (10).* A variation in the use of methods (9) or (10) would be to instruct the woman to do nothing for contraception but on the infrequent occasion of a fertile cycle, as evidenced by a delay in menstruation, to use a luteolytic method.

13 *Once-a-month contraprogestational pill.* To be taken regularly at the time of the expected menses, to interfere with luteal maintenance of

early decidua, and bring on endometrial sloughing whether or not the cycle had been fertile. Several compounds with a potential for this activity are available.

14 *Once-a-week mild contraprogestational pill.* To be taken regularly to cause a slight intermittent interference with preparation of endometrium for nidation, but not sufficiently potent to cause withdrawal bleeding. One candidate compound now in clinical trial.

15 *Precoital progestins.* At least one compound has been used that appears to cause a rapid change in the cervical mucus, making it impossible for sperm to penetrate.

16 *Postcoital estrogen or antiestrogen taken orally.* On the basis of limited human use, this procedure is claimed to be effective in preventing pregnancy following isolated exposures. Presumably, it affects the rate of ovum transport. Several compounds are active in animals.

17 *Postcoital antizygotic agent taken orally.* Work is required to seek compounds that would appear in the tubal fluid at adequate concentration to be toxic to the zygote without manifesting general toxicity.

18 *Immunization with sperm antigens.* The objective is to prevent deposited spermatozoa from achieving fertilizing capacity in the female tract. Success has been reported in animals, but considerable basic work, including safety studies, are needed before human trials are feasible.

19 *Injection of passively transferred antibodies to HCG.* This procedure would be employed on the occasion of a missed period, during the time that HCG stimulation of the corpus luteum is required to maintain a nascent implantation. Animal studies establish the feasibility of the approach.

20 *Immunization with steroid-binding proteins.* Recently described tissue-specific intracellular binding proteins can be derived from the uterus. Active antibody production would prevent estrogenic changes from occurring in the uterus without interfering with the ovarian cycle.

21 *Improved methods to detect ovulation.* This is feasible on the basis of simplification in the method to detect either LH or progesterone in urine, saliva, or blood (finger-prick sample).

22 *Release of premature* egg. This may be achieved by causing LH peaking early in the cycle, thus depleting the pituitary to prevent subsequent ovulation. May be achieved by use of LH releaser or by single dose of estrogen early in the cycle.

23 *Regularizing ovulation by the use of natural ovulation-inducing substances.* This line of investigation becomes a distinct possibility with the advent of synthetic LH releaser.

24 *Reversible tubal occlusion.* Instead of sectioning the fallopian tube, a removable plug would be introduced, either transcervically or through an abdominal approach.

25 *Simplification of tubal ligation operation.* Now being tested by several surgeons, the tube is sectioned or electrocoagulated by a peritoneoscopic or culdoscopic instrument.

26 *Intrauterine infusion of cytotoxins.* Intrauterine infusion of cytotoxins has been reported as a means of inducing sterility. The safety and permanency of the procedure need further study. The objective is to occlude the intramyometrial portion of the fallopian tube.

27 *Oral or parenteral preparation to assure multiple births at will.* Purified human pituitary gonadotropin can stimulate multiple ovulations; at least one synthetic compound has similar activity. Fine adjustment of dosage, on the basis of ovarian function tests, would provide assurance of multiple births, if desired.

28 *Sex determination at will by immunization with Y-sperm antigen.* There is some evidence that specific antigens from Y sperm may be identifiable. If so, women could be immunized against this antigen in order to inactivate Y sperm and assure female sex determination. No similar approach to male sex determination can be envisaged.

29 *Sex determination at will by artificial insemination.* There have been occasional claims of success in separating X and Y spermatozoa in an ejaculate by physical means (centrifugation, electrophoresis, sizing, column diffusion). By using only Y-bearing gametes in artificial insemination, male zygotes could be assured. No confirmed procedure has yet been established, however.

For Use by the Male

1 *Subdermal implant to suppress spermatogenesis.* Release of an androgen from a capsule of silicone rubber can be achieved at a low and constant rate for several years. This can provide a basis for gonadotropin suppression at levels of androgen therapy that may be medically acceptable.

2 *Periodic injection of long-acting androgen.* Testosterone enanthate, or other esters, can suppress sperm production while providing androgen replacement therapy. Depot injections can remain active for 3 to 6 months.

3 *Subdermal implant of progestin.* Low doses of progestin, below the threshold for pituitary suppression, can prevent maturation of epididymal sperm in some animals. A similar activity in the human male would prevent fertility without suppressing spermatogenesis.

4 *Subdermal implant of antiandrogen to prevent epididymal maturation of sperm.* Cyproterone acetate has this activity in rats and is now being studied in primates, including man.

5 *Oral tablet of synthetic spermatogenesis inhibitor.* Several compounds that act directly on the testis to prevent spermatocyte maturation have been reported. In animals some compounds require continuous administration, whereas others can be given for a few days each month. A nontoxic compound of this type is being sought.

6 *Oral tablet to alter biochemical constitution of seminal fluid.* Although no specific compound has yet been identified that can influence fertility by this mechanism, the possibility of this type of action exists. The appearance of exogenously administered substances in seminal fluid has been reported.

7 *Immunization with testis or sperm antigens.* This procedure can cause aspermatogenesis in animals. Purification of antigens, control over reversibility, mode of immunization remain as problems to be investigated for human application.

8 *Reversible vas deferens occlusion.* The use of a liquid silicone rubber that vulcanizes into a pliable plug at body temperature has been attempted in animals.

9 *Reversible vas deferens ligation.* Procedures are being tested to modify the procedure for surgical vasoligation, in a manner that would improve prospects for reversibility.

For Use by Either the Male or the Female

1 *Use of antagonists against gonadotropin-releasing factors.* The isolation of these polypeptides, may provide a basis for specific LH or FSH suppression so that gamete production in either the male or the female could be prevented.

2 *Immunization with enzymes specific for normal reproductive function.* There is evidence that specific isoenzymes can be identified in gonadal or placental tissue. Although this work is at a preliminary stage of development, it could lead to a source of highly specific antigens.

3 *Oral administration of chemical inhibitors of releasing-factor production.* Monamine oxidase inhibitors and various biological amines interfere with reproductive function in either male or female. These substances may act by interfering with the function of neurosecretory cells in the CNS.

4 *Immunization with purified gonadotropins.* In either male or female animals, gonadotropin antibody formation leads to the expected result of gonadotropin deficiency. In some animals LH immunization prevents the maintenance of early pregnancy.

5 *Oral administration of antigonadotropic drugs.* Several antigonadotropic agents, either synthetic chemicals or natural plant products, have been reported. Taken regularly, an active and safe preparation of this type could prevent sperm production in the male or interfere with reproductive function in the female in several ways.

6 *Topical application of pheromones.* Laboratory experiments demonstrate that volatile agents produced by one animal can influence reproductive functions in another. The role of such substances in human reproductive physiology remains to be established.

ABORTION

methods

Abortion is believed to be the most common form of birth control in all parts of the world, even in countries where modern contraceptives are readily available. This is true in spite of the fact that abortion is illegal in many countries. Where it is legal, it is often permitted only under carefully defined conditions.

Abortion is the termination of pregnancy in the early stages. Two medically approved procedures are commonly used during the first 12 weeks of pregnancy. The first involves scraping the lining of the uterus, thereby dislodging the embryo or fetus, a procedure usually referred to in medical terms as dilation and curettage. The second, and newer, method uses a suction device to draw the fetus from the lining of the uterus. After the sixteenth week of pregnancy the procedure is considerably more complicated, in which case abortion is performed only if the pregnancy seriously endangers the mother's life. This more complicated and difficult procedure usually involves removing, with a needle, some of the fluid from the sac that surrounds the fetus and replacing that fluid with a strong saline solution which stimulates labor and results in expulsion of the fetus.

safety

When performed under appropriate medical circumstances by a competent physician, abortion is safer than a full-term pregnancy. If delayed beyond the twelfth week, however, the risks of complications, or death, increase considerably. The risks increase when the abortion is illegal, the amount of increase depending upon the circumstances. These may range from self-inducement with a knitting needle to the use of nonsterile conditions by untrained or semitrained people. Of course, some illegal abortions are performed by competent physicians under ideal conditions. In the United States, as in many parts of the world, illegal abortions are the greatest single cause of maternal deaths, accounting for about 45 percent of them.

Illegal abortions can probably be obtained in every country in the world. They are most prevalent in countries where laws are more restrictive. In Italy, where contraceptives are prohibited, the abortion rate is estimated to be almost equal to the birthrate. In Chile, where contraception is also illegal, some women are known to have had as many as 20 abortions. In China, Japan, and several eastern European countries, abortions are essentially available upon request, even though a woman must go through the legal procedure of applying for and receiving official sanction before she can have one. Rumania and Bulgaria, alarmed at their very low birthrate which they attribute in part to the ease of obtaining abortions, have recently tightened their regulations.

In the United States several states have liberalized their abortion laws to permit abortion in cases in which child bearing presents a risk to the mental or physical health of the mother, in which pregnancy is the result of incest or rape, and in which there is a substantial likelihood that the child will be physically or mentally defective.[3] To obtain a legal abortion, a woman must usually submit her case to a hospital reviewing board of physicians, a time-consuming and expensive process. Because hospital boards have been generally conservative in their interpretation of changes in the abortion laws, large numbers of illegal abortions continue to be performed each year. The estimated number of illegal abortions per year in the United States is usually given as one million. This amounts to more than one abortion for every four births.

[3] On January 22, 1973 the U.S. Supreme Court struck down essentially all state laws on abortion and made abortion during the first three months of pregnancy the decision of the woman and her doctor.

Attitudes toward abortion are highly variable. Many individuals and some religious groups consider abortion immoral; others feel that there is nothing immoral about it and that every woman should have the right of access to abortion. This is obviously a decision to be made by each individual. Regardless of attitudes, few people would claim that abortion is preferable to contraception. Many people feel that abortion is preferable to the birth of unwanted children, especially in an overpopulated world. Until a more effective form of contraception becomes available, abortion will continue as a common method of birth control.

IF?

If society refuses to accept the need for population control, and if family-planning programs prove ineffective in stabilizing the population size in underdeveloped countries, what are the alternatives? Numerous answers to this question have been proposed; many of them involve governmental actions. These kinds of programs may very well establish the principle that child bearing is a privilege, not a right. The alternatives then become "food for thought." Some of the possibilities are listed below. No arguments are made pro or con, but each young person should examine each of the alternatives carefully and after making his own judgment decide upon his contribution to a solution of the population problem.

1. Extension and legalization of free abortion services to the community as a whole.
2. Use of fertility control agents on a regional (population) basis.
3. Issuing of licenses to have children.
4. Temporary sterilization of women and/or men.
5. Compulsory sterilization of men and/or women having three or more living children.
6. Payments to married couples who refrain from having children during particular time intervals.
7. Bonuses for spacing children or for achieving nonpregnancy.
8. Imposition of a tax on births after a determined number.
9. Allocation of free schooling for two children only in each family.
10. Raising the minimum age for marriage.
11. Required participation of women in the labor market.
12. Minimizing the social position of the family in community life.
13. Instituting a second type of marriage, one of a more temporary nature designed for a childless relationship.

Nearly all organisms that are well adapted to their environment have built-in mechanisms that check population growth before the essentials for survival, food and habitat, are permanently destroyed. However, Nature's population control processes are unemotional, impartial, and truly ruthless—conditions man will surely wish to avoid.

Suggested Readings[1]

Anderson, P. K., Ed. *Omega: Murder of the Ecosystem and Suicide of Man.* Dubuque, Iowa: Brown, 1971.

Berelson, B. et al. *Family Planning and Population Programs.* Chicago: University of Chicago Press, 1966.

Boughey, A. S. *Man and the Environment.* New York: Macmillan, 1971.

Ehrlich, P. R. and A. H. Ehrlich. *Population, Resources, Environment.* San Francisco: Freeman, 1970.

Hardin, G., Ed. *Population, Evolution, and Birth Control,* 2nd ed. San Francisco: Freeman, 1969.

Hinrichs, N., Ed. *Population, Environment and People.* New York: McGraw-Hill, 1971.

[1] For a complete bibliography on contraception, see Segal, S. J. and C. Tietze, "Contraceptive Technology: Current and Prospective Methods" in A Report on Population and Family Planning, July, 1971, The Population Council, 245 Park Ave., New York, New York 10017.

Corn is one of the three most important food crops. The photograph shows the damage to the ends of some ears by the corn earworm.

The Carrying Capacity of Earth

THE CONCEPT OF CARRYING CAPACITY

carrying capacity

The previous chapters explored the problems of the human population and the mechanics that man could bring to bear on its control. We now ask how great a population the earth can support and what factors will establish the final population size.

Here we borrow a concept from wildlife management and attempt to apply it to man. Carrying capacity describes the ability of an ecosystem to sustain a level of production without damage to its essential life support systems (Fig. 14-1). These systems include not only the basic producer organisms but also essential mineral cycling, maintenance of the soil, diversity of plant and animal life, adequate water levels, and sufficient reserves to offset fluctuations (droughts, excessive rains, extremes of temperature). Carrying capacity is expressed in terms of population—the maximum population for long-term survival.

Carrying capacity can also be expressed on a per-person basis. In this case it is the smallest area into which the operations necessary to support one individual can be compressed. We might then say that carrying capacity is the number of people a square meter, a hectare, an acre, or a square mile can support on a continuing basis.

Fig. 14-1. When the number of deer exceeds the carrying capacity of the habitat, starvation results. *(Courtesy Michigan Department of Natural Resources.)*

In the few places remaining where societies operate without dependence on fossil fuels, we can determine the carrying capacity based purely on current solar energy. Granted, energy of fossil fuels traces back to the energy of sunlight stored during photosynthesis in times past. We are concerned here with utilization of the current input of solar energy subsidized or not subsidized by dipping into fossil fuel energy resources.

India

Parts of India surviving on primitive agriculture—with no use of fossil fuels—have populations that average 640 persons per square mile. The carrying capacity, then, is 1 person per acre. Coral atolls and stable tropical forests, also running on current solar energy input, might support 1 person per square mile.

tropics

effect of fossil fuel use

Evaluation of any scheme to feed more and more people involves the vital although sometimes overlooked contribution of fossil fuel energy. The "borrowing" of energy from the past has the effect of adding to the carrying capacity. The results in the United States can be seen in a comparison of 1880 agriculture efficiency—which involved no fossil fuel use—and current yields. Based on the harvest of organic matter, the efficiency of utilization of sunlight on farms in 1880 amounted to 0.03 percent. In 1964, with fossil fuel energy input into agriculture, the average efficiency for North American grains was 0.12 percent and that of United States rice production reached 0.25 percent. The energy in the harvested crop is still the energy of sunlight trapped during photosynthesis, but fossil fuels enable the farmer to remove factors that limit the processes of photosynthesis.

ARE WE STUCK WITH EARTH?

colonizing other worlds

Is there any possibility that world population problems might be solved by emigration to another planet? This proposal is sometimes voiced: When things get too bad here, we will simply transport the excess to a new unpolluted world. In all likelihood the proponents of such schemes have not actually explored the costs and may not even realize the number of people to be moved. To hold the earth's population to zero growth by exporting the excess, we would in 50 years produce Earth's present population density on our moon, Mercury, Mars, Venus, the 12 moons of Jupiter (Fig. 14-2), and the 10 moons of Saturn (Fig. 14-3). In 200 years we would have saturated the remainder of the planets in our solar system.

Fig. 14-2 (above). Jupiter, showing the large red spot (upper left). *(Courtesy Hale Observatory.)*

Fig. 14-3 (below). Saturn and its ring system. These and other remote planets offer no solution to Earth's population problem. *(Courtesy Hale Observatory.)*

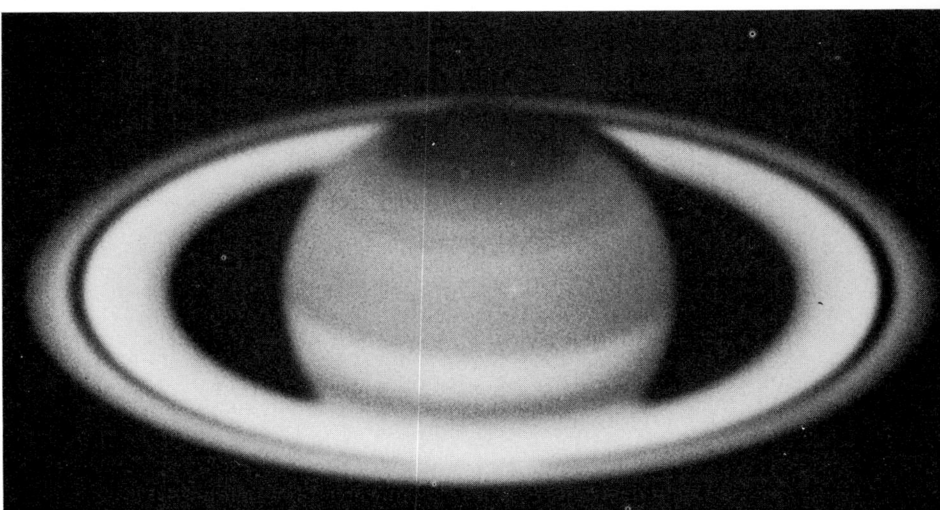

Just what would the cost be for such a venture? If we could transport 100 people to Mars for the same price as sending 3 to the moon, about 2000 space ships a day, each carrying 100 people, would be needed. This amounts roughly to 3 billion dollars a day. Three days of launchings would absorb the United States gross national product for 1 year.

However, to come down to earth with what might appear to be a more practical approach—Are there areas left on earth into which excess people might move? The answer here is a qualified "yes." The Congo might be farmed. More people could move into the Amazon Valley. The United States has some sparsely settled areas. The cost of clearing jungle areas has dropped from $350 an acre to about $65 per acre. Disregarding what this would do to the countless millions of nonhuman lives, we then ask whether such mass migrations are feasible. Could India, for example, solve its population increase this way, even with outside help? Each year in India 21 million children are born, and each year 9 million persons die. The annual net gain is 12 million. If there were places to go and we employed the United States long-range jet fleet of 600 planes each carrying 150 passengers twice a week, we could still handle only 75 percent of India's annual increase.

migrations

From this we may conclude that we cannot solve the population crisis by sending people to outer space. Furthermore, even if we had the space on earth, we are already beyond a practical solution through emigration. It is logistically impossible to move people fast enough, and even if we could the new areas would soon be filled also. The population problem will have to be solved with people remaining where they are (Fig. 14-4).

Fig. 14-4. The planet on which Earth's population problem will have to be solved. Earth as seen on an Apollo mission. *(NASA photograph courtesy of Hasselblad.)*

WHAT WILL SET THE POPULATION LIMIT?

principle of the minimum

A little more than 100 years ago, Justus von Leibig proposed what is now known as the *principle of the minimum*. In simple terms von Leibig reasoned that the size of a population is limited by whatever need is in shortest supply. An organism, and therefore a whole population of these same organisms, needs many things. A rose plant, for example, needs carbon dioxide, water, light, and chlorophyll in functioning leaf and stem cells if photosynthesis is to take place. The rate of photosynthesis at any one moment, however, is determined by the quantity of one of these needs. In low light intensity increasing carbon dioxide might not change the rate of photosynthesis because in this case light acts as a limiting factor. Only by increasing light could the rate of photosynthesis be changed. At moderate light intensities carbon dioxide could be the limiting factor.

Just as some environmental factor controls the population density of every other organism, the human population growth will be curtailed at some point. To attempt to predict that point is indeed risky and subject to error because of failure to recognize all the limiting factors. Our population could be controlled by the availability of energy (food and other uses), by exhaustion of nonrenewable resources, or by the availability of pure air and water. If the problems inherent in these crises are solved and no restraints placed on human reproduction, the population limits could be set at a very high level by such agents as heat and space. In all probability, several factors will combine to set population levels unless man decides to set them himself. Such synergistic reactions are difficult to predict, so we examine a few of them separately for possible consequences.

Malthus

The carrying capacity of the earth, as far as the human population is concerned, depends on what becomes the limiting factor. Whether this factor, whatever it turns out to be, results from man's choice, or is by default the operation of Malthus' laws of population control, is of crucial importance.

HEAT

The effect of man's heat production in its contribution to the total heat load of the planet could possibly become a limiting factor in population growth. If the present rate of energy use continues (a 5-percent increase each year), conditions could be critical within 100 years (Fig. 14-5).

energy input

The biosphere has several sources of energy input—the sun, sunlight reflected by the moon, the stars, and the earth's own internal and surface nuclear energy. Of these, the sun makes by far the major contribution. For the annual average temperature of the earth to remain constant, the

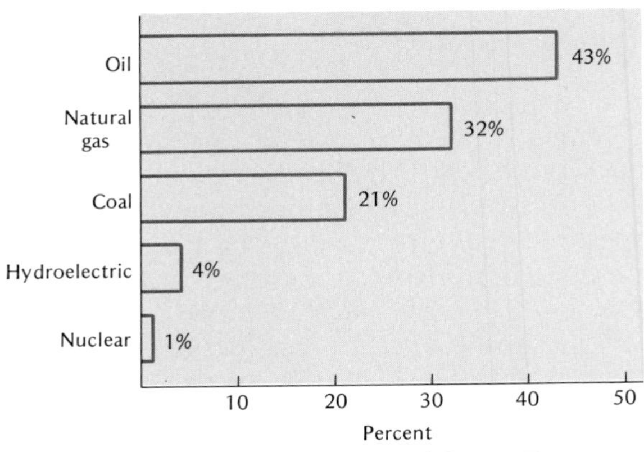

Fig. 14-5. Energy sources in the United States. Energy may become a limiting factor to further increase in standard of living.

energy input must be balanced by the amount of energy stored in a non-heat form plus that radiated back into space. Whether man's activities will result in a warmer or a cooler climate remains in doubt at the moment. More carbon dioxide in the air would have a warming effect; more particulate matter in the air reflecting the sun's rays would cool. One might offset the other.

polar ice pack

Bernt Balcher, a pilot and engineer who accompanied Roald Amundsen to the Arctic in 1925 and acted as a pilot for Richard Byrd on the first crossing of the South Pole in 1929, supports the case for a warming trend. According to Balcher, the Arctic Ocean ice pack measured 43 feet in depth in 1893. It is now only 6 to 8 feet thick over an area of 5 million square miles. Continuation of this trend would free the Arctic Ocean of ice by the turn of the century. Balcher feels that this would have relatively little effect on the ocean level but would change climates drastically. As the ice melts, storm tracks in the Northern Hemisphere would shift northward. While this might reduce tornado risk in Oklahoma, it would at the same time severely reduce the rainfall on the plains of North America, Europe, and Asia. The chief food-producing areas of the world would then turn to deserts.

Balcher further predicts an increase in temperature across the northern United States, but cooler winters in Florida. The more moist air in the Far North would increase the snowfall by 8 to 10 feet in polar land areas. This would change the ecological relationship of the tundra, a region that now receives relatively light annual snowfall.

Once melted, polar ice caps are not easily reformed. The loss of the ice removes one of the stabilizing conditions that formerly helped to keep the temperature constant. The ice reflected much of the sun's radiant energy even though so far from the equator. After the ice is gone, the watery area replacing it would absorb radiant energy. This would accelerate the temperature rise.

Total melting of the world's glaciers would certainly flood the United States coastal plain and similar elevations around the world. Where the coastal plain meets the rolling hills of the Piedmont near Rocky Mount, North Carolina, for example, the elevation is about 100 feet. The coastal plain is an old seabed, so the ocean has been there before. Some studies indicate a possible sea level rise of 200 feet or more if the Greenland and Antarctic glaciers melt.

In spite of man's increasing energy use, the planet Earth could enter a cooling stage. Increasing pollution of the air with particulate matter could accomplish this. (Regardless of what is done in the United States to reduce air pollution, will the developing countries, in their headlong plunge into industrialization, spend money for stack scrubbers and other emission controls?) Or an especially productive volcano could trigger the change. Once started, similar to overheating, the trend may become self-reinforcing and become irreversible at least as far as man's efforts are concerned. Some past experiences indicate possibilities.

volcanic eruptions

The coldest decades in England's long history of temperature recording coincide with volcanic eruptions. The 1781–1790 span included the eruption of Mount Skaptar (Iceland) and Mount Asama (Japan). Both eruptions occurred in 1783. In the middle of the 1811–1820 decade, Mount Tamboura on the island of Sumbawa in Indonesia coughed 150 cubic kilometers of ash into the air. The following year (1816), the northern United States had no summer and the average July temperature in England, based on records over 250 years, dropped 5 degrees. Finally, Mount Krakatoa, between Java and Sumatra, virtually obliterated the area by explosive eruptions in 1883 and lowered England's temperature. What could an era of volcanic activity do to world food production?

A lower average world temperature would also affect the glaciers. Snow accumulation would exceed melting, and the resulting increase in ice thickness would put more weight on the bottom ice. Increased pressure lowers the freezing point. The destabilized ice cap would then start to move. Massive slumping of huge ice blocks into the Antarctic Ocean would create global tidal waves which could eliminate a substantial portion of the terrestrial life on earth. With the spread of the ice northward from the Antarctic and southward from Greenland, more efficient reflection of the sun's heat would further lower the temperature. The Antarctic and Greenland together hold enough ice to cover the earth with a layer 50 yards thick. A 3°C drop in average world temperature is sufficient to start another ice age.

A sobering thought: Evidence indicates that the last two ice ages began with catastrophic inundations and extensive loss of life. Conditions now are very close to those that initiated these ice ages. In fact, the four ice ages during the last million years, all within the experience of man, have left evidence of man flooded out, frozen out, or driven out by drought. In the past these have been natural changes—risks that have always been present. However, man may be altering the climate just at a time of critical population pressure on food production. Any climatic change, either warmer or cooler, can be expected to reduce crop production, at least temporarily. As climate changes, agriculture must change too, but agricultural changes always lag behind.

FOOD

Estimates of the population that the world could support are somewhat irrelevant if we hope for a life above the level enjoyed by ants. We should, however, explore the possibility that food may not be the convenient limiting factor that we counted on to prevent Earth from becoming a giant anthill with only man and his domesticated plants and animals left.

Food production, the incorporation of energy into biologically usable materials, has a maximum potential limited by the sun's input of energy into the earth's biosphere. This is a known quantity, and its effective and total utilization is a matter of technology. Because the present world population growth, 2 percent each year, would in 650 years produce a population density of one person per square foot of surface area, we need to know the factors that can prevent such an absurdity.

earth's food production

Figures based on the total annual energy capture of the oceans (4 to 12×10^{16} kilocalories) combined with that of the land (4 to 6×10^{16} kilocalories), and an assumed individual need of 2200 kilocalories per day, indicate that the present population consumes only 1 percent of the annual food production. To suggest that this means a potential world population of 300 billion, however, is to assume man to be totally vegetarian and the only consumer of food. If one herbivore is inserted into the food chain and man is the only carnivore, the resulting energy waste, figured conservatively at 80 percent, would reduce the maximum population to 60 billion. Production is not uniform over the whole earth, however, and any conditions that eliminated many other life forms would likely also have eliminated man. Such scientific "doodling," as many population estimates are, serves to remind us of some of the principles of food chain relationships.

Changing Farm Practices

crop rotation

In the drive for more food for more people, many farm practices, long cherished in the United States as the only way to run a farm, are coming under question. Crop rotation is one of these. In a mixed agriculture in which cash crops include wheat, corn, soybeans, and clover (the last-mentioned for sale of seed), and in which relatively small farms can still be maintained, crop rotation may be practical. To improve the soil crops are rotated to insure clover on a field every third or fourth year. Thus corn is followed by wheat and then clover in a 3-year rotation. The trend is now toward big business farming, however. Small fields of 10 to 50 acres and diversified crops do not produce maximum profits.

With reasonable initial fertility, production can be maintained even with single-crop farming. Some areas of Puerto Rico have been producing sugarcane for 400 years and yet yields are higher today than in former times. The same holds for sugarcane in Hawaii over the past 100 years. Continuous planting of lima beans in California over 45 years has produced a steady yield. The current shift in the Midwest is to continuous corn on the soils best suited to corn.

sugarcane

small farms

A small, inefficient farm is obsolete. Today's trend in the developed countries is toward greater use of machinery and less manual labor. In the United States this was forced onto the farmer by the increased cost of

Fig. 14-6 (above) and 14-7 (below). Formerly, wheat and oats threshing was a community affair. Usually, 12 to 14 farmers organized a threshing ring which moved from farm to farm. The various jobs, pitching, bundle hauling, grain hauling, and operating the blower, were drawn by lot at an organizational meeting. Such a group could thresh 30 to 40 acres in a day. Now, one man with a self-propelled combine completes the harvesting of 20 acres in about 6 hours.

farm labor or, more often, the absence of persons willing to work for what the farmer could afford to pay (Figs. 14-6 and 14-7). The final blow to some areas came when the expiration of U.S. Public Law No. 78 stopped the seasonal importation of Mexican labor into the Southwest. Mechanical robots costing $20,000 or more replaced the *braceros*. In the 2 years that followed the cutoff of Mexican labor, the picking of tomatoes in California changed from 100 percent hand picked to 85 percent machine picked.

Actually, the move toward farm mechanization began much earlier. In the early part of the century, a farmer in the Ohio Valley, for example, might have 20 head of horses—a 4-horse team for himself and each of his four sons. With these they could farm 60 to 80 acres, sometimes 100 acres. In the course of a day, 1 man could plow 1 acre.

tractors

With the advent of the small tractor pulling at most a two-gang plow, the farmer with one tractor, three horses, and one son, now farmed his own 100 acres and rented 50 more. The tractor plowed 20 acres a day. This did not last long, however. By the late 1930s and early 1940s, the farm had two tractors and no horses. In one generation 12 million mules and horses were replaced by five million tractors.

mechanization

The dramatic increase in farm machinery and decrease in farm labor began after World War II. Between 1950 and 1965 the number of farm workers dropped nearly one-half and the value of farm machinery doubled. During this same period sprinkler irrigation increased from 100,000 acres to 5 million acres. Machines now pick cotton, harvest tobacco, shake nuts and fruits off trees, shuck and shell corn in the field, turn 6 furrows at a time, and plant 10 rows of corn each trip across a field. For machinery to be used economically, large fields are necessary. Small fields have been combined and the brushy fence rows that used to harbor quail and cottontails are gone. Airplanes sow rice, oats, and other small grains,

Fig. 14-8. Man-hours required to produce 100 bushels of corn. 1800, 350; 1950, 34; 1960, 3½; 1966,—1 man-hours.

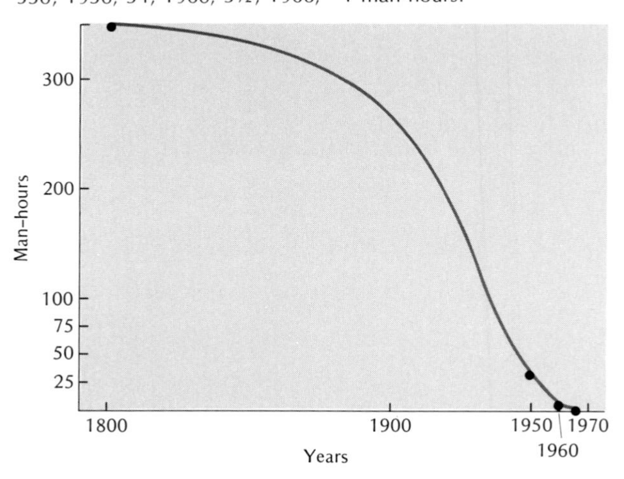

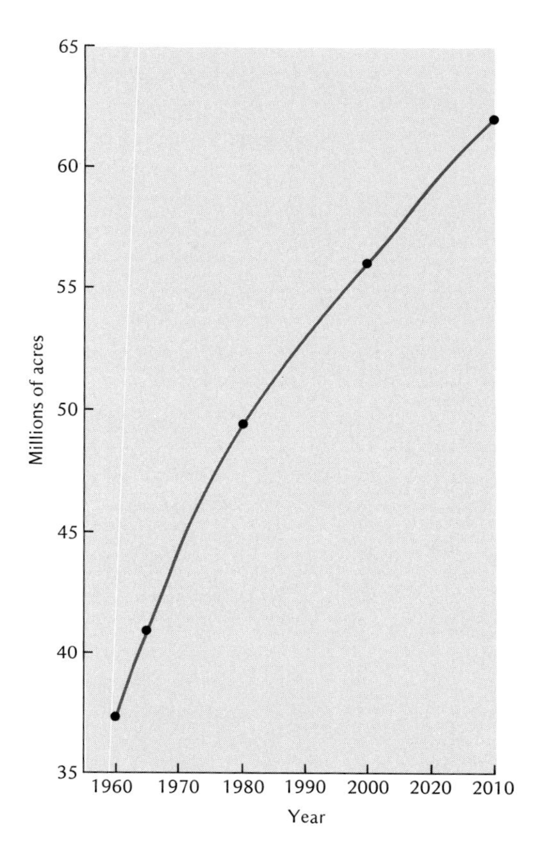

Fig. 14-9. Projection for irrigated land in the United States. 1960, 37.24; 1965, 41.84; 1980, 49.48; 2000, 56.55; 2020, 62.28 million acres. This may allow a higher population density.

and coat the good earth with poisons to control weeds and insects. The results: increased yields from fewer man-hours. Figure 14-8 illustrates the decrease in the time required to produce 100 bushels of corn. Since 1945 wheat and tobacco yields have doubled, or nearly so, and corn yields have tripled. The highest corn yields on experimental plots are now over 250 bushels per acre.

Less cultivating of the soil—plowing, rolling, disking, harrowing—represents a new trend in many areas. The advantages are numerous: less erosion, less compacting of the soil by machinery, and reduced costs. Machinery now in operation can prepare the soil, plant the seed, and apply the fertilizer and weed control chemicals in just one pass over a field.

Many earlier farm practices evolved as they did because of the space required to drive a team of horses. This was especially true of row crops such as corn. Forty inches between corn rows was traditional spacing. With the passing of the horse on the farm, the narrowing of rows became possible. Where rows are 30 inches apart, plants are less crowded in the direction of the row for the same number of plants per acre and may give 20 percent greater yield. Extended use of irrigation (Fig. 14-9) may further increase yields in many areas.

The Green Revolution

Borlaug

The awarding of a Nobel Peace Prize in 1970 to Norman E. Borlaug marks not only the recognition of his impact on the world's food problems but also the importance attached to solving those problems. Starting in 1944 at the International Corn and Wheat Improvement Center (Centro Internacional del Mejoramiento de Maíz y Trigo) in Mexico, Borlaug and his associates worked to increase the yields of varieties of dwarf wheat. More than a billion people—about 35 percent of the world population—depend on wheat as their principal food. Borlaug's work enabled Mexico not only to provide its own wheat but to have excess for export. Wheat production in Mexico increased sevenfold between 1945 and 1967. Similar results are anticipated in India, Pakistan, Turkey Tunisia, Nepal, Morocco, Iran, and Afghanistan. The same principles are now being applied to corn, millet, sorghum, and rice.

dwarf wheat

Wheat is not just wheat—there are many varieties. When Mark A. Carleton became director of the U.S. Department of Agriculture cereal crops investigation in 1894, top priority was placed on collecting wheat from all parts of the world. Today this collection contains more than 20,000 specimens and is the source of much of the genetic material used to develop disease-resistant strains. The successful development of wheat resistant to wheat rust was made possible by use of these stocks.

The development of high-producing varieties of wheat in Mexico illustrates the value of international cooperation. The United States supplied the Center with 13,000 lines of seed, but the backbone of the breeding program, Norin 10 wheat, had been brought to the United States previously from Japan (1946).

Norin 10

The acceptance of Borlaug's wheat is indicated by the heavy export of Mexican wheat seed. India, West Pakistan, and Turkey imported more than 82,000 tons for planting in 1967 and 1968. The new short-stemmed Mexican wheats yield well over a wide range of latitude. They are not sensitive to day length. Also, they mature in less time than most varieties they replace.

The success of the new wheat in Mexico prompted the Rockefeller and the Ford Foundations to launch a crash program to improve rice yields. Work began in 1962 at Los Baños in the Philippines. In 2 short years strains IR5 and IR8 were developed. As a result, the Philippines over a 5-year period evolved from a rice importer (1 million tons annually) to a self-sufficient provider and will soon have a surplus to export. Similar results have been attained elsewhere. Rice production in Sri Lanka increased 34 percent and in Pakistan 162 percent. The difference between the two countries was due primarily to the number of farmers using the new strains.

rice

Weed Killers and Food Production

herbicides

The vital role herbicides now play in agriculture warrants further consideration of these chemicals in relation to food production. The following discussion reemphasizes the technical nature of modern agriculture.

Herbicides fall into two general groups. Dinitro compounds, oils, and many arsenicals kill only the part of the plant they touch. These are *contact* herbicides. *Translocated* herbicides affect plants in places other than at the point of contact. They move from one part of the plant to another. 2,4-D (2,4-dichlorophenoxyacetic acid) is the best known member of the second group.

selectivity

Selectivity—action against one kind of plant but not against another—results from a variety of causes. When applied in high concentrations, 2,4-D kills or severely damages almost all plants. In low concentrations it kills broad-leaved weeds but not cereal grains and other grasses. In contrast, trichloroacetic acid (TCA) and dalapon, chlorinated acetic acids, protect dicots such as legumes against monocots (grassy weeds).

Physiological differences among plants account for the sensitivity of some species to a particular chemical and the resistance of other species. The exact reasons are not well understood. Herbicides may interfere with enzyme systems or affect metabolic processes in other ways. In other cases differences in cell membrane permeability render one species vulnerable and another immune. Some plants are immune because they have enzymes that break down the herbicide to nontoxic compounds. Ability or failure to transport the chemical from the point of entry to a site of action may produce selectivity. A chemical applied to the soil that interferes with some step in photosynthesis kills no weeds unless it moves through the plant to the chlorophyll-bearing cells.

The action of a particular chemical can be different at different stages of plant growth. The sensitivity of most grains to auxin herbicides is high during the seedling stage and again at flowering, but low during tillering (development of additional stems at the base) and setting of the fruit.

The value of herbicides lies in reducing the competition of crops with weeds for available water, nutrients, and even light. Formerly, laborious plowing at a shallow depth reduced the weed crop in corn and in soybeans when the latter were planted in rows spaced to permit such plowing. Crops planted with what was commonly referred to as a wheat drill (wheat, oats, rye, barley, clover) had rows too close together for any further cultivation except for the use of a heavy roller (to break the surface crust after hard rains) and what was somewhat hopefully called a "weeder." The latter, a multitude of flexible curved spring teeth, could be used after wheat or corn had reached a height of 3 or 4 inches but no more. Its effectiveness was questionable. Herbicides, selected for a crop and for specific weeds, can be applied before, during, or after planting and thus greatly reduce the amount of cultivation. This reduces erosion, lowers overhead expenses and, because it is more effective than former methods of weed control, increases yields. The technology is somewhat complicated, however, as the following discussion of a particular herbicide indicates.

Herbicides used for weed control in cornfields are likely to kill many broad-leaved plants. Atrazine, manufactured under the trade name AAtrex, is registered for selective control of broad-leaved and grassy weeds in fields of corn, sorghum, sorghum-Sudan hybrids, perennial ryegrass, sugarcane, and for turf grasses such as St. Augustine, centipede, and zoysia. It can be effective in controlling weeds in commercial plantings of firs (grand, noble, white, and Douglas) and pines (Scotch, lodgepole, and ponderosa). By using low applications, 1 to 4 pounds per acre, more than 60 specific weeds can be controlled, including such problem species as quackgrass, ragweed, Russian thistle, dog fennel, several kinds of foxtail grass, cocklebur, crabgrass, and nutgrass. A heavier application, 10 to 60 pounds per acre, is used for roadside weed control.

atrazine

Atrazine enters plants primarily through the roots and passes upward with water through the xylem. In prolonged cool, wet weather, atrazine may be absorbed through the leaves. Its effect on the plant occurs in the chloroplasts of the leaf and stem cells. Unlike some weed killers which are simply growth hormones, atrazine kills by blocking some step in the photosynthetic process. The exact step inhibited may vary for different susceptible species, but the primary mechanism is the ability to block noncyclic photophosphorylation. Some plants are immune to the action of atrazine because they metabolize it rapidly to nontoxic compounds.

FDA

The U.S. Federal Food and Drug Administration has set tolerance levels for atrazine at 0.02 part per million in eggs, milk, and meat products of poultry, cattle, sheep, hogs, goats, and horses. Somewhat higher amounts, 0.25 part per million, are tolerated in corn grains, fruits, and nuts. The 15 parts per million allowed in corn and sorghum fodder appears to be a safe limit. Cattle, dogs, horses, and rats fed a diet containing 25 parts per million atrazine over extended periods showed no ill effects. Most of the atrazine consumed by animals is metabolized and the products excreted through the urine within 24 hours. This was demonstrated by incorporating radioactive carbon into the atrazine molecule.

The Perils of Plenty

food surplus

The results of a food shortage are easily predicted: people starve. The usual answer to starvation is seen as an effort to supply more food. Food surpluses, however, as they are now occurring, can be fraught with problems as severe as food shortages. The second meeting of the World Food Congress in The Hague, the Netherlands, in early July 1970, explored many of these matters. Addeke H. Boerma, Director General of the United Nations Food and Agriculture Organization (FAO) is reported by *Time* magazine (July 13, 1970) to have expressed the opinion that only careful management of the green revolution would avoid "a conflagration of violence that would sweep through millions of lives."

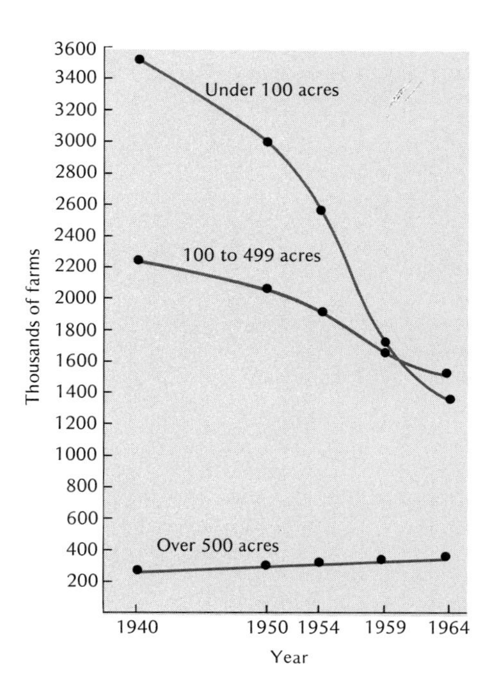

Fig. 14-10. Changes in farm size and number in the United States, 1940–1964. The increase in the number of farms over 500 acres occurred through the consolidation of small holdings under one ownership. (Data compiled from *United States Book of Facts, Statistics, and Information,* 1969.)

agribusiness

Several factors contribute to the problems of plenty, at least as the thrust toward solving food problems is now directed. Chief among these is the big business aspect of efficient food production. The new miracle varieties of wheat and rice require capital outlay for the seed itself, fertilizer, irrigation in many cases, insecticides, herbicides, and the expensive farm machinery needed to operate a modern farm. Furthermore, high yields with a reasonable return on investment are best achieved with large acreages. Figure 14-10 illustrates the decline in the number of small farms as farmers with money bought land from the owners of small tracts. The number of farms under 100 acres dropped from the 1940 level of more than 3.5 million to less than 1.5 million by 1964. A farmer in the north-central states now handles 350 to 500 acres; some farm more. In Utah a farm operation may involve 2000 to 6000 acres. Many peasants in India own 1 acre at one place and perhaps ½ acre a mile away.

In areas where both large and small land holdings exist, the green revolution serves primarily to widen the gap between the rich and the poor. With the drop in the prices of farm products due to oversupply, the small farm in the United States has largely been abandoned as a serious device for making a living. This occurred at a time when many older farmers were retiring and their children had already moved to the cities. Consequently, those who stayed in farming bought expensive machinery and rented additional land. The custom of owner and renter dividing the cost of seed, fertilizer, sprays, and hauling grain to market left very little

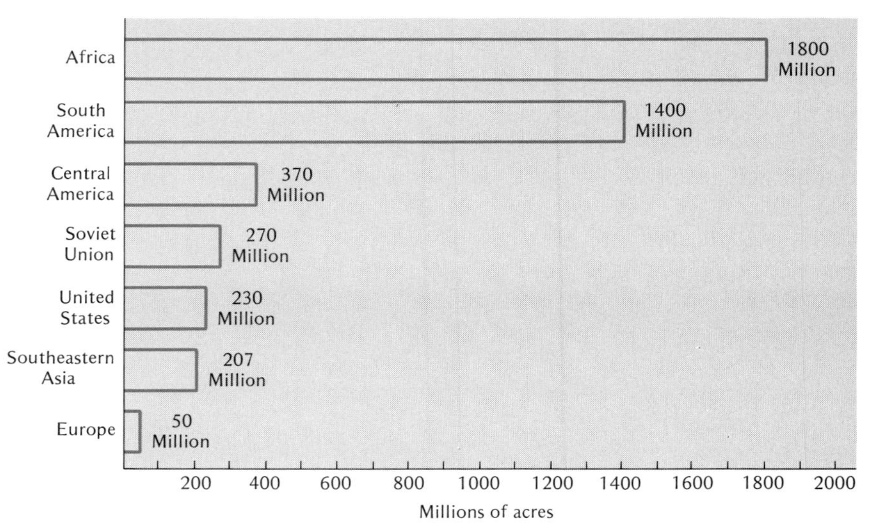

Fig. 14-11. The world's reserve farmland.

for the owner. A rented 140-acre farm in certain cases returned to the owner no more than $1500 per year.

In the United States these conditions were offset by Medicare and Social Security. In countries without such financial backup, the seeds of plenty are the seeds of revolution. Such countries simply are not prepared to cope with abundance. It is too new and usually happens suddenly.

arable land

Successful management of the green revolution becomes complex, particularly in developing countries. A clue to the problem lies in the fact that arable land still exists which is not now being farmed. Charles Kellogg of the U.S. Department of Agriculture noted that Africa, south of the Sahara, has about 1.8 billion more acres that could be farmed, South America 1.4 billion acres, North and Central America 600 million acres, and southeastern Asia 207 million acres (Fig. 14-11). The primary hindrance to farming these relatively good soils is the lack of transportation. In the United States the development of transportation followed a logical evolution: water, roads, railroads, more roads, air. In South America transport developed directly from canoes to flying machines. Because of adverse conditions of climate and terrain, roads and railroads for the most part were bypassed as means of transportation. As late as 1938, at only one point in the country of Honduras did two main rail lines cross.

Central America

The Standard Fruit Company port of La Ceiba on the Gulf coast of Honduras, a shipping point for bananas, oranges, coconut products, and mahogany lumber, had no automobile roads leading out of town. A narrow-gauge railroad connected with Tela 30 miles up the coast and with Los Coyoles across the Congrejal Mountains.

Further complications in guiding the green revolution involve the other services that must be provided in addition to just giving farmers better-yielding varieties. If maximum production is to be achieved, soils must be fertilized. This entails fertilizer production, shipping to distribution points, and machinery to apply the fertilizer to the soil. To avoid waste soils have to be tested to determine the kinds and quantities of nutrients needed. Mechanization of agriculture requires the production, distribution and, most import, maintenance of farm machinery. Also, the whole petroleum industry from tankers to delivery trucks immediately becomes involved. A tractor without gasoline and oil does not work. In the United States these components of modern agriculture developed over 60 years or more. To transport this complex social system to developing countries, as some suggest, would completely displace their own cultures. Few if any can cope with instant affluence.

The lack of facilities for handling a sudden increase in food presents a major problem in many countries. Pakistan had no elevators to store the mountainous crops. Twenty percent of the yearly crop is lost to rodents, insects, and theft. In 1968 India closed scores of schools and used them for storing surplus wheat. Even with the surplus, however, the starvation rate in India was little affected. The distribution and marketing system for supplying food to the people did not exist.

Other problems plague efforts to obtain more food. The new wheat imported into Turkey did not taste like the old. In keeping with man's tendency to classify anything different as bad, the new wheat sold for 10 percent less than local varieties. The new rice is a little more gluey than the old variety. The inhabitants of southern and southeastern Asia who eat rice with their fingers do not want the new high-yield types. Because of buyer resistance in the Philippines, the new rice sells for 30 percent less than the standard kinds. In fact, dried rice plants (fodder) are often fed to cattle. In Japan, a country that formerly imported rice, surpluses have reached such proportions that one company uses rice in air jets to clean fan blades.

The changes in farm size in the United States that occurred between 1940 and 1964 clearly predict what might happen on an even larger scale in developing countries. Figure 14-10 shows the consolidation of farms during that period. A further breakdown of the 1964 data shows that 4.6 percent of the farms, those of 1000 acres or more, contained 52.6 percent of all farmland and 24.3 percent of the cropland. About 70 percent (69.3) of the 1964 cropland was on farms of 260 acres or more.

farm size

The displaced peasant may become a major problem for the next 2 decades. Long-range planning is urgently needed, and some suggestions are being considered. Land reform that would give everyone a small parcel offers no solution. Small holdings have been the bottleneck that thwarted efficient agriculture in many places in the past. Farm cooperatives might be formed in which small land parcels are pooled. Since not all members of the cooperative would be needed to work the land, surplus labor might specialize in the other services normally associated with agriculture—fertilizer manufacture and distribution, gasoline refining, repair and maintenance, and so on. We should note, however, that there is already a labor surplus in the cities which could be absorbed into such new opportunities. In either case extensive educational programs and capital investment would be necessary. The question then becomes one of time. This may be more serious than finding a source of financing, particularly in countries that now have a population doubling time of 17 to 20 years. The simple, hard facts are: A few people with modern equipment and fossil fuel to spend can produce far more food than many people utilizing man and water buffalo power, given the same total acreage. In the United States 500,000 farmers could produce all the food we can use.

The present breakthrough in food production has triggered enough side effects to warn of the consequences of an all-out attempt to feed all the people that unregulated human reproduction will add to the fragile biosphere of this planet. What happened more slowly in the United States, putting the small farmer out of business, may now sweep across the developing countries. The owner of a few acres (often only 1 or 2 acres is a farm in India) cannot compete with the increased production of large, more efficiently operated land holdings. As their lands are engulfed by "agribusiness," the displaced farmers will migrate to already overcrowded cities.

unemployment

These cities are not industrialized to the point of being able to furnish employment for an influx of unskilled labor. The food problems of the 1960s may become the unemployment problems of the 1970s.

Suggested Readings

Aldrich, S. R. "Some Effects of Crop-Production Technology on Environmental Quality." *BioScience,* 22(2):90–95, 1971.

Allen, L. H., Jr., S. E. Jensen, and E. R. Lemon. "Plant Response to Carbon Dioxide Enrichment under Field Conditions: A Simulation." *Science,* 173(3993): 256–258, 1971.

Bazell, R. J. "Arid Land Agriculture: Shaikh up in Arizona Research." *Science,* 171(3975):989–990, 1971.

Boerma, A. H. "A World Agricultural Plan." *Scientific American,* 223(2):54–69, 1970.

Carter, L. J. "Land Use: Congress Taking up Conflict over Power Plants." *Science,* 170(3959):718–719, 1970.

Editor. "Note: New Plants by Induced Mutation." *American Biology Teacher,* 33(1):21, 1971.

Editor. "The Third World: Seeds of Revolution." *Time,* pp. 24, 27, July 13, 1970.

Flattau, E. "Tomorrow a Wasteland?" *Science Digest,* 70(5):22–26, 1971.

Galle, O. R., W. R. Gove, and J. M. McPherson. "Population Density and Pathology: What Are the Relations for Man? *Science,* 176(4030):23–30, 1972.

Holden, C. "Fish Flour: Protein Supplement Has Yet to Fulfill Expectations." *Science,* 173(3995):410–412, 1971.

King, K. W. "The Place of Vegetables in Meeting the Food Needs in Emerging Nations." *Economic Botany,* 25(1):6–11, 1971.

Pinchot, G. B. "Marine Farming." *Scientific American,* 223(6):14-21, 1970.

Sigurbjörnsson, B. "Induced Mutations in Plants." *Scientific American,* 224(1): 86–95, 1971.

Zobel, B. J. "The Genetic Improvement of Southern Pines." *Scientific American,* 225(5):94–103, 1971.

The wild yak is endangered as a distinct species because of frequent hybridization with domestic animals. The cow shown here, raised on the Standwood, Michigan, ranch of William Cummings, is being bred to a buffalo bull as part of a continuing genetic study of the buffalo.

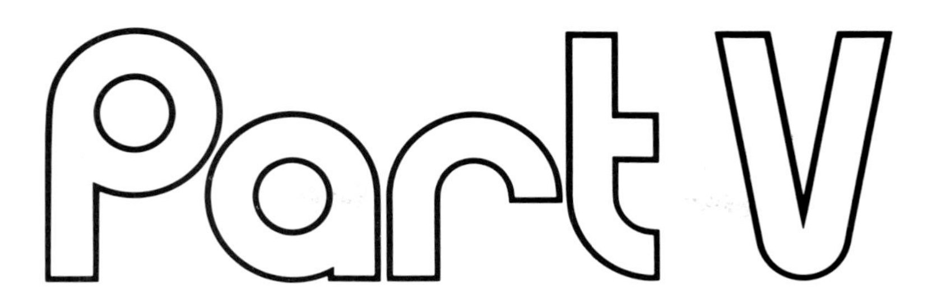

AFTERWORD

White-tailed deer fawn. *(Courtesy Michigan Conservation Department.)*

The Quest for an Environmental Ethic

This final chapter is directed to students and other individuals who would like to assume a more active role in solving some of our environmental problems and at the same time become less a part of the causes. A review of some of the problems is included in this chapter to help answer the question, "What can one person do?" The ultimate answer will be action-oriented and will involve attitude changes. The overall attempt must be directed at efforts to reduce the amount of stress that each of us places upon the biosphere. Survival is at stake.

PROGRESS

camas

The use of resources by a population is a function of that population's culture. An illustration of this concept can be found in the uprising of the Nez Perce Indians under the leadership of Chief Joseph. Camas, a tuberous plant of the western prairies, provided a staple in the diet of these Indians but meant nothing to the white settlers. Conflict over the use of this resource came as the settlers began to plow the prairies and destroy the camas

Fig. 15-1. Clovis-type fluted point, age 10,000 to 12,000 years, from Alamance County, North Carolina. These sharp, quartz spearheads greatly improved man's effectiveness as a hunter. The area from which the rock came that produced this tool is being exploited differently today because of cultural differences. (From the collection of William R. West. *Courtesy Carolina Biological Supply Company.*)

fields. In the culture of the Indians, the plowing not only was senseless, it was downright destructive. In the same way the flint ridges of eastern Ohio and northern Wyoming were important areas for the Indians but are of negligible value to the present occupants. Figure 15-1 illustrates one aspect of the limited environmental impact of the Indian culture. The coal and oil that lay beneath the surface of adjacent areas, so important in our own culture today, meant nothing to the Indians.

Kariba

These illustrations also show us that population alone is not the whole story. The pressure exerted by a population is a variable function of its culture. Some populations, because of their culture, exert a much lower per capita pressure on the available resources. Before the construction of the Kariba Dam, the 50,000 Tonga natives living along the Zambesi River in Africa were in equilibrium with their resources. Their primitive fishing methods provided a sustained harvest, and their spear hunts were no threat to native game animals. Somewhat the same situation occurred in parts of the United States during the depression of the 1930s when many unemployed families moved back to the hills of the Ozark, Cumberland, and Blue Ridge mountains. Although their tools were metal, these hill farmers reverted essentially to a Neolithic culture. Fortunately, during their short sojourn in the city, they had not forgotten how to live off the land. By taking a few more squirrels and rabbits, and picking a few more berries, their subsistence farming cushioned them against the economic disaster of the cities.

In contrast, the present culture in the United States and in other affluent nations, as well as in some of the developing countries, holds to two principles which must be reevaluated if we are to solve our environmental problems. The first can be stated simply: Bigger equals better.

progress

The commercial economy of the United States places a high value upon expansion. Unless the economy is continually expanding and the gross national product is greater each year, the country is suspected of being on the verge of bankruptcy. Unless a town is building more shopping centers, taller downtown buildings, and attracting more industry to pollute its waters and compound its solid waste disposal problems, it is dying. This philosophy was well illustrated by the frenzy in which Michigan's white pine forests were annihilated (Figs. 15-2 and 15-3).

Figs. 15-2 and 15-3. Prior to 1900, the economy of Michigan was based on the lumber industry. Summer (above) and winter (below) aspects are shown. The "inexhaustible" supply of white pine was completely harvested to build the expanding towns and farm buildings of the midwestern and prairie states. *(Courtesy Michigan Department of Natural Resources.)*

need versus demand

A second possible fallacy by which we guide our destiny equates demand with need. When power companies seeking a license for thermal pollution threaten that "the *need* for electrical power will double in the next ten years," they are really predicting that the *demand* will double. Considering the possible effects of thermal pollution and the further depletion of natural resources (coal, oil, uranium), we have a right to ask whether this demand represents an actual need. A West Coast used car dealer advertised for people to visit his car lot "under a thousand lights." And it was. A thousand light bulbs burned day and night to help people make the proper choice of a second or a third automobile. Does the fact that the dealer can pay the light bill justify such use of a limited resource? Does public interest demand that extravagant use of any nonrenewable resource be considered antisocial behavior? Or, more to the point: What is the relation between population density and personal freedom?

ENVIRONMENTAL DETERIORATION

We generally accept the idea that man is now facing an environmental crisis and that some corrective measures must be found if we are to prevent a deterioration of the environment to the point of completely changing our life-style as well as our standard of living. Many people assume that the fundamental cause of environmental deterioration is population growth and affluence. Although these factors contribute to the problem, the basic cause is incompatability between the environmental system and the type of technology upon which much of our industrial and agricultural production is

Fig. 15-4. One of the numerous early farm homes abandoned in Michigan as a result of soil depletion. *(Courtesy Michigan Conservation Department.)*

dependent, particularly in the United States. Figure 15-4 shows the inevitable results of misuse of the land.

The environment is a complex, dynamic system. It is characterized by numerous changes and events, all of which ultimately result from the use of energy derived from sunlight. This energy is transmitted through a series of physical, chemical, and biological processes. The last-mentioned are crucial; they represent a vast network of interrelated actions among all living things and among them and their physical and chemical environments. In all instances the processes are cyclic. These cyclic processes accomplish the self-purification of the environmental system. Wastes produced by one step in the cycle become the necessary raw material for a subsequent step in the cycle.

cycles

The U.S. Environmental Protection Agency

The U.S. Environmental Protection Agency (EPA) was established December 2, 1970, bringing together for the first time in a single agency the major environmental control programs of the federal government. EPA is charged with mounting an integrated, coordinated attack on environmental problems concerning air and water pollution, solid wastes management, pesticides, radiation, and noise. To insure that the Agency is responsive to environmental needs in every part of the country, it has established regional offices in 10 major cities. Each office is staffed by specialists in each program area and is headed by a regional administrator possessing broad authority to act for EPA in matters within his jurisdiction.

Several on-going Federal programs were transferred to EPA under the President's Reorganization Plan No. 3. These programs included the functions of:

The Federal Water Quality Administration
The National Air Pollution Control Administration
The Bureau of Solid Waste Management
The Bureau of Water Hygiene
The Bureau of Radiological Health (environmental radiation programs)
The Federal Radiation Council

Also transferred to EPA were responsibilities and authority for:

Establishing standards for environmental chemicals
Establishing tolerances for pesticide chemicals
Registration and labeling of pesticides
Conducting research on pesticides
Conducting research on ecological systems

In addition, the Noise Abatement and Control Act of 1970 assigned to EPA the responsibility for studying the problem of noise and making recommendations for control.

When cyclic processes are stressed by adding to their individual components, the response is usually a change in the system sufficient to cause the cycle to break down. If water is overloaded with animal waste, the oxygen needed to support the bacteria that normally decompose these wastes may be reduced to zero; this phase of the cycle stops, thus the self-purification cycle as a whole is halted and the water becomes an unfit habitat for the support of organisms normally present. The natural, cyclic environmental system is balanced and self-sustained. If overstressed, it collapses; if permitted to function, it is capable of maintaining its properties, but only as long as it remains in cyclic balance. Since the environment is irreplaceable and essential to all human activities, changes must be compatible with environmental processes.

Unlike all other organisms, man is capable of exerting environmental influences far beyond those that he would exert simply as an organism in a natural environment. For example, the body energy expended by an individual amounts to about 1000 kilowatt-hours per year, yet in the United States about 15,000 kilowatt-hours of power are expended per capita per year. At the same time, man's activities in the more highly developed countries of the world have introduced into the environment not only intense stresses due to natural agents, but also new substances not encountered in the natural environment: artificial radioisotopes, detergents, pesticides, plastics, toxic metals and gases, and a host of man-made synthetics.

These human intrusions on the natural environment have thrown major segments of the ecosystem out of balance. Environmental pollution is a result of this breakdown. Figures 15-5 and 15-6 illustrate the symptoms of a wide-spread disease.

The fouling of surface waters is basically the result of overloading the natural aquatic ecosystem with nutrients, either directly by dumping organic matter in the form of sewage and industrial wastes, or indirectly by the release of nutrients produced by waste treatment or leached from overfertilized soil. Water pollution indicates that a limited, natural, self-purifying cycle has broken down. Air pollution is a sign that man's activities have overloaded the self-cleansing capacity of the weather system to a point where natural winds, rain, and snow are no longer capable of cleaning the air. Soil deterioration is a sign that organic matter, in the form of food, is being extracted from the cycle at a rate that exceeds soil rebuilding. Crop yield can be restored by loading the soil with inorganic fertilizer but at the expense of increasing water and air pollution. Add to these sources of

Fig. 15-5. An indication of a lack of environmental concern. Public beach along Clear Lake, Louisiana. *(Courtesy U.S. Department of Agriculture, Soil Conservation Service.)*

Fig. 15-6. Vandalism to picnic tables at Anna Ruby Falls recreation area, Chatahoochee National Forest, Georgia. (*Courtesy U.S. Forest Service.*)

pollution the man-made synthetics (pesticides, detergents, and plastics) and the dissemination of materials not naturally a significant part of the ecosystem (lead, mercury, and radioisotopes), and it becomes less difficult to understand why the self-purifying capabilities of the natural ecosystem lose their efficiency. Environmental pollution is a warning that there is something dangerously wrong in the way man reacts to his environment.

We should ask ourselves the question, What has gone wrong, why, and what can we do about it?

One proposed answer to the question what has gone wrong and why is based upon the notion that man, unlike other animals, lacks the instinct or the desire to "clean his own nest"; thus environmental pollution increases as population size increases. This is not an acceptable explanation, for the cleanliness of most animals in nature is not the result of their own sanitary activities. No animal lives in its own wastes, and wastes do not accumulate in natural nonhuman situations. They are utilized by other organisms. They become incorporated into a cycle, and as long as this cycle is unbroken no waste can accumulate. Man is no less sanitary than other animals. He pollutes his environment only because he has broken out of a closed, cyclic system.

waste in nature

Making a Community Inventory

Before a community can act intelligently to protect its environment, it needs to know its problems. You can help by taking part in a community-wide inventory, documenting your findings with photographs and statistics. Here are some things to look for.

Water pollution: Is your city's water supply taken from protected sources, or does it depend on water that has been used before by upstream cities or industries? Is your water supply adequately protected from encroachment, or do zoning authorities permit building of new sources of pollution within the watershed? Does your city live up to its obligation to

continued

downstream neighbors by properly treating sewage? Which polluters are lax in their treatment of plant wastes? What are their attitudes toward pollution, and what are their plans for future pollution abatement?

Air pollution: Does your community have a program for reducing air pollution? Is the program enforced? Does the city permit open burning of trash, garbage, or industrial wastes? Does the community require automobiles to have effective emission control devices? Are they checked periodically? Who are the polluters who contribute measurably to air pollution? What are their plans for abatement? Can you identify apartments and office buildings whose heating plants send up clouds of smoke?

Agricultural pollution: Check with your county extension agent for a listing of pesticides and fertilizers used by local farmers. Where are livestock feedlots located and what facilities are provided to prevent harmful runoff into water resources? Is an educational program provided to assist farmers in properly using pesticides and fertilizers to protect themselves, their employees, consumers, and wildlife?

Land pollution: Does your community have adequate zoning ordinances to encourage good land planning and good land use? Are industries permitted to destroy future productivity of land by dumping industrial wastes or mine tailings? Does the city have adequate litter laws, coupled with enough litter receptacles to encourage people to dispose of trash properly? How are abandoned cars disposed of in your community; and at whose expense? Are abandoned buildings allowed to stand indefinitely, attracting litter, rats, and other unwanted tenants? How does your community dispose of its garbage? Do you have a community dump, open at reasonable hours to encourage its use? What recycling possibilities are being explored?

Noise pollution: Does your community require that automobile and motorcycle exhausts be properly muffled, and is the law enforced? Is construction work confined to reasonable hours? Are low-flying aircraft prohibited? Are loudspeakers and other noise makers regulated?

Environmental education: Does the public school system use its resources for environmental education? Are there outdoor classrooms and are they being utilized? Are teachers required to have training in environmental education? Are inner city children being neglected in outdoor education programs? Are adult or extension courses being offered to help create an informed body of citizens?

Ecological awareness: Has the community taken steps to preserve unique natural environments such as swamps and marshes, stream banks and forests, and rare and endangered species? Are land developers permitted to alter watersheds at will, creating erosion or causing downstream flooding by increasing runoff of surface water from paved lots and rooftops? Are there adequate safeguards for the use of pesticides and herbicides? Are there dangers of agricultural pesticides entering the community water supply? Are dairy products tested for residual insecticides? Are power line rights-of-way granted regardless of scenic value?

These are just some of the questions that might be asked in your community environmental inventory; there are others that are equally important in your locality.

After you have completed your inventory, discuss alternative solutions to the problems you have uncovered with government officials, civic groups, and the press.

Concerned, informed citizens must—and can—obtain results. It is everybody's world and everybody's responsibility to make it better.

population

Another proposed explanation for the environmental crisis is overpopulation. The argument holds that people are polluters, thus the more people, the more pollution. An examination of the available statistical data indicates that this is not the whole story. In the United States, most of the major pollution problems have developed since World War II (1946). The intensification of the problems since this time is in the range of 10- to 20-fold or 1000 to 2000 percent. The use of mercury for chlorine production (a major source of mercury pollution) has increased 2100 percent; the use of inorganic nitrogen fertilizer (a source of water pollution as a result of leaching) has increased by 1400 percent; the amount of phosphate discharged into municipal sewage has increased by 500 percent, and that of nitrate by 260 percent. Lead deposited in glaciers has increased by about 400 percent.

The above changes in pollution levels cannot be accounted for simply by an increased population size in the United States, for during the same period population size increased by only about 45 percent. It is possible that an increase of 40 to 50 percent in population size may be the cause of a much larger increase in pollution intensity. For example, the need for additional food, clothing, and shelter for a growing population might very well intensify production, and therefore the resulting pollution, particularly if the efficiency with which production supplied the needs of the population were to decline. There is some indication that efficiency has declined (mass transportation, power generation, and so on), but it seems highly unlikely that these changes are sufficient to account for our environmental crisis.

city growth

Another factor that contributes to the environmental crisis is population distribution (Fig. 15-7). The rapid growth of large cities with internal crowding and deteriorating social conditions lead to a worsening of pollution problems. It is true that the size and population density of cities have disproportionately large effects on the amount of pollution produced per individual, but the effect is too small to explain the total observed increases in pollution intensity. Problems due primarily to radioactive fallout, fertilizers, pesticides, mercury, and other industrial pollutants are not caused by urbanization.

Fig. 15-7. Housing development occupying prime agricultural land in California. (*Courtesy U.S. Department of Agriculture, Soil Conservation Service.*)

Fig. 15-8. Intersection of routes 287 and 78 in New Jersey. Note large area required for intersection. *(Courtesy U.S. Department of Agriculture, Soil Conservation Service.)*

The intensification of environmental problems associated with organization is due not so much to the size of the population as is to the maldistribution of the living and working places in metropolitan areas. Minority groups have been concentrated in urban ghettoes and more affluent social groups in the suburbs. Homes and places of work have become separated for both groups. Those suburbanites who work in the city, but are unwilling to live there, must commute; ghetto dwellers who work in outlying industries, but are unable to live in the suburbs, must also commute. This helps to explain why automobile vehicle-miles traveled within metropolitan areas increased from 1050 miles per capita in 1946 to 1790 miles in 1966. This increased transportation has added to the burden of pollution (Fig. 15-8).

distribution of people

The shift in population from rural areas to cities is itself associated with technological advances which have contributed directly to environmental deterioration. The intensification of agricultural production has reduced the need for farm labor, while at the same time it has increased the use of fertilizers, pesticides, and fossil fuels.

There appears to be no reasonable explanation for the rapid increase in pollution levels since 1946 based solely upon growth in population size, urbanization, or a supposed decrease in production efficiency. The explanation lies elsewhere.

Since 1946 there has been a sharp increase in the amount of pollution produced per person. We have become more affluent; we use more of the world's goods and therefore produce more of the world's wastes than any other nation. With only 6 percent of the world's population, we use 40 to 50 percent of the world's goods. Figure 15-9, a recent development, contrasts sharply with the more common method (Fig. 15-10) of handling the wastes resulting from our high standard of living.

Fig. 15-9. Recycling of junk automobiles. *(Courtesy of Michigan Department of Natural Resources.)*

per capita uses

There was essentially no change in the per capita consumption of goods required for food, clothing, or shelter in the United States between 1946 and 1966. Total calories consumed per person actually dropped during this period—from 3380 per day per person to about 3170 per day per person. Protein consumption increased by about 3 percent, whereas the so-called accessory foods (minerals and vitamins) decreased by about 6 percent during this period. These figures are reflected in the overall agricultural production data for the United States. Total grain production per capita decreased by about 8 percent during the 20-year period. At the same time meat and lard production decreased by about 6 percent. In general, then, food consumption per capita has remained substantially unchanged since 1946. The same is true of clothing. The amount of clothing produced per

Fig. 15-10. Roadside dump. *(Courtesy U.S. Department of Agriculture, Soil Conservation Service.)*

capita has not increased significantly. Statistical data reveal essentially the same picture for housing. Housing units occupied in 1946 were 0.272 per capita; in 1966 the figure was 0.295. These figures do not take into account the quality of housing, but the same situation is reflected in the production figures for housing materials.

The above paragraph indicates that there was no significant change in the per capita consumption of goods required for food, clothing, and shelter between 1946 and 1966. At the same time, however, the per capita gross national product (goods and services for the support of one individual) increased by 50 percent—from $2222 in 1946 to $3354 in 1966 (corrected for inflation). Clearly, the increases in available goods and services have been in areas other than food, clothing, and shelter. Some of these are easily identified. Apart from increased services, some of them are: automobiles and trucks (an increase of 96 percent in vehicles per capita), certain household appliances—dryers, air conditioners, and dishwashers—(an increase of about 15 percent from 1960 to 1966), and wood pulp products (an increase of 150 percent from 1946 to 1966). In addition, we have seen marked increases in the per capita utilization of fuel-generated electric power, gasoline, and many "extras" such as television sets, radios, electric can openers, corn poppers, hair dryers, boats, snowmobiles, and so on. The availability of these kinds of goods tells us that we are indeed an affluent society, but these items are too small a part of the nation's overall production to account for the increase in pollution level.

the extras

The major problems of pollution cannot be blamed on overpopulation and affluence alone. The answer, in large part, involves not technology that has allowed us to become more affluent, but that part of technology that has allowed us to go far beyond simply being an affluent society. This technology has made possible the replacement of natural materials (cotton, wool, silk, and wood) with man-made plastics; there has been a marked increase in the amounts and varieties of other man-made synthetics (detergents, pesticides, herbicides, and so on); automobile engines have been redesigned to operate at increasingly higher compression ratios; electric power has increasingly replaced home heating directly by fuel; the use of materials such as aluminum and certain chemicals that require the consumption of increased amounts of power for manufacturing has become a "necessity"; at the same time there have been striking changes in agricultural practices, especially the increased tendency to feed cattle away from pastures, reduced crop rotation, increased use of inorganic fertilizers, and the massive introduction of synthetic pesticides and herbicides.

nonbiodegradables

The technology that resulted in the kinds of changes cited in the above discussion has been developed since World War II, during the period that coincided with rising pollution. These technologies are unsuited for accommodation by natural environmental processes. The substitution of plastics for natural fibers requires the use of fuel-generated power (with its attendant pollution) in place of sunlight absorbed by plants and transmitted by natural cyclic processes. Synthetics such as detergents and pesticides are outside (and therefore incompatible with) the coordinated system of biochemical processes that living things have evolved. They are therefore not assimilated by natural environmental cycles; they act as pollutants. The high-compression gasoline engine, with its attendant high temperature, causes oxygen and nitrogen to combine as nitrogen oxide, a compound otherwise rare in nature and not readily assimilated by natural environmental processes. Smog is a result. Electric power-generating plants produce several by-products not compatible with natural environmental cycles. These include sulfur dioxide, nitrogen oxides, and radioisotopes. Newer agricultural techniques have disrupted soil cycles so that natural soil fertility is reduced and fertilizers leach into groundwaters. Pesticides disrupt the balance between pests and their natural predators with the resultant appearance, increasingly, of insecticide-induced outbreaks of insect pests, and the accumulation of insecticides in wildlife and in man.

The most logical explanation for the environmental crisis man faces is the massive introduction into the developed countries of the world of new technologies that are strikingly incompatible with the natural, balanced processes that sustain the environmental system.

GLIMMERS OF HOPE

limits to growth

A 1-day symposium entitled *Limits to Growth*[1] held at the Woodrow Wilson Center in Washington, D.C., March 2, 1972, drew a wide audience, including senators and ambassadors as well as representatives of the news media. Meetings of this type indicate a growing concern for our environmental problems. The principal paper presented at the symposium was a computer-based study of the combined effects of population growth, pollution, food production, industrialization, and use of nonrenewable resources. The conclusions could be anticipated—continuation at current rates will bring major disaster within 50 years.

In his evaluation of these predictions, Abelson (1972) reminds us that conclusions based on treating Earth as a single system have limited application. Population growth is not uniform over the world. Rates differ by as much as a factor of six. Also, in an effort to reduce the number of variable factors, different kinds of pollution that are not at all alike in toxicity were grouped together. More important, however, a shift in thinking may produce action that will avert some or most of the calamity. The rate of population growth in the United States is now decreasing. Responsible people in the middle and upper-middle income groups are talking of a no-growth society.

[1] Based on a study made at Massachusetts Institute of Technology, Cambridge, Massachusetts.

Fig. 15-11. This homestead in Cades Cove, Smoky Mountain National Park, indicates that man can still live in equilibrium with the environment. This requires a low population density, however.

U.S. Fertility Rate Goes Down

Data compiled for the first half of 1972 suggest that the United States has reached a fertility level that would eventually result in zero population growth (ZPG). Two independent studies* show that the average number of children per family in the 18–24 year age group for married women is 2.3. When this figure is corrected to include all women of child-bearing age, the fertility rate becomes 2.1 children per female. This is the replacement rate—the number of offspring per female to maintain a population at a constant level.

The number of births for the first half of 1972 was 9 percent below that of a similar period in 1971 in spite of a 3 percent increase in the number of women of child-bearing age. On the basis of births per year per 1000 women aged 15 to 44, the data for the first half of 1972 projected for the year gives a fertility rate of 73.1. This falls below the previous low record of 1936 (75.8) and certainly far below the 1957 high of 122.9.

Even if a ZPG fertility rate is maintained from 1972 on, the population of the United States can be expected to increase for the next 70 years. This results from the increasing number of girls that reach child-bearing age each year as a result of past high birth rates. Twenty years ago, the number of girls 18 years of age was only 1.1 million; today there are 1.9 million. And so, although births dropped 151,000 in 1971, the number born (1.6 million) was 600,000 greater than the number of

* Census study Series P-20, number 240 (available from the U.S. Government Printing Office, Washington) and the monthly *Vital Statistics Report* Volume 21, number 6.

deaths for that period. The leveling of the growth curve to a stable population will occur gradually. The population by the year 2000, assuming that the 1972 fertility rates persist, would be about 270 million—some 60 million more people than at present. However, this is a far more hopeful prospect than the realization of previous estimates, many of which exceeded 300 million.

A milestone in the history of man has been reached. The United States, along with a few other countries, is demonstrating that man can regulate his population density and thus escape to a large extent those factors which in the past have controlled population numbers—disease, famine, and war.

Organizations That Can Help

Numerous resources are available to individuals concerned with improving the environment. National and local service organizations, social clubs, churches, and business groups may be actively concerned with environmental problems. Local private industry, private utility companies, and local divisions of national corporations are possible sources of help. Many already have active community improvement and incentive programs. Several national organizations with state, regional, or local chapters can be contacted directly to be of service in home communities. Some of these are listed below. All are nonprofit.

Appalachian Trail Conference, 1718 N Street, N.W., Washington, D.C. Coordinates volunteer maintenance of the 2000-mile Appalachian Trail from Maine to Georgia. Can provide publications and other guidance to other groups interested in establishing and maintaining trail systems.

Citizens for Clean Air, 40 West 57th Street, New York, New York. Citizen group working for public education on health, esthetic, and economic effects of air pollution.

Desert Protective Council, P.O. Box 33, Banning, California. Works to safeguard desert areas of scientific, scenic, historical, and recreational value, and to promote understanding of desert resources.

Ducks, Unlimited, P.O. Box 8923, Chicago, Illinois. Membership organization to perpetuate wild waterfowl principally by preservation and rehabilitation of wetlands in the United States and Canada.

Garden Club of America, 598 Madison Avenue, New York, New York. Organization of local member clubs which promotes knowledge and appreciation of horticulture, landscape design, and natural resource conservation.

General Federation of Women's Clubs, 1734 N Street N.W., Washington, D.C. An organization of 51 state federations of local women's clubs. Supports study and action programs for community betterment.

continued

Izaak Walton League of America, 1326 Waukegan Road, Glenview, Illinois. Membership organization with local chapters and state divisions. Promotes conservation of natural resources; development, protection, and enjoyment of high-quality outdoor recreation and natural beauty resources; and public education in these concerns. Publishes a monthly magazine. Cosponsors books and other educational materials.

National Association of Soil and Water Conservation Districts, 1025 Vermont Avenue, N.W., Washington, D.C. Membership organization of 3000 local districts and 50 state associations working to conserve and develop land, water, and related natural resources.

National Audubon Society, 1130 5th Avenue, New York, New York. Membership organization with state chapters. Works for conservation of all natural resources and conservation education. The Society publishes a monthly magazine, manuals, bulletins and teaching aids, and a publications list. Offers films and speaker services.

National Trust for Historic Preservation, Decatur House, 748 Jackson Place, N.W., Washington, D.C. Membership organization made up of individuals and groups. Provides advice and technical assistance on preservation and restoration of buildings or sites significant in American history and culture. Publishes leaflets on such subjects as preservation law and restoration techniques, a quarterly journal, and a monthly newspaper.

National Wildlife Federation, 1412 16th Street, N.W., Washington, D.C. Membership organization with affiliated state organizations. Dedicated to encourage wise use and management of natural resources. Sponsors annual National Wildlife Week. Publishes booklets, newsletters, a bimonthly magazine of general interest, and a monthly nature magazine for children. Distributes television and radio materials.

The Nature Conservancy, 1522 K Street, N.W., Washington, D.C. Membership organization with primary purpose of acquiring land to help preserve the country's natural heritage. Publishes quarterly *News,* and pamphlets on scientific, educational, and legal aspects of natural area and open space preservation.

Sierra Club, 1050 Mills Tower, 220 Bush Street, San Francisco, California. Membership organization devoted to exploring, enjoying, and protecting natural scenic resources. Produces conservation films, exhibits, and manuals; sponsors conferences on wilderness and natural science; and publishes books on wilderness and other scenic resources, guidebooks, a monthly bulletin, and other conservation-education materials.

Urban America, 1717 Massachusetts Avenue, N.W., Washington, D.C. Educational organization seeking to improve total quality of life in cities. Publications include a bimonthly, *City;* and brochures and other materials.

The Wilderness Society, 729 15th Street, N.W., Washington, D.C. Membership organization dedicated to increasing knowledge and appreciation of wilderness, and to see established policies and programs for its protection and use. Publishes quarterly magazine, *The Living Wilderness.*

A *Conservation Directory* which lists national, regional and state citizen and professional organizations and officials in natural resource and related fields is published annually. It is available ($1 per copy) from the National Wildlife Federation, 1412 16th Street, Washington, D.C.

Oslo pact

One of the hopeful signs that Earth may not have passed the point of no return to a quality environment can be found in the international scope of treaties concerned with these problems. Early in 1972, 12 western European nations meeting in Oslo signed an agreement to ban ocean dumping of highly toxic materials such as cadmium and mercury, carcinogenic chemicals, and certain indestructible plastics. The disposal of less harmful substances would be controlled by individual governments. This agreement covered only the North Sea, the northeastern Atlantic, and parts of the Arctic Ocean. Discharge into the ocean from land pipelines and drainage streams was not covered, nor were oil spills.

Iran convention

In 1971, at an international wildlife conference in Ramsur, Iran, several nations signed a convention protecting wetlands of international importance to waterfowl.

Stockholm conference

Mrs. Inga Thorsson, a Swedish diplomat, suggested in 1967 that the United Nations tackle the question of environmental pollution. Through the prodding of the Swedish Ambassador to the United Nations, the U.N. General Assembly passed a resolution the following year which called for an environmental conference. This conference, involving 114 nations, 140 nongovernment conservation organizations, and 1400 journalists, was held during June 1972 in Stockholm, Sweden.

The preliminary work that preceded the conference may have been as important as the conference itself. Each participating nation prepared a critical review of its environmental problems. For some, this was the first time such problems were recognized. Nearly 7000 pages of documentation were received by Maurice Strong, Secretary-General for the conference. These were distilled into an 800-page report for the use of the delegates.

The conference was hampered by politics and cold-war issues. The emerging nations could not accept the suggestion of restraint in their bid to industrialize. Industrialization is their hope to achieve a high standard of living. Debate on the consequences of wide-scale development of industry in terms of pollution and exhaustion of nonrenewable resources was viewed by the delegates of "have-not" nations simply as an effort to hold them back.

Although nothing binding was signed by any delegation, several significant decisions were reached. Of most importance was the agreement to establish a new, permanent organization within the United Nations endowed with a $100-million fund to help finance efforts to deal with global environmental problems over the next 5 years. Plans were made for a subsequent conference to finalize an international agreement on the restriction of ocean dumping of toxic materials. This conference has now been held.

The population problem was not supposed to be on the agenda but it could not be kept out of discussions. The Conference did decide to increase emphasis on population policy and the dissemination of family planning information.

A recommendation to the International Whaling Commission calls for a 10-year moratorium on commercial whaling. Plans were made for an Earthwatch Program to coordinate and expand pollution monitoring systems. In all, 186 recommendations were brought before the conference.

The work at Stockholm was a vital beginning of a global effort to assure all a life worth living. What many people consider "a life worth living" may be in for radical change, particularly among people of the more affluent nations when they fully realize the long-range effects of current short-term goals. Dr. Jacques Piccard, marine scientist, stated at the conference that he felt that living styles and our philosophy of what constitutes success must be changed completely. In a memorable address at the conference, Lady Barbara Ward Jackson, co-author of *Only One Earth,* very aptly stated man's obligations:

> To act without rapacity, to use knowledge with wisdom, to respect interdependence, to operate without greed—these are not simply moral imperatives. They are an accurate scientific description of the means of survival.

Suggested Readings

Abelson, P. H. "Editorial." *Science,* 175(4027):1197, 1972.

Adler, I. *How Life Began.* New York: Signet Science Library, 1957.

Benarde, M. A. *Our Precarious Habitat.* New York: Norton, 1970.

Bresler, J. B. *Environments of Man.* Reading, Mass.: Addison-Wesley, 1968.

Clegg, E. J. *The Study of Man.* New York: American Elsevier, 1968.

Commoner, B. *The Closing Circle: Nature, Man and Technology.* New York: Alfred A. Knopf, 1971.

De Bell, G., Ed. *The Environmental Handbook.* New York: Ballantine Books, 1970.

Editor. "Picking Up the Pieces from Stockholm." *Conservation News,* 37(2):2–4, 1972.

Ehrlich, P. *The Population Bomb.* New York: Ballantine Books, 1968.

Holdren, J. P. and P. R. Ehrlich, Eds. *Global Ecology.* New York: Harcourt, 1971.

Jackson, W. *Man and the Environment.* Dubuque, Iowa: Brown, 1971.

Johnson, C. E. and M. M. MacDonald, Eds. *Society and the Environment.* New York: Van Nostrand Reinhold, 1971.

Johnson, H. D. *No Deposit—No Return.* Reading, Mass.: Addison-Wesley, 1969.

Knobloch, I. W., Ed. *Readings in Biological Science.* New York: Meredith, Company, 1967.

Odum, H. T. *Environment and Society.* New York: Wiley, 1971.

Southwick, C. H. *Ecology and the Quality of Our Environment.* New York: Van Nostrand Reinhold, 1972.

Strohm, J. and J. Hess. "It all happened at Stockholm." *International Wildlife,* 2(5):34–35, 1972.

APPENDIX
The Metric System

Over the next 10 to 15 years, the United States will gradually convert much of its system of measurements to the metric system. This is necessary in order to survive in international trade because all other major world powers use the metric system or are in the process of converting to it.

The metric system involves a standard unit for length, for mass, and for volume, whose magnitude is set arbitrarily as one. Thus the standard unit of length is the meter, of mass is the gram,[1] and of volume is the liter. Other units, based on powers of 10, fall readily into place because of the consistent use of prefixes. "Kilo-" always means 1000 of the standard units. A kilometer equals 1000 meters. Table 1 gives the values of the common prefixes used in the metric system.

TABLE 1
Prefixes Used in the Metric System of Measurement

Prefix	*Symbol*	*Multiple*	*Prefix*	*Symbol*	*Multiple*
Tera-	T	10^{12}	Deci-	d	10^{-1}
Giga-	G	10^{9}	Centi-	c	10^{-2}
Mega-	M	10^{6}	Milli-	m	10^{-3}
Myria-	my	10^{4}	Micro-	μ	10^{-6}
Kilo-	k	10^{3}	Nano-	n	10^{-9}
Hecto-	h	10^{2}	Pico-	p	10^{-12}
Deka-	da	10	Femto-	f	10^{-15}
			Atto-	a	10^{-18}

[1] The kilogram is sometimes listed as the standard unit of mass. The International Bureau of Weights and Measures, Paris, France, keeps a 1-kilogram block made of platinum-iridium as its official reference. However, the scale of metric units of mass uses standard prefixes coupled to *gram*. Thus reference to centigrams implies hundredths of a gram, not hundredths of a kilogram.

The following tables give the values to convert from metric to United States customary units and the reverse. Only the conversion factors of practical use are given. Normally, there would be little occasion to convert millimeters to yards or decimeters to statute miles.

To use these tables to convert from United States customary units to metric, divide the value of the United States units by the factor found at the intersection of that column and the appropriate horizontal column. For example, 50 inches divided by 0.3937 equals 127 centimeters. For the opposite conversion, metric to United States customary, multiply the metric value by the intersected factor. Twenty meters times 3.2808 equals 65.616 feet. This degree of accuracy may be greater than necessary for most uses. The values in the tables are carried to several decimal places primarily to illustrate the manipulation of the decimal point in the metric system.

TABLE 2
Length

		Divide				
		Inches	*Feet*	*Yards*	*Statute miles*[a]	*International nautical miles*[b]
Multiply	Millimeters	0.03937	0.00328			
	Centimeters	0.39370	0.03281	0.0109		
	Decimeters	3.93701	0.32808	0.1094		
	Meters	39.3701	3.28084	1.0936	0.000621	0.000539
	Kilometers		3280.84	1093.61	0.6214	0.5399
	Myriameters		32808.4	10936.1	6.214	5.399

[a] 5280 feet.
[b] 6076.1155 feet.

TABLE 3
Area

		Divide				
		Square inches	*Square feet*	*Square yards*	*Acres*	*Sections*[a]
Multiply	Square millimeters	0.00155				
	Square centimeters	0.15500	0.00107			
	Square decimeters	15.50	0.10764			
	Square meters	1550.003	10.7639	1.1959	0.000247	
	Ares[b]		1076.391	119.59	0.02471	0.0000386
	Hectares[c]		107639.1	11959.9	2.47105	0.003861
	Square kilometers			1195990.0	247.105	0.3861

[a] A section equals 1 square mile (statute) or 640 acres.
[b] An are equals 100 square meters or 0.01 hectare.
[c] A hectare equals 10,000 square meters or 100 ares.

TABLE 4
Volume

		Divide				
		Cubic inches	*Cubic feet*	*U.S. fluid ounces*	*U.S. liquid quarts*	*U.S. liquid gallons*
Multiply	Milliliter[a]	0.061025	0.000035	0.033815	0.0010567	0.000264
	Liter	61.02545	0.035315	33.81497	1.0567176	0.264179
	Kiloliter	61025.45	35.31566	33814.97	1056.7176	264.1794

[a] One milliliter is now considered equivalent to 1 cubic centimeter. One kiloliter, then, equals one stere (cubic meter).

TABLE 5
Mass[a]

		Divide				
		Grains	*Avoirdupois ounces*	*Avoirdupois pounds*	*Short tons[b]*	*Long tons[c]*
Multiply	Milligram	0.015432	3.5×10^{-5}	2.2×10^{-6}		
	Gram	15.43235	0.03527	0.002204		
	Kilogram	15432.35	35.2739	2.2046	0.0011023	0.000984
	Metric ton[d]			2204.6	1.1023	0.9842

[a] Mass is the measure of the matter in an object; mass varies with velocity; mass is the measure of a body's resistance to acceleration and is independent of its position in space. Weight is the gravitational force exerted by the earth or other celestial body; weight is the product of an object's mass and the local value of gravitational acceleration.
[b] One short ton equals 2000 pounds.
[c] One long ton equals 2240 pounds.
[d] One metric ton equals 1000 kilograms.

Index

Numbers in italic indicate pages on which complete references are listed. Numbers in boldface indicate pages on which illustrations appear.

A

B